Work 280

食物交易者
Food Traders

Gunter Pauli

[比]冈特·鲍利 著
[哥伦]凯瑟琳娜·巴赫 绘
朱溪 译

上海远东出版社

丛书编委会

主　任：贾　峰
副主任：何家振　闫世东　林　玉
委　员：李原原　祝真旭　牛玲娟　梁雅丽　任泽林
　　　　王　岢　陈　卫　郑循如　吴建民　彭　勇
　　　　王梦雨　戴　虹　翟致信　靳增江　孟　蝶

特别感谢以下热心人士对童书工作的支持：

匡志强	宋小华	解　东	厉　云	李　婧	陈　果
刘　丹	熊彩虹	罗淑怡	旷　婉	杨　荣	刘学振
何圣霖	廖清州	谭燕宁	韦小宏	李　杰	欧　亮
陈强林	王　征	张林霞	寿颖慧	罗　佳	傅　俊
胡海朋	白永喆	冯家宝

目录

食物交易者	4
你知道吗？	22
想一想	26
自己动手！	27
学科知识	28
情感智慧	29
艺术	29
思维拓展	30
动手能力	30
故事灵感来自	31

Contents

Food Traders	4
Did You Know?	22
Think about It	26
Do It Yourself!	27
Academic Knowledge	28
Emotional Intelligence	29
The Arts	29
Systems: Making the Connections	30
Capacity to Implement	30
This Fable Is Inspired by	31

真菌朝着土地深处挖掘，寻找树根。每当遇到一株植物，真菌都会用细线和他连接起来。他热衷于做生意，对一株植物说道：

"我能帮上忙吗？我发现你身边没太多食物哦。"

"你能为我提供什么呢？寻常的食物吗？"

A fungus is digging deep into the ground, looking for roots. Each time it comes across a plant, it connects to it with a tiny wire, and being interested in doing business, he asks the one plant,

"Can I be of any help? I see you do not have a lot of food around."

"What do you have to offer me? The usual foodstuff?"

真菌朝着土地深处挖掘……

A fungus is digging deep into the ground ...

……当我还是种子时被带到这里……

... my seed was dropped here ...

"不,今天我有些好东西给你,比如碳元素、氮元素和磷元素。"

"我的确需要这些物质。"植物回答。"当我还是种子时被鸟儿的排泄物带到这里,其实我并不太走运,因为这个地方的土壤太贫瘠了。在这里生存太难,更别说要成长壮大了。"

"No, today I have some good things for you, like carbon, nitrogen and phosphorous."

"I can surely do with those," Plant replies. "When my seed was dropped here in some bird droppings I wasn't very lucky, as it landed in an area where the soil is poor. It is tough to survive, let alone thrive, here."

"所以我们要把你需要的送来,把它们直接送上门,即使你需要的东西也许来自森林的另一头。"

"太感谢了。不过价格呢?你为我带来了那么多,需要多少钱呢?"

"That is why we get you what you need, delivered right to your doorstep, even if it has to come from the other side of the forest."

"And I am grateful. But the price! How much do you want for such a lot?"

……直接送上门……

... delivered right to your doorstep ...

……我们来达成公平的交易。

... we will strike a fair deal.

"让我们开诚布公地聊聊吧。你知道的,当周围有充足的食物供给,那么食物既容易获取也便宜。而当食物匮乏时,食物获取就变得困难,自然也更贵了。"

"瞧,我就知道你需要我的糖分和脂肪。我只是担心你从我这里拿走的太多了,以至于我自己都不够用。"

"我知道你也需要这些,植物女士,所以我们来达成公平的交易。"

"Let's be open about this, and fair. You know that when there is a lot of food around, it is easy and cheap. And when there isn't, it is more difficult – and more expensive."

"Look, I know you want my sugars and fat. I'm afraid that you will ask so much from me that I do not have enough for my own use."

"I know you are in need, Mrs Plant, so we will strike a fair deal."

"这么说,你打算做一次与众不同的交易吗?"

"在这个世界,你总能发现很多人乞讨、借贷、盗窃或欺骗!但是也有很多人值得有机会过上更美好的生活。"

"我们中的数百万人的确应得到更好的待遇。我不是在乞求,只是要求你也考虑我们的实际情况。毕竟,你投资建造了这个巨大的网络,获取糖分并交付碳元素,你理应借此生存。"

"So, you are ready to make a deal different to your other deals?"

"In this world, you'll find those who beg, borrow, steal and cheat! But there are also those who really deserve a better chance in life."

"Millions of us do deserve better. I am not begging, just asking that you consider our reality as well. You deserve making a living too, after all you've invested in this great network, taking sugars and delivering carbons."

……很多人乞讨、借贷、盗窃或欺骗!

... those who beg, borrow, steal and cheat!

……我们真菌构造了可以运输食物的高速通道。

... we fungi build the highways to deliver food.

"是的，我们真菌构造了可以运输食物的高速通道，我们也需要储藏的空间、运输的站点来获取你的糖分、脂肪，并卸下你所需要的碳元素。"

"可你们真菌是生意人呀，"植物评价道。"你们这么做并不是为了慈善。"

"1克土壤里就有我们大约1千米长的细线，连接着一切。在我们的'木维网'里，一切都会被观察到。"

"Yes, we fungi build the highways to deliver food, and we need storage space, and delivering stations, to take your sugars and fat, and offload the carbon you need."

"You fungi are traders," Plant remarks. "You are not in this for charity."

"One gram of soil has about a kilometre of our tiny wires, connecting everything. In our 'wood wide web', nothing goes unnoticed."

"人类来到我们的森林研究植物和动物,却从来没想起关注脚下的'生命之网',这不是很奇怪吗?"

"是啊,如果人类真的想了解大自然是怎样运作的,他们就必须理解我们在森林里作为食物交易者的角色。"

"Is it not strange that people come to our forests to study plants and animals, but do not think of looking at 'The Web of Life' under their feet?"

"Yes, if people really want to understand how Nature works, they must understand our role as the food traders of the forest."

脚下的"生命之网"。

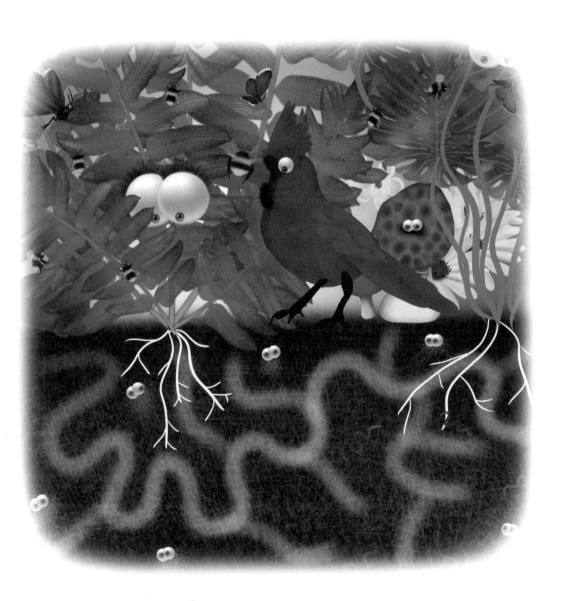

"The Web of Life" under their feet.

······慷慨解囊······

...will you be generous...

"但是你会对老的、幼的和贫困的对象慷慨解囊吗？"

"我的目标是整片森林蓬勃发展，而不只是服务于最好最强壮的那些个体。关键不是最大程度地利用一项交易，而是要确保每个个体都拥有发展所需的一切。"真菌骄傲地说道。

"But will you be generous with the old and the needy, and the young seedlings?"

"My aim is to see the whole forest thrive, and not only to be of service to the best and strongest. The key is not to make the most on one deal, but to ensure that everyone has all they need to thrive," Fungus says proudly.

"所以你不只是用智慧来交易,还用了你的诚心?"

"正是如此!如果做生意只是一门心思考虑自己的利益,那么他很快就会发现没人愿意和他合作了!"

……这仅仅是开始!……

"So you are trading not only with your mind, but also with your heart?"

"Exactly! Someone who trades with only his own interest in mind, will soon find he has no one to trade with!

... AND IT HAS ONLY JUST BEGUN!...

……这仅仅是开始！……

... AND IT HAS ONLY JUST BEGUN! ...

由真菌在土壤中建立的"木维网"的密度比现实中的万维网高出几百万倍。通过固定生物线进行的通信密度超过任何其他设想或计划的通信网络。

The Wood Wide Web established by fungi in the soil is millions times more dense that the World Wide Web. The density of communications through fixed biological wires outperforms any other communication network imagined or planned.

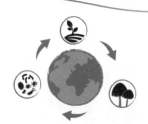

All information in a forest circulates in the local ecosystem. This communication network is self-sufficient in energy, and operates as an interface between plants, fungi, and microorganisms. The system works and trades 24 hours a day.

森林中的所有信息都在当地的生态系统中流通。该通信网络在能量上达到自给自足,并且充当了植物、真菌和微生物之间的接口。这个系统每天24小时工作并进行交易。

The Wood Wide Web, with physical connections thinner than a silken thread, and a length measuring 1km in just 1gr of soil, connects and communicates information about available resources, imminent threats, nutrient balance and water density.

木维网具有比丝线更细的物理连接，在1克土壤中可长达1千米，连接和传达信息，包括可用的资源，迫近的威胁，养分平衡和水的密度。

Fungi and plants engage in trading, whereby fungi deliver mainly carbon. The fungi ultimately decide how many plants, and which plants, pay what price – expressed in an amount of sugar and fats.

真菌和植物进行交易，其中真菌主要传递了碳。真菌最终决定了交易中有多少植物，有哪些植物该支付多少——通过糖和脂肪的数量表现出来。

Some plants have to pay a higher price for acquiring carbon and other trace minerals that are desperately needed. This higher cost limits their growth. However, this extra income for the fungi is partly redistributed to the needy.

一些植物必须支付更高的"价格"才能获得急需的碳和其他微量矿物质。较高的成本限制了它们的生长。但是，真菌的额外收入部分重新分配给了有需要的植物。

Fungi have a sense of taxing and redistribution of wealth. Part of the extra price is invested to rebuild the fungal network, which is regularly destroyed by tilling.

真菌具有征税和财富再分配的意识。多支付的"费用"部分被投资于重建真菌网络，该网络经常被耕作破坏。

The fungal-flora network secures information and nutrition that is supportive to all life. Fungi learnt over millions of years that the young, old and weak share the wisdom and the knowledge that go beyond the mere utilitarian goals of trade.

真菌－植物网络确保了对所有生命体的信息和营养支持。真菌经历了几百万年的学习，让年幼、年长或虚弱者都共享智慧和知识，这些远超交易的纯功利主义目标。

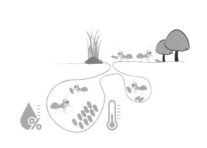

The fungal network services cover health and safety measures, and even emergency calls. All exchange of information is local, processed, stored, shared and acted upon locally. This leads to the term "data farming" as opposed to data mining.

真菌网络服务涵盖了健康和安全措施，甚至包括紧急呼叫。所有信息交换都是本地处理、存储、共享并起作用的。这产生了术语"数据耕耘"，而不是数据挖掘。

How is it that the internet of the fungus is more performant than our internet?

真菌的互联网和我们的互联网相比性能如何？

How could all communications be achieved with only biological wires, without anything wireless?

如何仅用生物线，而不使用无线的方式来实现所有通信？

Would you determine the price based on the needs of the young and old?

你会根据年轻人和老年人的不同需求来制定价格吗？

Would you ask for information overseas when it is locally available?

你会从海外获取本地信息吗？

Do It Yourself! 自己动手!

Find out what the internet density is in your area? The proposal is to increase connectivity to one million communication points per one square kilometre, building a network of antennae, putting in fibre-optic networks from house to house, and having a minimum of 7 to 10 Wi-Fi points per household. Now make a comparison with the density of one kilometre of wires per cubic centimetre of soil. Can you calculate an estimate of how much more dense the Wood Wide Web is, compared to the best of the internet that people have imagined to date?

你所在地区的互联网密度是多少?建议将连接程度提高到每平方千米100万个通信点,建立天线网络,每个家庭联通光纤,并且每个家庭至少拥有7—10个Wi-Fi点。现在与每立方厘米土壤含有1千米线路的密度进行比较。对比人们迄今所能想象的最佳互联网络,你能估算出木维网的密度会高出多少吗?

TEACHER AND PARENT GUIDE

学科知识
Academic Knowledge

生物学	菌根网络是连接植物以输送水、碳、氮等物质的地下菌丝网络；丛枝菌根网络（150—200种）和外生菌根网络有1万多种；不能合成因此依赖真菌的真菌异养植物；已被研究的植物科中有92%存在菌根。
化　学	碳、磷、氮、水、化感物质和防御化合物的运输。
物　理	液压升降机可帮助根深蒂固的树木和幼苗之间几百米距离的水体传输。
工程学	由菌根真菌连接的植物利用这些地下网络产生并接收警告信号。
经济学	感染了菌根的新苗可加快其在社群里的扎根发展，并提高林业效率；植物和树荫表面区域间碳的源−汇梯度通过作为交易者的真菌网络，调节了植物之间的碳转移；投资回报，投资于网络、存储和交付，但必须维护并进行再投资。
伦理学	真菌考虑到资源利用率相对较高的条件（强光或高氮环境）将碳或养分转移到条件较差的植物上；真菌网络确保了更快的生长速度和幼苗的存活；贸易不是慈善，但贸易可以是公平贸易；需要花一些时间来了解事物的运作方式。
历　史	自5.41亿年前的寒武纪革命以来，真菌网络就开始运行；1984年，哥伦比亚必查达的拉丝加维奥塔斯首次大规模恢复了8000公顷的森林，使树苗接种了菌根。
地　理	如在温带和北方森林中看到的那样，当次生演替发生时，菌根网络有助于幼苗的再生；在烧毁和施救的森林中，一种外生菌根真菌网传输了氮和磷等物质，有助于幼苗生长。
数　学	源−汇模型用于研究如何计划和调节土壤中的养分；在某一时刻最大化一次交易的结果，或者随着时间的推移为每个人优化结果。
生活方式	在有能力时尽可能慷慨；任何被高度需求的东西都被认为是昂贵的。
社会学	认识到我们不能绝对独立地生活；乞讨、借贷、偷窃和作弊；我们都关注相同的东西，而不关注其他人不看的地方。
心理学	应激压力；即使一开始的条件不是最好的，也要使生活变得美好。
系统论	木维网。

教师与家长指南

情感智慧
Emotional Intelligence

真菌

真菌注意到植物缺少食物，并愿意帮助她。他与她洽谈交易时保持透明，并陈述了交易条件。真菌在了解了情况后打算公平交易，并且考虑到植物从中应得的利益，提出与她做生意。真菌证明了所收取的价格合理，并向植物解释双方交易的实际情况。对于人类还不了解真菌网络的运作方式，尤其是其中交易的运作原理，他表示认同。他耐心、明确地告诉植物，他的商业核心是符合职业道德的，并且希望为所有人服务。真菌意识到只为自己的利益而努力，最终只会自欺欺人的事实。

植物

起初植物不是很友好，认为真菌只会提供寻常的食物，并且担心真菌因为了解她的处境而索要超出其承受能力的条件。植物惊讶地发现真菌提出的交易方式非常特殊，并且认为像她这样的成千上万的人也应该得到更好的待遇。她认识到真菌为了实现交易进行了大量投资，是交易者也不经营慈善组织，理应获得丰厚的收益。植物转移了思考方向，想知道为什么人类不去寻找土壤中生命网的更多信息，但随后又回到核心问题，直截了当地询问真菌会不会对有需要的人慷慨解囊，通过其智慧和真诚进行交易。

艺术
The Arts

画一棵树时，我们总是从雄伟的树干和树枝开始，然后添加树叶、花朵、种子和果实。现在的挑战是完整画出我们在森林土壤深处的东西。同样的现实，只是我们从通常不会用到的角度来观察。这有助于你开阔思路，让你发现甚至没有意识到其存在的各种关系和联系。请清楚地描绘出它们所形成的比万维网和物联网更好的通信系统。

TEACHER AND PARENT GUIDE

思维拓展
Systems: Making the Connections

世界已经认识到数据就是新的石油资源。然而，主要的挑战是数据产生的几乎所有收入都是由全球范围内的极少数参与者赚取的。我们应该在了解人类设计的、人类至上的互联网的细节之前，对自然界进行交流的方式有所了解，以此进行自我教育，并从中寻找灵感。我们能效仿自然界出奇高效的数据交换和交易信息系统吗？如果能，那么我们在社交网络上的聊天就可以类比为树木的树干和树枝，有牢固的根系支持，有菌根真菌的强化以确保将本地网络连接到连贯的系统中。树叶和灌木丛中的微生物滋养土壤，提供使所有物种都能发挥作用的能量，并且结合自然界的五个王国（细菌、藻类、真菌、植物和动物）中每个王国的独特贡献。生态系统的综合体——大自然是迄今为止最大的通信网络。虽然5G网络或3万颗卫星的性能可能会给我们留下深刻的印象，但自然界的信息和通信系统已被证实具有更高的性能和韧性。我们往往只观察到我们在破坏生态系统中所起的作用，以及对生境和动植物生命的影响，但很少深入了解正在迅速发展的自然网络，即土壤、水和空气中的真菌以及其他微生物的网络。

动手能力
Capacity to Implement

让我们设计一个数据场。就像肥沃的土地——微生物、蘑菇、苔藓、昆虫和植物与水、空气和土壤之间相互作用，数字农业让人们对家庭和社区内部的相互作用有了深刻的理解。人们不仅希望访问数据，还希望控制和分析数据，然后建立数据网络，将各个网络相互连接。再加上感觉和灵敏度，这将是物联网所永远无法比拟的。首要的是数据是本地的。这使得这个"物联网"可以演变为民有民享的物联网。最终，我们希望生命互联网出现，不仅包括人类，也涵盖生态系统中的所有其他物种。你可以制定一个计划，将其应用到你的家和学校吗？

教师与家长指南

故事灵感来自
This Fable Is Inspired by

托比·基尔斯
Toby Kiers

托比·基尔斯于1976年出生于纽约。她在美国缅因州的鲍登学院（Bowdoin College）学习生物学，并获得学士学位。2005年，她到加州大学戴维斯分校继续接受教育，并获得生态与进化博士学位。她与日本京都市的日本科学促进会合作，深入开展了一些相关研究，并在巴拿马的史密森尼热带研究所度过了一段时间。托比是马萨诸塞大学安姆斯特分校的达尔文研究员。她还是阿姆斯特丹弗里耶大学的进化生态学教授和牛津大学的高级研究员。托比致力于了解植物、动物和微生物之间如何形成共生伙伴关系，以及导致这些伙伴关系破裂的原因。她阐述了植物根和真菌如何形成类似于人类市场的复杂的地下贸易网络。

图书在版编目（CIP）数据

冈特生态童书.第八辑：全36册：汉英对照/
（比）冈特·鲍利著；（哥伦）凯瑟琳娜·巴赫绘；
何家振等译.—上海：上海远东出版社，2021
ISBN 978-7-5476-1773-1

Ⅰ.①冈… Ⅱ.①冈…②凯…③何… Ⅲ.①生态环境-环境保护—儿童读物—汉、英 Ⅳ.①X171.1-49

中国版本图书馆CIP数据核字（2021）第249940号

策　　划	张　蓉
责任编辑	程云琦
封面设计	魏　来　李　廉

冈特生态童书
食物交易者
［比］冈特·鲍利　著
［哥伦］凯瑟琳娜·巴赫　绘
朱溪　译

记得要和身边的小朋友分享环保知识哦！
八喜冰淇淋祝你成为环保小使者！

Work 279

耐久的蓝色
Long-lasting Blue

Gunter Pauli

［比］冈特·鲍利 著
［哥伦］凯瑟琳娜·巴赫 绘
朱溪 译

上海远东出版社

丛书编委会

主　任：贾　峰
副主任：何家振　闫世东　林　玉
委　员：李原原　祝真旭　牛玲娟　梁雅丽　任泽林
　　　　王　岢　陈　卫　郑循如　吴建民　彭　勇
　　　　王梦雨　戴　虹　翟致信　靳增江　孟　蝶

特别感谢以下热心人士对童书工作的支持：

匡志强　宋小华　解　东　厉　云　李　婧　陈　果
刘　丹　熊彩虹　罗淑怡　旷　婉　杨　荣　刘学振
何圣霖　廖清州　谭燕宁　韦小宏　李　杰　欧　亮
陈强林　王　征　张林霞　寿颖慧　罗　佳　傅　俊
胡海朋　白永喆　冯家宝

目录

耐久的蓝色	4
你知道吗?	22
想一想	26
自己动手!	27
学科知识	28
情感智慧	29
艺术	29
思维拓展	30
动手能力	30
故事灵感来自	31

Contents

Long-lasting Blue	4
Did You Know?	22
Think about It	26
Do It Yourself!	27
Academic Knowledge	28
Emotional Intelligence	29
The Arts	29
Systems: Making the Connections	30
Capacity to Implement	30
This Fable Is Inspired by	31

一株小小的、不爱出风头的靛蓝植物安静地生长在墨西哥的田野里。这时一只路过的软体动物注意到了她，便开口说道：

"这么说，就是你生产出了可以用来印刷书本的漂亮蓝色墨水，以及几百年来一直用于制衣的染料。"

A small, unassuming indigo plant is quietly growing in a field in Mexico when a visiting mollusc notices her and comments:
"So, you are the one responsible for this beautiful blue ink used to illustrate books, and the dye that has been used for clothing for centuries."

一株小小的、不爱出风头的靛蓝植物……

A small, unassuming indigo plant...

……有的人类也有着美丽的蓝色眼睛……

... some people do have beautiful blue eyes ...

"是的,看来我们都有为世界提供蓝色染料的殊荣。"

"所有人都喜欢蓝色——天空是蓝色的,水是蓝色的,甚至有的人类也有着美丽的蓝色眼睛。"

"是啊,的确如此。不过你我产生的是一种耐久的、真正的蓝色。"

"Yes, it seems that we both have had the honour of supplying the world with blue pigments."

"Everyone loves blue – the sky is blue, the water is blue, and then some people do have beautiful blue eyes."

"Yes, that is true. But you and I make a true blue, one that lasts."

"天空的颜色会在日落时从蓝色变成粉色、红色和橙色。没有办法能够让蓝色固定在天空中。这只是一种错觉罢了。"

"你说的对。确实不存在真正的蓝色,这其实都是因为光线和水滴的折射,还有你大脑的提示造成的。但是我们的蓝色,不管有没有阳光都能保留下来。"

"你知道我的蓝色其实是来源于黄色吗?"

"The colour of the sky changes from blue to pink, red and orange at sunset. There is no way to fix that blue in the sky. It is an illusion."

"You are right. There is no real blue, it is all about light and water droplets, and what your brain tells you to see. But our blue is here to stay, with or without the sun."

"Did you know that my blue is made from yellow?"

……这只是一种错觉罢了……

... It is an illusion ...

……用黄色来制造蓝色……

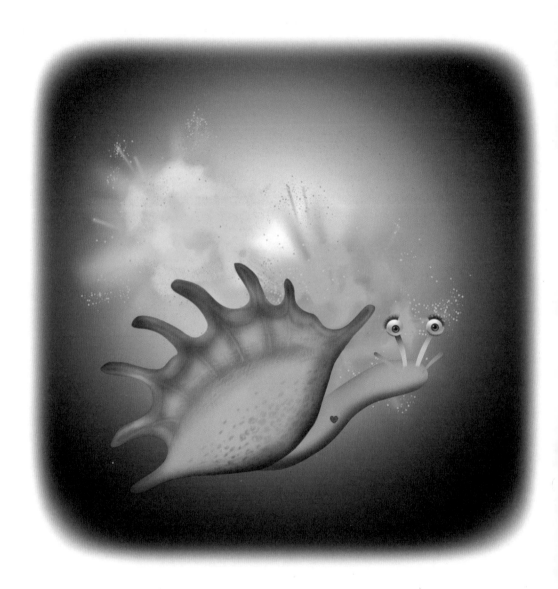

... to make blue from yellow ...

"你还知道怎么用黄色来制造蓝色？"靛蓝植物问道。

"我能产生一种深黄色的胶状物。当人类把它抹到布上，并挂在亮蓝色的天空下让阳光照射，它就会变成蓝色。你晒得越久，颜色就会变得越深。"

"像魔法一样？"

"You know how to make blue from yellow?" Indigo Plant asks.

"I make this jelly that has a strong yellow colour. When people put it on cloth and hang it in the sun under a bright blue sky, it turns blue. The longer you keep it in the sun, the stronger the colour becomes."

"Like magic?"

"很长一段时间里人类都以为这是魔法。现在他们已经知道,我的胶质能把空气中的氧气去除,让颜色变蓝。"

"空气里有许多氧气呢。"靛蓝植物评论道。

"问题不在于怎么做到,而是想染一件衣服的话,得把我成千上万个兄弟姐妹的胶质挤出来。"

"For a long time, people did think it was magic. Now they know that my gel takes oxygen from the air to make it blue."

"There is plenty of oxygen in the air," Indigo Plant remarks.

"The problem is not how to do it, but that to dye one garment you need to squeeze gel out of ten thousand of my brothers and sisters."

……我的胶质能把空气中的氧气去除……

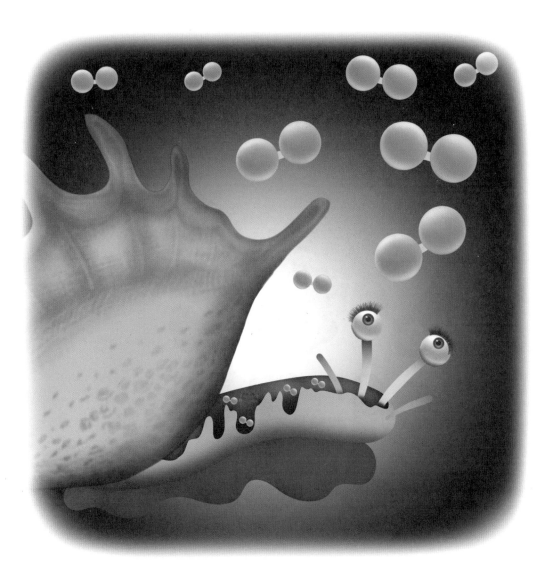

... my gel takes oxygen from the air ...

……用到许多尿液……

... use a lot of pee ...

"那的确是太多了,特别是如果你的再生还不够快的话。也许这也是利用我的蓝色色素比用你的更普遍的原因了。你知道我产生的蓝色来源于绿色吗?"

"你能把绿色变为蓝色?也是利用氧气和阳光吗?"

"不是的,要想把我的色素从绿色变为蓝色,人们通常会用到许多尿液。"

"That's a lot, especially if you do not reproduce fast enough. Perhaps that is why using my blue pigment became more popular than yours. Did you know I make blue from green?"

"You can change green to blue? Also using oxygen and the sun?"

"No, to change my pigment from green to blue, people used to use a lot of pee."

"噢，天哪！好吧，尿液的确每天都会产生，所以它的用量不成问题。只不过它到底是怎么把绿色变为蓝色的呢？"

"这都多亏了能消耗氧气的细菌创造了这美妙的蓝色。"靛蓝植物说道。

"Oh dear! Well, a lot of that is produced every day, so it should not be a problem. But what exactly makes it turn from green into blue?"

"It is all thanks to the bacteria that eat away all the oxygen, to create this wonderful blue," Indigo Plant says.

……能消耗氧气的细菌……

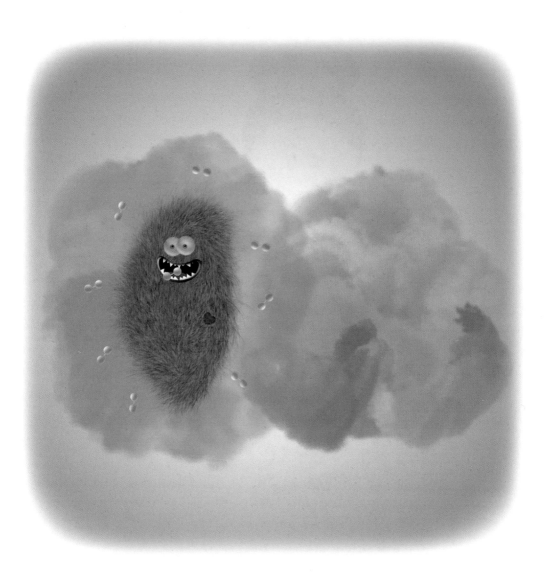

... bacteria that eat away all the oxygen ...

蓝色服装从此变得流行起来……

Blue garments were popular even then ...

"你在告诉我人类要想变得时髦,身着优雅的蓝色服装,得把布料浸泡在尿液里,以便让细菌消耗掉所有氧气?"

"是的!蓝色服装从此变得流行起来,但是很少有人会意识到产生蓝色的过程利用到了什么,还有怎样才能避免让它褪色。"

"Are you telling me that to be fashionable, in an elegant blue garment, people had their clothes soaked in pee, so the bacteria can eat away all the oxygen?"
"Yes! Blue garments were popular even then, but few people realised what went into the process of making them, and not having them fade."

"太阳晒和经常洗涤能让一切颜色褪色,蓝色也不例外。但即使这样,如今褪色的衣服也变得时尚。"

"再想到把衣服染成蓝色是从利用人类的尿液开始……"

……这仅仅是开始!……

"Everything fades in the sun and with lots of washing, blue will start fading too. But then, even faded garments are now very fashionable."

"And to think that dyeing cloth blue started with using people's urine…"

... AND IT HAS ONLY JUST BEGUN!...

……这仅仅是开始！……

...AND IT HAS ONLY JUST BEGUN!...

Did You Know? 你知道吗?

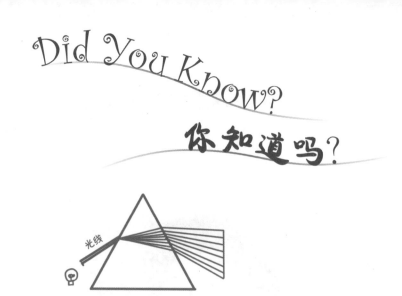

Of the 280,000 flowering plants, less than 10% are capable of producing the colour blue. The colour blue is therefore often a refraction of the light rather than a real blue pigment.

在28万种开花植物中,只有不到10%能够产生蓝色。蓝色通常是光的折射造成的,而不是真实呈现的蓝色色素。

Light-emitting diodes were developed by Nick Holonyak in 1962. It had red, yellow and green solid-state diodes. Blue diodes were invented in 1993 by Shuji Nakamura, who was awarded a Nobel Prize.

发光二极管是由尼克·何伦亚克于1962年研发的。有红色、黄色和绿色的固态二极管。蓝光二极管由诺贝尔奖获得者中村修二于1993年发明。

India was historically the world's largest supplier of blue pigments. India's association with indigo is reflected in the Greek word (Ἰνδικόν), for the dye, *indikón*, meaning "from India". The Romans Latinised it to *indicum*.

印度历来是世界上最大的蓝色染料供应国。印度与靛蓝的关联体现在希腊字 Ἰνδικόν 中，染料 *indikón* 的意思是"来自印度"。罗马人将其拉丁化为 *indicum*。

Indigo formed the foundation of centuries-old textile traditions throughout West Africa amongst the Touareg nomads, the Yoruba and Hausa of Nigeria, and the Mandinka traditions of Mali.

靛蓝构成了整个西非几百年纺织传统的基础，涉及图瓦雷克游牧民族，尼日利亚的约鲁巴和豪萨，以及马里的曼丁卡。

Japan prohibited commoners from wearing silk, leading to the cultivation of cotton, since this was one of the few fibres that could be dyed with indigo.

日本禁止平民穿着丝织品，结果促进了棉花的大量种植，因为这是少数可以用靛蓝染色的纤维之一。

In North America, indigo became the second-most important cash crop, after rice, for the colonies. As a major export crop, indigo supported the slave trade.

在北美，靛蓝成为殖民地仅次于水稻的第二种重要经济作物。作为主要的出口作物，靛蓝为奴隶贸易提供了支持。

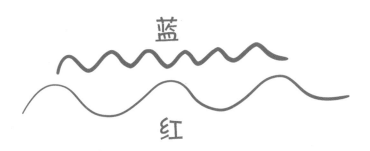

The colour blue has a short wavelength. Blue is more visible than any other colour. At sunset, sunrays fall at an angle to the Earth's surface, and since red light waves are longer, it turns the sky red.

蓝色的波长较短。蓝色比其他任何颜色都更显眼。日落时，阳光照射地球表面形成夹角，并且由于红光波长较长，因此使天空变红。

Making the blue pigment traditionally required soaking indigo plant leaves in human urine for a week. The fabric was soaked in the dye for a day and then put out in the sun, where it dried and turned bright blue.

传统的制作蓝色色素的过程通常需要将靛蓝植物的叶子在人类尿液中浸泡一周。织物在染料中浸泡一天后，再放在阳光下晒干，就会变成亮蓝色。

Do you like the colour blue?

你喜欢蓝色吗?

What to do when there is not enough for everyone?

资源不足时该怎么办呢?

Is a pair of faded blue jeans your favourite fashion item?

褪色的蓝色牛仔裤是你喜欢的时尚品吗?

Would you wear clothing dyed with urine?

你愿意穿用尿液染色的衣服吗?

Do It Yourself! 自己动手！

Clothing has become subject to fast fashion, meaning that garments are not made to last, and that the industry quickly shifts to new styles and colours. This has led to the fashion industry becoming one of the most polluting. Have a look at what is in your wardrobe, and how old your garments are. Do you wear clothing that was worn by an older sibling, or even by one of your parents? See if the colours of your older garments are still vivid and bright? Look at the labels to see what information you can find on which colours are natural, and which ones are synthetic. As you will discover, we do not know much about this subject. Become a pioneer, do some research, and then share your findings with friends and family members.

服装受制于快时尚，这意味着服装生产不是为了经久耐用，并且服装业会迅速转向新的风格和颜色。这也导致时装业成为污染最严重的行业之一。看看你衣柜里的衣服，这些衣服穿了多少年？你穿过你哥哥或姐姐，甚至爸爸或妈妈的衣服吗？这些旧衣服是否仍然鲜艳明亮？看一下衣服标签，你能找到什么信息，哪些颜色是自然色，哪些又是合成色。你会发现，我们对此知之甚少。做一名先驱者，研究一番，然后与亲朋好友分享你的发现。

TEACHER AND PARENT GUIDE

学科知识
Academic Knowledge

生物学	蚊子容易被深色尤其是蓝色所吸引；光的波长对人体有生物活性，会影响人体细胞的功能；骨螺能产生蓝色色素；原产于热带地区的木蓝是蓝色色素的起源。
化 学	火星的红色来自氧化铁；花青素，蝶蓝素；骨螺的蓝色色素是一种稀有的动物产生的有机溴化合物；尿液中的氨有助于将颜色从绿色变为蓝色。
物 理	红色是婴儿能看到的第一种颜色；丁达尔散射会产生漫反射的蓝色；天然色是基于入射白光的光谱选择性吸收呈现的；靛蓝不溶于水；在人眼看来呈白色的阳光是彩虹所有颜色的混合体；太阳光是一种电磁波，会引起空气分子内部的带电粒子（电子和质子）上下振荡；当阳光照射穿过空气时，其中蓝色部分会使带电粒子的振荡速度快于红色部分；振荡越快，产生的散射光就越多，因此蓝色的散射比红色更强烈。
工程学	蓝光波长为450—495纳米，红光为620—750纳米；帽贝使用矿化的结构组织来呈现蓝色的光学效果；一条蓝色牛仔裤染色只需用3克靛蓝染料。
经济学	染料工业是价值数十亿美元的行业；大多数色素是合成的，提取自石油；在古代近东和古代墨西哥，皇家用的紫色染料价值高于黄金，被称为蓝金。
伦理学	为了生产一种被过度需求的产品，行业甚至会为此破坏原本可持续的供给状况；当供应减少价格上涨时，会给生产者带来错误的激励。
历 史	在秘鲁发现了可追溯到6 000年前的靛蓝；日光疗法最早是由埃及人使用的；古代希腊人、中国人和印度人都会使用光疗法；泰尔红紫色或宝蓝色是腓尼基人制造的；亚里士多德在他的著作《动物志》中命名的骨螺是科学界仍在使用、最古老的经典贝壳名称；自从12世纪的艺术品中描绘圣母玛利亚的蓝色斗篷以来，蓝色便在欧洲流行；在中世纪，绿色代表背叛和不可靠。
地 理	骨螺种群生活在潮间带或浅潮下带的岩石和珊瑚中，几乎遍布世界各地。黎巴嫩的提尔是骨螺蓝色染料的起源地；琉球群岛出产靛蓝。
数 学	如何计算大自然的承载能力：在不改变生态系统的情况下，可以从多少公顷土地中提取多少量的彩色颜料。
生活方式	蓝色是全世界最受欢迎的颜色；红色与美丽和激情相关联。
社会学	蓝色与工人阶级的颜色息息相关。在德国，蓝色意味着醉酒；在中国戏曲中，蓝脸是反面人物的代表之一。
心理学	人们会将色彩与童年时期的物件和情绪联系起来；粉色是令人放松的颜色；彩色比黑白更容易令人记住；颜色会影响对食物和服饰的品味。
系统论	植物和动物不仅是重要的食物来源，还是染料颜色的重要来源。许多天然色素是可食用的，从同一种的农作物中能获得多种收益。

教师与家长指南

情感智慧
Emotional Intelligence

软体动物

软体动物对靛蓝植物表示赞赏。他也是靛蓝的提供者。他支持靛蓝植物的说法，认同大多数蓝色只是一种幻觉。他毫无压力并自信地测试了靛蓝植物的学识，并向她解释将黄色转化为蓝色的能力。他公开分享自己的"商业机密"，打破了魔法创造颜色的神话，并提供了简单、科学的解释。他非常清楚他提供的颜色是独特的并且有强烈需求，但是为了染一件衣服就需要牺牲自己成千上万的同胞，对此他感到痛苦。靛蓝植物提供了替代的染料来源，他想知道颜色创造的过程是否相同。当听到要利用尿液时，他感到惊讶。他谈到褪色的服装现在很流行，这表现出了他的洞察力。

靛蓝植物

靛蓝植物十分谦虚，为能够提供蓝色色素而感到荣幸。她对自己的角色以及所提供颜色的质量充满信心。当得知软体动物的黄色凝胶可以转变为稳定的蓝色染料时，她感到很惊讶。她指出了一个事实，即软体动物面临着必须牺牲许多同类才能提供足够的染料。靛蓝植物分享了使用人类尿液能把绿色转变为蓝色的"商业秘密"，并愿意为公之于众承担责任。她以能够为所有人提供流行的色彩而感到自豪。

艺术
The Arts

蓝色是一种原色，不能通过混合其他颜色来获得。但你可以创造不同深浅和色调的蓝色。让我们学习如何混合颜色来实现目标。从深蓝色开始，添加白色来创造浅一些的蓝色。接下来从中度蓝色开始，添加一点橙色来创建灰调的蓝色。当我们从绿色开始添加紫罗兰色时，我们会得到另一种蓝色。你能创造至少10种不同深浅和色调的蓝色吗？记录下创建这些颜色的详细信息，以便你能复制这些颜色。

TEACHER AND PARENT GUIDE

思维拓展
Systems: Making the Connections

从一开始，人类就有使用色彩的渴望，并且成为人类间差异化的重要方式。在罗马时期，公务员穿着蓝色制服。紫色曾被认为与皇室、权力和财富有关。黄色和橙色是婚礼仪式上服饰的颜色。人们经常忽略的事实是颜色在产业和经济中所扮演的角色。像香料一样，颜料自古以来就因为与等级息息相关而被往来贸易。要获得稳定的色彩需要扎实的科学和工程技术，因此彩色颜料变得像金银这样的金属一样有价值，也就不足为奇了。现代工业每年会选出大约5 000种不同的主导颜色，应用于从化妆品到时装的各个领域。不幸的是，这些颜色中的大多数不仅是合成的，而且被设计为不可降解的，原因是易褪色的产品被认为是劣质的。这些色素颗粒的尺寸很小，很容易越过过滤系统，溶解在水中，并随时间累积在各处，从而成为自然环境中最持久的污染物之一。天然产品可以代替合成产品使用，因为大自然就能够生产成千上万种颜色。吃剩的水果和蔬菜可以轻易地成为彩色染料的来源。这确实需要使用不同的工业生产模式。但是鉴于全新的、因地制宜的经济，现有的物资流如果能产生更多价值，那么这里就有机会建立一种可持续的、令人赏心悦目的经济活动。

动手能力
Capacity to Implement

让我们来玩一玩花园、公园和森林里发现的植物汁液吧！确保在摘取植物前获得了允许。还可以在果蔬店找找可能产生彩色颜料的植物。比如橙色的木瓜，红色的西红柿和西瓜以及粉红色的番石榴。对于蓝色，请试试黑莓和蓝莓，不过它们可能会产生难以去除的甜菜根状的污渍。实验一下，尽情享受，看看哪种植物原料能够留下最难去除的印记。与朋友和家人分享你的发现。

教师与家长指南

故事灵感来自
This Fable Is Inspired by

玛丽亚·乔奥·梅洛
Maria João Melo

玛丽亚·乔奥·梅洛于1995年在葡萄牙新里斯本大学获得物理化学博士学位。随后，她于1996年至1998年在意大利完成了博士后研究，研究方向是应用科技修复与保存艺术品。玛丽亚在新里斯本大学科技学院创立了保护与修复系。她将化学应用于对文化遗产的保护，包括保护艺术品中的自然色彩，历史染料和聚合物。她还是光降解研究的专家。

图书在版编目(CIP)数据

冈特生态童书.第八辑:全36册:汉英对照/
(比)冈特·鲍利著;(哥伦)凯瑟琳娜·巴赫绘;
何家振等译.—上海:上海远东出版社,2021
ISBN 978-7-5476-1773-1

Ⅰ.①冈… Ⅱ.①冈… ②凯… ③何… Ⅲ.①生态环境–环境保护–儿童读物—汉、英 Ⅳ.①X171.1-49

中国版本图书馆CIP数据核字(2021)第249940号

策　　划　张　蓉
责任编辑　程云琦
封面设计　魏　来 李　廉

冈特生态童书
耐久的蓝色
[比] 冈特·鲍利　著
[哥伦] 凯瑟琳娜·巴赫　绘
朱溪　译

记得要和身边的小朋友分享环保知识哦!
八喜冰淇淋祝你成为环保小使者!

Work 281

哥伦布之蛋
The Egg of Columbus

Gunter Pauli

[比]冈特·鲍利 著
[哥伦]凯瑟琳娜·巴赫 绘
朱溪 译

上海远东出版社

丛书编委会

主　任：贾　峰
副主任：何家振　闫世东　林　玉
委　员：李原原　祝真旭　牛玲娟　梁雅丽　任泽林
　　　　王　岢　陈　卫　郑循如　吴建民　彭　勇
　　　　王梦雨　戴　虹　翟致信　靳增江　孟　蝶

特别感谢以下热心人士对童书工作的支持：

匡志强　宋小华　解　东　厉　云　李　婧　陈　果
刘　丹　熊彩虹　罗淑怡　旷　婉　杨　荣　刘学振
何圣霖　廖清州　谭燕宁　韦小宏　李　杰　欧　亮
陈强林　王　征　张林霞　寿颖慧　罗　佳　傅　俊
胡海朋　白永喆　冯家宝

目录

哥伦布之蛋	4
你知道吗？	22
想一想	26
自己动手！	27
学科知识	28
情感智慧	29
艺术	29
思维拓展	30
动手能力	30
故事灵感来自	31

Contents

The Egg of Columbus	4
Did You Know?	22
Think about It	26
Do It Yourself!	27
Academic Knowledge	28
Emotional Intelligence	29
The Arts	29
Systems: Making the Connections	30
Capacity to Implement	30
This Fable Is Inspired by	31

一位父亲想考验自己的儿子，于是从鸡笼里取出了一枚刚下的鸡蛋，问道：

"你有足够的耐心和稳固的双手让这枚鸡蛋竖立起来吗？"

"爸爸，这根本不可能！鸡蛋是弯弯的，要想让它在平滑的表面上竖立起来是不可能的。"

A father wishes to challenge his son, so he gives him a freshly laid egg from the chicken coop, and asks,
"Do you have the patience, and the steady hand, needed to balance this egg upright?"
"Dad, that's impossible! An egg is curved, and it's impossible to put it down standing straight up on a flat surface."

一位父亲想考验自己的儿子……

A father wishes to challenge his son ...

……你只要试一试

... you should simply try

"啊,你只要试一试,第一次失败之后,再试一次。反复尝试直到成功。要有创意一些,跳出思维定势。"

"努力尝试,再发散性思考。说起来容易做起来难啊,爸爸。我在想到底该不该考虑重力的影响……"

"Ah, you should simply try, and when it fails the first time around, then you should try again. Try until you succeed. Get creative, think out of the box."

"Trying harder and thinking out of the box. Easier said than done, Dad. I wonder if I shouldn't use the influence of gravity…"

"重力？真的吗？你的呼吸造成的干扰可比地球或者月亮的磁力都要大哦。"

"也许等月圆之时这就能成功呢，或者等中国人过春节的时候……"

"听着，甚至你的心跳都会比天空中太阳或者月亮的位置对鸡蛋的影响更大。"

"Gravity? Really? The disturbance created by your breathing will be greater than any magnetic force from the Earth, or the Moon."

"Maybe this will work when it is full moon, or when the Chinese celebrate spring…"

"Look, even your heartbeat will have a bigger influence than the position of the Sun or Moon in the sky."

也许等月圆之时这就能成功呢。

Maybe this will work when it is full moon.

……中国人设法把鸡蛋平衡竖立起来……

... the Chinese do manage to balance their eggs ...

"可是中国人每年都能几次把鸡蛋平衡竖立起来呀,那我们肯定也可以!"

"我希望你能控制鸡蛋。你要做的是找到鸡蛋底部的小疙瘩或者小'酒窝',你就会成为高手。"

"可是爸爸,我没办法装作自己是鸡蛋平衡艺术的高手呀。"

"But the Chinese do manage to balance their eggs on a few occasions a year, and so should we!"
"I would like you to be able to control the egg. All you need, is to find a few small bumps or dimples on the bottom of the egg, and you will be a master."
"But I never pretended to be a master of the art of egg balancing, Dad."

"差不多是时候去发现你的能力了,要敢于挑战从前不敢想象的事。"

"那我可以挑战你吗,爸爸?"

"你当然可以啦!"

"那你可以把两个鸡蛋系在一起吗?"

"It is about time you discovered what you can do, and be challenged to do things you never imagined possible before."
"May I challenge you as well then, Dad?"
"Of course you may!"
"Can you tie two eggs together in a knot?"

那你可以把两个鸡蛋系在一起吗?

Can you tie two eggs together in a knot?

那你展示给我看看要怎么做吧!

Then show me how you do it!

"噢，那倒简单！"

"那你展示给我看看要怎么做吧！"

"阐述事物的原理并不是学习的最佳方式。"

"那你想让我怎么来学习呢，爸爸？我都不知道从哪儿开始！我想最好的老师肯定能给出最佳的回答。"

"Oh, that's easy!"
"Then show me how you do it!"
"Demonstrating how something works is not the best way to learn."
"So how do you want me to learn then, Dad? I don't even know where to start! And I believe that best teacher can give the best answers."

"如果你想做的是从前已经做到的事情,那么你可以依照范例来学习。但是,如果你想改变世界,让所有人的生活更美好,就需要跳出思维定势,提出从前不曾想出的观点。"

"老实说,我不认为我是能改变世界的人。"

"If you want to do what has been done before, then learn by example. But, if you want to transform the world and make life better for all, then you need to think out of the box and come up with ideas never imagined before."

"I honestly don't think I can be the person to transform the world."

如果你想改变世界……

if you want to transform the world ...

那还在等什么呢,爸爸?

So what are you waiting for, Dad?

"你刚刚问了一个问题,却不知道问题的答案。所以你会充满好奇地去寻找答案。然而,我也问了你一个我自己都没有答案的问题。"

"那还在等什么呢,爸爸?请教教我怎样才能想出这些问题的答案吧。"

"You asked a question, and you did not know the answer. So you are curious to find out. However, I asked you a question to which I do not have the answer."

"So what are you waiting for, Dad? Please teach me how to go about figuring out the answers to these questions."

"为充满未知、唯有变化是永恒的生活做好准备，不能只去复制别人已做过的，而应该想象别人没有做到过的。"

"所以，我可以问一些你从来没遇到过的问题吗？"

"当然！一定要确保你爸爸对答案毫无头绪哦！"

……这仅仅是开始！……

"Getting ready for life with all its uncertainties, where change is the only constant, is not about copying what others have done – it is about imagining what no one has done."

"So, may I ask the questions that have not yet been asked?"

"Just make sure your dad has no clue of the answers!"

... AND IT HAS ONLY JUST BEGUN!...

……这仅仅是开始！……

... AND IT HAS ONLY JUST BEGUN! ...

Did You Know? 你知道吗？

The expression "Columbus's egg" refers to a brilliant idea or discovery that seems simple or easy after the fact, one that had seemed an impossible task until the solution had been demonstrated.

"哥伦布之蛋"一词指的是一个绝妙的想法或发现，在解决方案被显示出来之前，它似乎是一项不可能完成的任务，而事后看起来却很简单或显而易见。

Christopher Columbus, when told finding a new trade route was no great accomplishment, challenged his critics to making an egg stand. When his challengers gave up, he tapped the egg on the table to flatten its tip.

克里斯托弗·哥伦布在被告知找到一条新的贸易路线并非什么了不起的成就时，向批评者发起了挑战，要求对方把鸡蛋竖起来。当对方放弃时，他轻敲桌子上的鸡蛋压平了它的尖端。

The story of the Columbus's egg is often alluded to when discussing creativity. It shows that anything can be done by anyone with the right mind-set and skills-set, however, not everyone knows how to do it.

在讨论创造力时，经常会提到哥伦布之蛋的故事。这表明，有"志"与"技"者事竟成，但是，并非所有人都知道如何去做。

Egg balancing is a traditional Chinese practice. Eggs typically have some imperfections, so the vast majority can be balanced on their broad ends with minimal effort.

竖鸡蛋使之平衡是中国的传统习俗。鸡蛋通常有一些不平整，绝大多数鸡蛋可以非常轻巧地在宽的一端竖立平衡。

Sprinkling salt onto a surface making the flat surface uneven, allows the egg to balance on a few crystals. Once the egg is balanced, the remaining salt can be gently blown away.

将盐撒在平面上会使表面不平整，这样鸡蛋就能在少数晶体上保持平衡。一旦鸡蛋保持平衡，再把其余的盐轻轻地吹走。

At the 1893 World Exposition in Chicago, Nikolas Tesla displayed his version of Columbus's Egg by using a rotating magnetic field that drove an alternating current induction engine to spin a copper egg on end.

在 1893 年芝加哥世博会上，尼古拉斯·特斯拉展示了他的哥伦布之蛋——使用旋转的磁场驱动交流电感应发动机持续旋转铜蛋。

In 1507, the German cartographer Martin Waldseemüller drew a new map of the world, based on the travel stories of the Italian explorer Amerigo Vespucci, who had travelled to the New World in 1499 and again in 1502.

1507 年，德国制图师马丁·瓦尔德西米勒根据意大利探险家阿美利哥·韦斯普奇的旅行故事绘制了一张新的世界地图，后者曾于 1499 年和 1502 年两次前往新大陆。

Christopher Columbus did not publish anything about his voyages. Vespucci's travel accounts, published in 1502 and 1504, were widely read, announcing the discovery of a continent unknown to Europeans of the time.

哥伦布没有发表过任何有关航行的文章。韦斯普奇于 1502 年和 1504 年出版了广为人知的旅行记述，并宣称发现了一块当时不为欧洲人所知的大陆。

If you do not talk about your invention, then someone else gets the credit?

如果不和别人谈论你的发明,那么别人会不会把荣誉抢走?

Do you have the patience to place an egg upright?

你有立起鸡蛋的耐心吗?

Do you have the courage to challenge your father? Will your father like it?

你敢挑战你的父亲吗?你的父亲会有什么反应?

Do you like questions that cannot be answered, or posing questions that have never been asked?

你喜欢未解之谜吗?你喜欢提出难以回答的问题吗?

Try it for yourself. See if you can place an egg upright. It is not that difficult, and with a bit of patience and perhaps by choosing an egg that has a few uneven spots on its surface, not impossible. Use a timer to see how long it takes you, and also record the number of attempts you have made before you succeeded. This dispels the story that you have to crack the thin end of the egg to make it stand on end, as Columbus did. Take care not to make a mess, so work on a tray, or use a boiled egg. When you succeed you will impress friends and family.

试试看能不能把鸡蛋立起来。这并不难，需要一点耐心。你可以选一枚表面不太平整的鸡蛋，把它立起来并非不可能。使用计时器，看你需要多长时间，并记录尝试了多少次。不必像哥伦布那样，敲破鸡蛋的一端才能竖起鸡蛋。注意不要弄得一团糟，最好在托盘上尝试，或者用熟鸡蛋。一旦你取得成功，会给朋友和家人留下深刻印象。

TEACHER AND PARENT GUIDE

学科知识
Academic Knowledge

生物学	在生物学中,"卵"指的是卵细胞;细胞质包括蛋黄营养物质,蛋黄被蛋清等物质包裹;鸡蛋为卵形,梨形蛋常见于海雀和海鸦等海鸟中,海龟蛋为圆形,鸵鸟蛋为椭圆形;鸡蛋由蛋白(白色)、蛋黄(黄色)、连接蛋黄的卵带、作为鸟类肢体起源的胚胎盘、气室和壳膜组成。
化　学	钙元素对子宫内蛋壳的形成起了作用。
物　理	鸡蛋的质心包含在三个接触点所构成的三角形内,这是平衡任何物体的条件;"特斯拉蛋"与带有四个线圈的环形铁芯定子一同运转。
工程学	蛋形消化室在水密性和气密性方面性能优越;汽车的流线形,前端相当于鸡蛋的尖端,后端像鸡蛋的底部;蛋形扬声器抑制其内部的振荡和回声,使蛋形设备产生原本的声音。
经济学	创新思维,"跳出思维框架"的思考是经济进步的关键。
伦理学	关键并不在于知道所有问题的答案,而是明白该问哪个问题,能够让至今尚未解决的问题得以解决。
历　史	列夫·托尔斯泰在他的《战争与和平》一书中提到了哥伦布之蛋;意大利天文学家卡西尼(1625—1712)应路易十四之邀到巴黎,成为天文台的首任台长,描绘了行星的椭圆轨道;捷克共和国的剧作家、前总统瓦克拉夫·哈维尔(Vaclav Havel)表示,学习就是看到两种你从未想过会有相关性的现象之间的联系。
地　理	寻求欧洲到印度的新航线;找到陆地并不自动等同于发现新大陆。
数　学	鸡蛋的形状近似于长球体的"长"半边与粗略的球形椭圆体的"短"半边相连;椭圆形是指像鸡蛋或椭圆的形状;笛卡尔和卡西尼的研究方法将行星轨道描述为椭圆形曲线,后来被牛顿修正为椭圆轨道。
生活方式	生活方式的改变决定了成功妊娠的可能性;影响因素包括家庭、营养、体重、运动、心理压力、环境和职业接触、吸烟、违禁药物、酒精和咖啡因的使用。
社会学	鸡蛋是椭圆形的,我们称其为"蛋形",但由于椭圆形可以改变,因此我们对鸡蛋的真实形状仍描述不清。
心理学	最好的解决方案通常是最简单的,而且往往不是最明显的,因为我们会被现有的知识和过往的经验蒙蔽双眼。
系统论	鸡蛋作为生命的象征,却也有精确的数学框架作为进化的证据。

教师与家长指南

情感智慧
Emotional Intelligence

父 亲　　父亲考验儿子，质疑他有没有耐心（精神力量）和稳定的双手（身体力量）来竖起鸡蛋。父亲不接受儿子拒绝挑战的行为。当儿子试图跳出思维定势时，父亲嘲笑儿子说："真的吗？"他坚持认为，心跳和呼吸对竖立鸡蛋的影响比其他许多因素加起来都重要。父亲不准备进行详细而复杂的争论，他只是想让儿子了解如何平衡鸡蛋。当儿子想挑战父亲并征求他的同意时，父亲以要求儿子提供帮助、改变世界为由，应对儿子的挑战。父亲没有正面回答问题，而是描述了怎样才能发现生活中产生改变的途径。

儿 子　　儿子在尝试之前就放弃了。当父亲坚持时，儿子声称说起来容易做起来难。儿子开始寻找需要跳出惯性思维的解决方案。当父亲不理会时，他又提出第二个"古怪"的可能性，虽然也被无视了。儿子以中国人在春节期间立鸡蛋的习俗为例，用事实论证。儿子向父亲明确表示，他并不想成为竖立鸡蛋的专家。他充满信心也十分礼貌地询问是否可以挑战他的父亲。当父亲声称把鸡蛋系在一起很容易，但拒绝向儿子展示方法时，儿子才明白只有找到最好的老师，自己才会想去学习新事物。儿子感受到父亲传承给他的责任的重要性。但是最终，他付诸提出从未提过的问题。

艺术
The Arts

我们将利用数学来绘制鸡蛋的确切形状。只要稍微改变椭圆的方程式就可以实现。将 y 或 y^2 乘以合适的项 $T(x)$，这样 y 轴越往右侧，则 y 值越大，越往左侧则 y 值越小。试试看，并了解数学怎样成为你的艺术伙伴。

TEACHER AND PARENT GUIDE

思维拓展
Systems: Making the Connections

蛋的形状可以分为四组：卵形、梨形、圆形和椭圆形。鸡蛋是卵形的。梨形蛋在海鸟中很常见，其尖端和底部之间的差异大于卵形。海龟产圆形蛋，而鸵鸟产椭圆形蛋。鸡在平坦的地方产卵，所以不用担心鸡蛋滚开或破碎，因此鸡蛋是卵形的。海龟会在沙滩上产卵，用沙子埋起来，鸵鸟会在草原上产卵，这两个地方都比较平坦，可以免于蛋滚开的危机。海鸟比如海雀和海鸦，在狭窄的岩壁上产卵。这些岩架通常是倾斜的，因此它们下梨形蛋也很合乎道理。海鸥和塘鹅也是海鸟，但由于它们的蛋不是梨形的，会有滚动的危险，因此它们会筑巢，将蛋固定在稳固的空间里。观察到这一点，我们就明白蛋有不同形状的原因了！大自然要形成这些形状很容易，而人类也认识到了其中的高效，但这样的认知从未成为主流。有蛋形污泥消化室，在水密性和气密性方面优于扁平设计。有蛋形排水沟，流水的断面呈蛋形，其排水性据说较好。有蛋形车身，其流线形已在汽车中发展起来，前端相当于鸡蛋的尖端，后端则像鸡蛋的底部。有蛋形扬声器，可以抑制扬声器内部发生的振荡和回声。所以蛋形被用来追求原声再现，而在圆顶形天花板下播放音乐也成了流行。一切可以理解为蛋的形状不仅与生物学有关，而且对我们的生活也有着重大的影响。

动手能力
Capacity to Implement

我们必须学会构思没有任何答案的问题。最好的问题是那些连我们的老师、大师和祖父都没有答案的问题。然而关键并非知晓答案，而是学会弄清楚引导我们找出答案的问题是什么。为什么事物会这样，我们怎么从观察中学习，而不是单纯地得到答案？当你研究背景情况时，思考能力将得到提升，就像在建立联系的过程中，因为那才是真正的学习。你的任务是描述为什么鸟蛋会有四种不同的形状。与朋友和家人分享这些信息，你会发现在思考中你的语言会更加精练。一旦发现了关联，你就会得到最符合当下的解决方案。

教师与家长指南

故事灵感来自
This Fable Is Inspired by

西山丰
Yutaka Nishiyama

西山丰于1967年至1971年在京都大学理学院学习数学。1971年至1985年，他在IBM公司担任系统工程师，开始了自己的职业生涯。后来他转入大阪经济大学，做了10年信息数学讲师。1995年，他被任命为正教授。2005年至2006年，他在剑桥大学做访问学者。他继续教授数学在日常生活中的应用。他最新的书讨论了"五之谜"，解释了为什么很多花卉都有五朵花瓣。他还写了大量关于鸡蛋形状的文章。他被称为"回旋镖教授"，因为他详细解释了为什么回旋镖会再次返回，以及为什么数独在益智游戏中取得了新的成功。

图书在版编目(CIP)数据

冈特生态童书.第八辑:全36册:汉英对照/
(比)冈特·鲍利著;(哥伦)凯瑟琳娜·巴赫绘;
何家振等译.—上海:上海远东出版社,2021
ISBN 978-7-5476-1773-1

Ⅰ.①冈… Ⅱ.①冈…②凯…③何… Ⅲ.①生态环境—环境保护—儿童读物—汉、英 Ⅳ.①X171.1-49

中国版本图书馆CIP数据核字(2021)第249940号

策　　划　张　蓉
责任编辑　程云琦
封面设计　魏　来　李　廉

冈特生态童书
哥伦布之蛋
[比]冈特·鲍利　著
[哥伦]凯瑟琳娜·巴赫　绘
朱溪　译

记得要和身边的小朋友分享环保知识哦!
八喜冰淇淋祝你成为环保小使者!

Work 282

搜寻酵母

Hunting Yeast

Gunter Pauli

[比] 冈特·鲍利 著
[哥伦] 凯瑟琳娜·巴赫 绘
廖铭诗 译

上海远东出版社

丛书编委会

主　任：贾　峰
副主任：何家振　闫世东　林　玉
委　员：李原原　祝真旭　牛玲娟　梁雅丽　任泽林
　　　　王　岢　陈　卫　郑循如　吴建民　彭　勇
　　　　王梦雨　戴　虹　翟致信　靳增江　孟　蝶

特别感谢以下热心人士对童书工作的支持：

匡志强　宋小华　解　东　厉　云　李　婧　陈　果
刘　丹　熊彩虹　罗淑怡　旷　婉　杨　荣　刘学振
何圣霖　廖清州　谭燕宁　韦小宏　李　杰　欧　亮
陈强林　王　征　张林霞　寿颖慧　罗　佳　傅　俊
胡海朋　白永喆　冯家宝

目录

搜寻酵母	4
你知道吗？	22
想一想	26
自己动手！	27
学科知识	28
情感智慧	29
艺术	29
思维拓展	30
动手能力	30
故事灵感来自	31

Contents

Hunting Yeast	4
Did You Know?	22
Think about It	26
Do It Yourself!	27
Academic Knowledge	28
Emotional Intelligence	29
The Arts	29
Systems: Making the Connections	30
Capacity to Implement	30
This Fable Is Inspired by	31

一只狗被主人用皮带牵着穿过森林，这时他们邂逅了一只狐狸。狗因为被束缚着，感到很难为情，并且项圈太紧了，以至于当狐狸问话时，他回答得很费劲。

"那么，你和你的主人今天在搜寻什么？"

A dog is walking through the forest, on a leash held by his owner, when they encounter a fox. The dog is embarrassed about being restrained, and with its collar so tight can hardly reply when the fox asks,

"So, what are you and your owner hunting today?"

……一只狗正穿过森林……

... dog is walking through the forest ...

……下次我们去猎狐……

... next time we go fox hunting ...

"不要来打扰我,否则下次我们去猎狐的时候我会先去找你的!"

"你很清楚,猎狐的日子已经过去了。我们现在有很大的机会逃脱,并且你知道我们有多快多聪明……"

"Don't you mess with me, or next time we go fox hunting I will go for you first!"

"You know very well that the days of fox hunting are over now. We now get a fair chance of escape, and you know how fast and how smart we are…"

"我知道人们对你们不公平,放了一大群狗去追你们中的一个。每只狐狸都寡不敌众,跑不动了,然后被捕杀。这是一种残忍的打猎方式,对此我深表歉意。"

"我接受你的道歉。但这一次你似乎在追寻一种与狐狸截然不同的东西。你的主人拿着罐头和平底锅,甚至还有一些柠檬汁!"

"I know that people have been unfair to you, releasing large packs of dogs to chase one of you down. Each fox was outnumbered, ran down and killed. It was a cruel way to hunt, and I would like to apologise for that."

"Apology accepted. But it seems that this time you are after something very different to a fox. Your owner is carrying cans and pans, and even some lemon juice!"

你的主人拿着罐头……

Your owner is carrying cans...

……让面包发酵的微小真菌吗?

... little fungi that make bread rise?

"你没看见他在搜寻酵母吗?"

"搜寻什么?不会是那些能让面包发酵的微小真菌吗?"

"就是这样!"狗叫道。

"Can't you see he is hunting yeast?"
"Hunting what? Surely not those tiny little fungi that make bread rise?"
"That is exactly it!" Dog exclaims.

"我已经好多年没有在森林里遇到寻找酵母的人了。你必须向我解释一下:一个人是如何把那些微小的酵母细胞弄到手的呢?"

"好吧,像今天这样的秋天,他会找一个潮湿的地方,长满了熟透果实的树。他宁愿周围没有什么动物。"

"It has been years since I've come across anyone looking for yeast in the forest. You need to explain this to me: How does a person get his hands around those minute little yeast cells?"

"Well, on an autumn day like today, he will look for a moist spot, one with trees full of overripe fruits. He prefers that there are few animals around."

……像今天这样的秋天……

… on an autumn day like today …

……当他收集酵母时……

...when he collects the yeast...

"你是说这个人在森林里打猎,想找一个没有动物的地方?他不是想猎杀我们,而是想捕捉那些微小到连肉眼都看不见的酵母细胞?"

"是的,当他收集酵母时,他也会得到一些尖尖的霉菌。"

"You mean this man, hunting in a forest, wants to find a spot without any animals? And instead of hunting us, he wants to catch yeast cells so tiny you cannot even see them with the naked eye?"

"Yes, and when he collects the yeast, he will also get some spiky moulds."

"他知道如何区分霉菌和酵母吗?"

"我听他说过,霉菌是绿色或蓝色的,长着毛。而酵母更白、更黏滑。"

狐狸问道:"1千克酵母中有成百万的小真菌,这是真的吗?"

"And does he know how to tell the difference between moulds and yeast?"
"I heard him say that moulds are green or blue, and grow hair. And that yeast is rather white and slimy."
"It is true that there are millions and millions of little fungi in a kilogram of yeast?" Fox asks.

……区分霉菌和酵母……

… difference between moulds and yeast …

……1克就有200亿个……

... twenty billion in just one gram ...

"实际上仅1克就有200亿个！10亿是1后面有12个零。"

"他想用几十亿个真菌做什么？烤面包？"

"不，他打算用来酿造自己的特色啤酒。"

"Actually there are twenty billion in just one gram! And a billion is a one with twelve zeros."

"What does he want to do with billions of fungi? Bake bread?"

"No, he is planning to brew his own special beer with it."

"我知道了,在凉爽的秋日,用在黑暗森林里搜寻到的野生酵母酿造啤酒。让我在一场公平的狩猎中尽情奔跑吧;这会使我快乐。"

……这仅仅是开始!……

"Brewing beer, with wild yeast hunted down in a dark forest, on a cool autumn day, I see. Let me just have a good run in a fair hunt; that is what gives me joy."

... AND IT HAS ONLY JUST BEGUN!...

……这仅仅是开始！……

... AND IT HAS ONLY JUST BEGUN! ...

Fox attacks on people are extremely rare since foxes have become tamer in an urbanised world. In fact, the chance of being bitten by another human is 60 times more likely than being bitten by a fox.

狐狸袭击人类是极为罕见的，在城市化的世界里，狐狸变得更加驯服了。事实上，被另一个人咬的概率是被狐狸咬的60倍。

Yeast produces around 80% of a beer's flavour. The yeast determines whether beer has notes of banana, freshly cut grass or something more like grapefruit. Different yeasts create different kinds of beer.

酵母能生产出啤酒80%左右的风味。酵母决定了啤酒是否有香蕉、新鲜割草或更像葡萄柚的味道。不同的酵母生产出不同种类的啤酒。

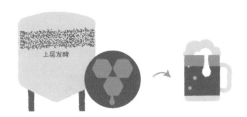

For thousands of years, top-fermenting yeasts were responsible for converting malt sugar into alcohol. About 500 years ago, uncultivated yeast arrived. This formed the bottom-fermenting lager yeast.

几千年来，上层发酵酵母负责将麦芽糖转化为酒精。大约 500 年前，未经培养的酵母出现了。这就形成了底部发酵的拉格酵母。

Wild yeast is the first domesticated living organism. It was used in Egypt to make wine 9,000, and bread 4,000 years ago. All wine yeast strains around the world are the same, and have followed human migration.

野生酵母是第一个被驯化的活微生物。在埃及，早在 9 000 年前它就被用来酿造葡萄酒，在 4 000 年前它被用来制作面包。全世界所有的葡萄酒酵母菌株都是相同的，并且它们跟随着人类迁徙。

In the modern era yeast is used to turn sugar into ethanol, as a substitute for diesel fuel. The process is nearly identical to the one used to make beer and wine.

在现代，酵母被用来将糖转化为乙醇，作为柴油的替代品。这个过程几乎与酿造啤酒和葡萄酒的过程一样。

Wild yeast helps process coffee and cacao in popular drinks and in chocolate. Yeast stops the germination process of the fruit and helps convert the taste into one we enjoy.

野生酵母有助于在流行饮料和巧克力中加工咖啡和可可。酵母阻止了水果的发芽过程，并将味道变得令人沉醉。

Wet cacao beans are piled up, covered with leaves and left for less than a week. The yeast from the surrounding environment, such as the soil and the trees, will grow in a natural way, without any chemical controls.

把湿可可豆堆起来，盖上叶子，放置不到一周的时间。来自周围环境例如土壤和树木的酵母，将以天然的方式生长，没有任何化学物质的控制。

The wild yeasts associated with coffee and cacao are more diverse than those of wine. Perhaps we derive more joy out of yeast than keeping a cat or a dog.

与咖啡和可可相关的野生酵母比与葡萄酒相关的野生酵母更加多样化。也许我们从酵母中获得的快乐更甚于养猫或养狗。

Hunting yeast you cannot even see?   你会搜寻你甚至看不见的酵母吗?

Twenty billion yeast cells in one gram?   在1克中有200亿个酵母细胞吗?

Is a fair hunt to be allowed?   公平的狩猎是被允许的吗?

Is it wise to embarrass someone who has lost?   让输了的人难堪是明智的吗?

Ask around and find out if there are any of your friends or family members who are in favour of hunting animals with dogs. Or do they enjoy watching a bullfight? Is hunting with dogs or bullfighting a sport? Enter into a debate on the subject, and make up your own mind. Now formulate your opinion on the matter. Make sure that you can argue the point and take a position in favour and against these activities.

四处打听一下，看看有没有你的朋友或家人赞成带狗狩猎动物。或者，他们喜欢看斗牛吗？带狗狩猎或者斗牛是一项运动吗？针对这个问题进行辩论，然后明确自己的观点。现在阐述你对这件事的看法。确保你能论证这一点，并坚持立场赞成或反对这些活动。

TEACHER AND PARENT GUIDE

学科知识
Academic Knowledge

生物学	酵母是一种单细胞微生物，估计有800种工业用酵母；酿酒酵母用于酿制葡萄酒；可可树用来发酵可可；咖啡和可可发酵是由特定地理区域的当地酵母群体进行的，并且似乎是独立产生的；当地酵母菌群给该地区的巧克力和咖啡带来了独特的风味；富含酵母的食物可以提供蛋白质和B族维生素；酵母中还含有线粒体，这是一种负责生产能量的细胞器。
化 学	一系列的酵母菌、乳酸菌和醋酸菌消化含果胶的果肉，并引发将风味和颜色赋予可可豆的生物化学变化；酵母通过发酵将糖和淀粉转化为二氧化碳和酒精；一个葡萄糖分子分解成两个酒精分子和两个二氧化碳分子；一个酵母家族可以产生600种不同的影响风味和香味的化合物。
物 理	酵母需要温暖和水分才能茁壮成长；二氧化碳使生面团鼓胀。
工程学	用酵母酿造啤酒可能是人类的第一个工程项目。
经济学	据估计，地球上只有5%的酵母菌被发现，在这些被发现的酵母菌中，有七八十种具有很高的生物技术价值，因此有足够的空间将野生酵母菌转化为当地经济的引擎。
伦理学	一只狐狸被一群狗猎杀并撕碎，这样的猎狐是不公平和残忍的；捕杀哪些动物以及采取何种方式猎杀，应得到社会的接受。
历 史	用嗅觉猎犬来追踪猎物可以追溯到公元前2000–3000年，起源于亚述和古埃及；第一次使用专门饲养用于猎狐的狗群是在17世纪末的英格兰。
地 理	猎狐的国家包括美国、澳大利亚、爱尔兰和加拿大；用于发酵可可的野生酵母起源于哥伦比亚和委内瑞拉的亚马逊和奥里诺科盆地，并且被运送到世界各地。
数 学	200亿个酵母细胞重量为1克。
生活方式	没有酵母就没有现代饮食：酸奶、啤酒、葡萄酒、咖啡、可可、面包；白色念珠菌是一种能引起人类口腔和胃肠道感染的酵母。
社会学	在16世纪，农民开始用狗猎狐以控制有害动物；猎狐，猎杀动物，这种与传统和上层社会强烈联系的追求造成了长期而深刻的争议。
心理学	动物保护与把动物开发为食物相悖；情感使我们按照某一种行动方针前行，例如寻求动物福利；悲伤是一种因动物生存受到威胁而表达的情感；由于宗教、神秘主义和哲学的原因，人类对动物感到同情、钦佩和尊重。
系统论	关于酵母与昆虫、无脊椎动物和深海鱼类，生活在热带森林或极端环境中的生物的关系，还有许多有待研究；酵母是单细胞微生物，与霉菌和蘑菇一起被归类为真菌王国的成员。

教师与家长指南

情感智慧
Emotional Intelligence

狐狸

狐狸有勇气接近被皮带牵住的猎狗，并且与之交谈。这种交流颇具优势。野蛮的狩猎方式已经被禁止。当狗道歉时，狐狸接受了道歉，表现出对如今实行的新奇打猎方式的兴趣和好奇，以此来转移对痛苦过往的注意力。比起捕猎，人类竟然更乐意在没有动物的情形下狩猎，对此狐狸很怀疑。狐狸揶揄这个猎人怎样才能将那些极小的细胞弄到手，还嘲笑猎人在追求肉眼看不见的东西。狐狸质疑猎人的智力，问猎人知不知道酵母和霉菌的区别。狐狸很享受这种崭新的优越感。

狗

狗很难为情，因为他现在被拴在皮带上，不能像以前那样猎狐。他有同情心，为过去不公平的狩猎道歉。当被问及新的狩猎时，狗又恢复了姿态，并且和狐狸来回争论。狗大方地解释了猎人是如何搜寻酵母。作为一个搜寻酵母的新手，狗为狐狸提供了许多让他耳目一新的知识。狗知道猎人的喜好，并与狐狸分享这些信息，告诉狐狸酵母的用处，甚至教他10亿是多少。他与狐狸交谈，第一次认为狐狸是和他平等的，而非猎物，想要弥补百年来的不幸。

艺术
The Arts

在这个故事里，传统的角色被逆转了。狐狸不是被一群一百多只狗追捕，反而取笑狗主人在森林里搜寻酵母。请你用漫画的方式，艺术地表达这个故事。不是每个人都擅长绘画，因此组建一个团队来构思和规划你的漫画，然后把绘画交给其中有艺术才能的人，用有趣的方式来表达这种对现实的反转。

TEACHER AND PARENT GUIDE

思维拓展
Systems: Making the Connections

自从人类把自己组织成一个社会团体，并且把自己确立为猎人以来，生态系统中的平衡一直在改变。放150只狗去扑倒一只狐狸，其实与控制有害动物的需要，或维持狐狸种群的愿望是无关的。很难说这是一项"运动"。随着人们对动物的人道待遇越来越坚定，有趣的是可以看到新的"狩猎"形式正在出现。其中之一就是搜寻酵母。这种看不见的单细胞微生物漂浮在空气中，从熟透的水果中寻找糖分，在近1万年前首次被驯化。我们利用这些微小生命形式为我们提供天然健康的食品和饮料，但很少有人意识到，迄今为止，自然界中野生酵母有所研究的还不到5%。而在这一小部分中，只有七八十种被用于工业化。更进一步的探索和新经济活动有着难以置信的发展空间，这些活动将利用还未经利用的生物多样性。寻找酵母是穿过森林的一次简单步行，只需要花费你的时间。有成千上万种不同的酵母有待发现，而且酵母菌有不同的表现水平和不同的生命力。有些酵母被称为母发酵剂，已经生产了几百年，让它们的"猎人"和驯化者得以生存。因此，不仅找到酵母是重要的，还要找到一种有实力提供服务很长一段时间的酵母。这就提供了一个通往新文化的窗口，一个有尊严的文化，谦逊并且尊重自然。

动手能力
Capacity to Implement

我们来培养一些野生酵母吧。用来烘焙的面粉里已经有野生酵母了。把面粉和水混合，并在温暖的地方放几天。两天后，酵头中开始形成气泡，表明野生酵母正在变得活跃和繁殖。为了保持酵母生长，在接下来的几天里，加入新鲜面粉和水，直到起泡。一旦到达了丰富多泡的阶段，酵头就可以使用了。这比大家想的容易多了！

教师与家长指南

故事灵感来自
This Fable Is Inspired by

珍妮弗·莫利内特
Jennifer Molinet

珍妮弗·莫利内特最初研究的是生物技术与工程科学，2012年在智利圣地亚哥大学获得本科学位。2013年她获得微生物学工程学位，并于2018年在这所大学获得微生物学博士学位。她的博士后研究对象是野生酵母。她强调需要通过收获野生菌株来探索新的遗传资源，这些野生菌株具有比想象中更高的遗传和表型多样性，表明生产新的发酵饮料是可行的。

图书在版编目（CIP）数据

冈特生态童书.第八辑：全36册：汉英对照/
（比）冈特·鲍利著；（哥伦）凯瑟琳娜·巴赫绘；
何家振等译.—上海：上海远东出版社，2021
ISBN 978-7-5476-1773-1

Ⅰ.①冈… Ⅱ.①冈…②凯…③何… Ⅲ.①生态环境-环境保护-儿童读物—汉、英 Ⅳ.①X171.1-49

中国版本图书馆CIP数据核字（2021）第249940号

策　　划　张　蓉
责任编辑　程云琦
封面设计　魏　来　李　廉

冈特生态童书
搜寻酵母
［比］冈特·鲍利　著
［哥伦］凯瑟琳娜·巴赫　绘
廖铭诗　译

记得要和身边的小朋友分享环保知识哦！
八喜冰淇淋祝你成为环保小使者！

Work 283

更高更强

Growing Strong and Tall

Gunter Pauli

［比］冈特·鲍利 著
［哥伦］凯瑟琳娜·巴赫 绘
周徽音 译

上海远东出版社

丛书编委会

主　任：贾　峰

副主任：何家振　闫世东　林　玉

委　员：李原原　祝真旭　牛玲娟　梁雅丽　任泽林
　　　　王　岢　陈　卫　郑循如　吴建民　彭　勇
　　　　王梦雨　戴　虹　翟致信　靳增江　孟　蝶

特别感谢以下热心人士对童书工作的支持：

匡志强　宋小华　解　东　厉　云　李　婧　陈　果
刘　丹　熊彩虹　罗淑怡　旷　婉　杨　荣　刘学振
何圣霖　廖清州　谭燕宁　韦小宏　李　杰　欧　亮
陈强林　王　征　张林霞　寿颖慧　罗　佳　傅　俊
胡海朋　白永喆　冯家宝

目录

更高更强	4
你知道吗?	22
想一想	26
自己动手!	27
学科知识	28
情感智慧	29
艺术	29
思维拓展	30
动手能力	30
故事灵感来自	31

Contents

Growing Strong and Tall	4
Did You Know?	22
Think about It	26
Do It Yourself!	27
Academic Knowledge	28
Emotional Intelligence	29
The Arts	29
Systems: Making the Connections	30
Capacity to Implement	30
This Fable Is Inspired by	31

松鼠一家安居在一棵杨树上。随着一代又一代松鼠在这里生活又离去，杨树也长得越来越高。

"我还能记得这棵树很小的时候呢。"松鼠爷爷告诉他的孙子。

"它长得有多快呀，爷爷？"

A squirrel family had set up home in a poplar tree. Generations of squirrels come and go, and the tree keeps on growing.

"I remember when this tree was still small," the grandfather of the squirrels tells his grandson.

"And how fast did it grow, Grandpa?"

……安居在一棵杨树上。

... set up home in a poplar tree.

……至少长高了1米吧……

... grew by at least a metre ...

"唔,我估计这棵树每年至少长高了1米吧!"

"那它以后还会长高多少呢?"

"嗯,看看周围的杨树,它可能不会再继续长高了。"

"Well, I would say that this tree grew by at least a metre every year."

"And how much higher will the tree grow?"

"Well, looking at the other poplars around, it seems that it won't grow anymore."

"你的意思是它会死掉吗?"

"一棵树不再长高并不意味着它会死去。相反,它会生长得越来越强壮。"

"但我听别人说,如果经济不增长的话,就会出现危机。"

"You mean it will die?"

"When a tree doesn't grow any taller, it doesn't mean it will die. On the contrary, it will grow stronger."

"But I hear people say that if the economy is not growing anymore, then there is a crisis."

你的意思是它会死掉吗?

You mean it will die?

我想变得更高、更强、更明智！

I want to be taller, stronger and wiser!

"那你告诉我,你是想要长得比你爸爸更高,还是比他更有智慧、更强壮?"

"我想变得更高、更强、更明智!"

"那你觉得当一棵树长得有其他树两倍高的时候,会发生什么呢?"

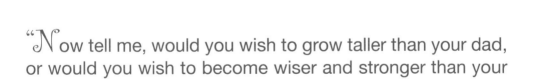

"Now tell me, would you wish to grow taller than your dad, or would you wish to become wiser and stronger than your dad?"

"I want to be taller, stronger and wiser!"

"What do you think will happen when a tree grows double the size of all other trees?"

"它会得到更多的阳光。"小松鼠答道。

"没错,小聪明!但是当风暴来临的时候,哪棵树最受到大风的影响?"

"当然是最高的那棵树啦。"

"那你认为其他的树能够保护那棵高高的树,让它不被风刮倒吗?"

"It will get more sun," the young squirrel replies.

"Correct, you wise boy! But when there is a storm, with a lot of wind, which tree will get most of it?"

"The tallest tree, of course."

"And do you think the other trees are able to protect the tall one, to keep it from blowing over?"

哪棵树最受到大风的影响?

which tree will get most of it?

……它会变得脆弱。

... it will become vulnerable.

"我希望是这样……但我不觉得事实如此。"

"如果一棵树一直不断地长高,它会变得脆弱。但当一棵树长宽、长出壮实的树干和强劲的树根,它就能变得坚韧,去抵挡最糟糕的天气。"

"那么为什么人们想让经济不断增长,而不是让它变得更加强壮,从而抵御风暴呢?"

"I would hope so… but I do not think so."
"If a tree keeps on growing taller and taller, then it will become vulnerable. When a tree grows wider, with a solid trunk and strong roots, it can withstand the worst weather. It becomes resilient."
"So, why do people want to grow their economy bigger and bigger? Why don't they want it to rather be more robust? To withstand the storms?"

"一个经济体就像一棵树。它需要深深地扎根，使用本地已有的资源，回应所有人的基本需求，建立社区。它需要变得更加坚韧，以便回击最糟糕的情况。"

"那为什么大公司把所有东西运来运去，在所有地方生产同样的商品，让所有东西贬值，给人们越来越低的工资甚至偷税漏税呢？到最后，谁会在这种生活方式中存活下来呢？"

"An economy is like a tree. It needs deep roots, using what is locally available, responding to the basic needs of all, and building up the community. It needs to be resilient, for when the worst times come around."

"And why do companies ship everything around the world, producing the same everywhere, making everything cheap, paying people less and less, and even evading taxes. Who will, in the end, survive such a way of life?"

它需要深深地扎根……

It needs deep roots ...

……每个人都贡献……

... everyone contributes to ...

"那些有钱的人——有钱能使鬼推磨。"

"当人们都没有工作的时候,谁会来赚钱呢?"

"在大自然中,每个人都贡献自己力所能及的部分,且只运用本地已有的资源。每个人都有工作!没有人只是为了钱而工作。"

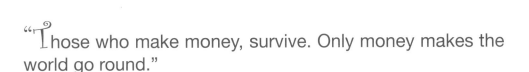

"Those who make money, survive. Only money makes the world go round."

"Who will be earning money when they are all out of jobs?"

"In Nature, everyone contributes to the best of their ability, and we only use what is local. Everyone has a job! No one does it only for the money."

"但是,爷爷,如果人们为了金钱开始冒险,做各种事情,他们的社区有可能变得快乐吗?"

"或许我们该问问那棵刚刚倒下的树!"

……这仅仅是开始!……

"But Grandpa, if people take risks, and do everything because of money, could their community ever be happy?"

"Maybe we should ask that tree that has just fallen over!"

... AND IT HAS ONLY JUST BEGUN!...

……这仅仅是开始！……

... AND IT HAS ONLY JUST BEGUN! ...

A recession is a widespread economic decline that lasts for several months. A depression is a severe economic and financial downturn that lasts for years. "The Great Depression" in the 20th century, lasted for a decade.

经济衰退是一种常见的经济下滑现象，通常仅持续数月。经济萧条是非常严重的经济金融低迷，持续数年。比如，20世纪的"大萧条"持续了整整10年。

A crash of the stock market, a spike in inflation, unemployment, or a series of bank failures can lead to an economic crisis but do not necessarily lead to a recession.

股市大跌、通货膨胀、失业率上升或一系列银行破产等都可能导致经济危机，但不一定会达到经济衰退。

In a recession the Gross Domestic Product (GDP) contracts for at least two quarters. This is accompanied by a drop in four critical economic indicators: income, employment, manufacturing, and retail sales.

经济衰退时，国内生产总值（GDP）会缩水至少 2 个季度。随之而来的则是 4 项关键经济指标的下降：收入、就业、生产和销售。

Stagflation is a combination of stagnant economic growth, high unemployment and high inflation. This exceptional situation is caused when politics disrupt the market causing inflation in a slow economy.

滞胀的特征是经济增长停滞、高失业率和高通货膨胀率同时出现。当政治因素扰乱市场，导致经济体增长缓慢且出现通货膨胀时，就会导致这种特殊现象。

When governments print money and banks create credit, while at the same time governments increase taxes, or banks increase interest rates, it would cause inflation and crisis and prevent companies from producing more.

当货币超发，银行扩张信贷，同时提高税率或者利率，就可能导致通货膨胀和经济危机，阻碍公司的生产和运营。

When assets witness a steep decline in value, businesses and consumers are unable to pay their debts, and financial institutions experience liquidity shortages, then there is a financial crisis.

当资产价值急速下滑时，企业和消费者无力偿付债务，金融机构将出现流动性短缺，这时就会发生金融危机。

Buyers who purchase an asset based solely on the expectation that they can later resell it at a higher price, rather than calculating the income it will generate in the future, are often the cause of a bubble and a crash.

一些买家购买资产时，并没有考量这些资产未来能产生多少收入，而是想当然地认为能够以更高的价格出售。这种投机行为是造成泡沫经济和泡沫破裂的主要原因。

"Herd behaviour" causes prices to spiral up far above the true value of the assets. If, for any reason, the price briefly falls, then investors realise that further gains are not assured, and then the spiral may go into reverse.

羊群效应会导致价格螺旋式上升，远高于资产真实价值。由于某种原因，当价格短暂下跌时，投资者会意识到无法保障更多的收益，价格就有可能反过来螺旋式下跌。

Can a tree safely grow double its normal size?

一棵树能安然无恙地长到正常高度的两倍吗?

Choose one: Would you rather be wiser, taller or stronger?

三选一:你想变得更有智慧、更高还是更强壮?

Where is the ultimate strength of the economy: local or global?

经济体强壮的根基应该在本地社区还是全球?

More profit (sun) always implies more risk (wind)?

得到更多的利润(阳光)一定意味着需要承担更多的风险(大风)吗?

Do It Yourself! 自己动手!

How much risk are you prepared to take? It is key that you know what the implications of your decisions are, and that you are making up your own mind, and not "follow the herd", taking risks that you do not understand. Reflect on this, and decide whether you have a healthy risk aversion: that means that you want to be sure that there is enough money for your family, and that should something happen, you will have a good reserve.

你准备好承担多少风险呢？只有了解不同的选项可能会有怎样的影响，才能做出对自己负责的决定，而不是随波逐流，承担你自己都无法理解的风险。由此反思，你是否有健康的风险预防习惯呢？风险预防意味着你需要保证家里有足够的存款，足够好的储备，以便安然应对各种突发情况。

TEACHER AND PARENT GUIDE

学科知识
Academic Knowledge

生物学	树长高得越快，枯死得越快；柳树和杨树长得比一般的树要快；当一棵树不再长高时，它的树干会越来越粗，至少长100年的年轮；松鼠能活到18岁；刚出生的小松鼠没有皮毛，就像刚出生的兔子一样。
化 学	碳在森林和大气之间的流动单位叫作碳通量；因为大气中的二氧化碳越来越多，世界各地树木的生长速度也越来越快。
物 理	狐尾松能活5 000年，但一株生长快速的轻木树却活不过40年；树平均能活200到300年；气候、水分、养分和土壤都会影响到树的生长。
工程学	只要修改杨树的2个基因，我们不仅能让它的生长速度提高1倍，还能让它长得更高、更粗、更加枝繁叶茂。但由于授粉的原因，转基因逃逸的风险很高，因此基因编辑严格受限。
经济学	郁金香狂热是泡沫经济的一个先例，起因是投机性的金融投资；经济增长缓慢会阻碍生活水平提高，减少税收，提高失业率（如果经济增长速度不能创造足够多的新工作岗位来补充被新科技所取代的岗位）；供给侧导致的经济增速减慢可能是生产力增速下滑造成的，需求侧的原因可能是总需求疲软。
伦理学	人们为什么不切实际地期待经济不断增长下去，即使我们的资源如此有限？人的选择不能仅仅用数学来解释；勤俭节约或许是对人类和地球最好的选择。
历 史	1637年郁金香狂热巅峰时期，郁金香的单价被炒到熟练手工艺人年收入的10倍以上；巴洛克流派的佛兰德画家小扬·布鲁盖尔（1601—1678）曾以猴子为主题创作了一幅油画，批判了这场经济危机；用数学进行经济社会分析的尝试始于17世纪；19世纪中期之后的150年内，工业化国家的平均身高增长了10厘米；简·丁伯根（1969年诺贝尔经济学奖得主）是经济计量学的创始人之一。
地 理	赤道区域的树在10到20年之内就能达到成熟期；在纬度靠北的区域，树通常一年至多生长1到2米，所以加拿大或欧洲北部的寒温带针叶林需要80—120年才能长成。
数 学	经济理论一般用数学模型来展示，用一系列的数学关系来阐述预设和可能的影响；博弈论拓宽了数学在经济学中的应用；经济计量学是运用统计学分析经济数据，从而为揭示经济关系和经济理论提供实证证据。
生活方式	有一些人盲目相信经济永远会不断增长，所有事情也会永远向好的方向发展。
社会学	体型偏小的人有可能血压偏低、患癌症的可能性偏低；人的身高可以作为营养质量和身体健康这两个重要福利要素的指标。
心理学	羊群效应：仅仅通过道听途说和其他人的选择来做决定；经济泡沫是跨金融、经济和心理学的交叉性问题。
系统论	二氧化碳把太阳的热量裹在地球表面从而导致全球变暖，而树吸收空气中的二氧化碳来滋养树叶、树干和根系，降低空气中二氧化碳的浓度。

教师与家长指南

情感智慧
Emotional Intelligence

松鼠爷爷

怀旧的心情涌上松鼠爷爷的心头,让他回忆起杨树还小的时候。他客观评判了树的生长速度,并对树的生长趋势做出了合理猜测。松鼠爷爷的提问让他能够和孙子分享他的立场。松鼠爷爷循循善诱小松鼠建立起自己的认知逻辑,批判性地观察现实,用一系列提问来引导小松鼠去反思现状。当小松鼠对经济增长发出疑问时,松鼠爷爷提供必要的信息来引发更多疑问。小松鼠问到收入时,爷爷以更有哲学性的方式为他答疑解惑。他戏剧性地回应了最后一个关于快乐的问题,暗示了最终的结论。

松鼠孙子

小松鼠渴望了解树的生长情况。当他意识到树不会继续长高时,他感到伤心甚至害怕。松鼠爷爷想要安抚他,小松鼠则把树和经济做类比。小松鼠愿意去理解爷爷的逻辑,用自己清晰的论述来回应问题。他的内心转折来自深深的疑惑:为什么人们想要滥用资源、让经济永远增长下去?小松鼠为了得到清晰的答案,对爷爷提出了更多疑问。经过这些思想挣扎后,他意识到生活核心的问题是:人们以后该如何过上满足的生活?

艺术
The Arts

1637年的郁金香狂热造成了最初的泡沫经济。佛兰德画家小扬·布鲁盖尔曾以猴子为主题创作了一幅油画,批判了这场经济危机。现代人的消费力远超地球所能承受的范围——你如何描述这一经济泡沫现象?艺术家是社会的敏感神经,当逻辑不能改变思想的时候,艺术是可行的另类表达方式。你如何艺术地描述和批判唯利是图、滥用资源的投机行为?

TEACHER AND PARENT GUIDE

思维拓展
Systems: Making the Connections

金融危机不仅仅在现代才发生。早在公元前33年的古罗马，货币短缺和信用破裂就导致许多公民面临破产。当今人类在以不可持续的速度消耗资源。我们的经济体系和消费欲望显然与现实脱节。我们最大的挑战之一是盲目相信经济会不断增长、不断产生利润、更多的财富会被分配给更多的人，更何况我们已经面临着人口过剩的问题。这也是为什么用树的生长作比喻格外重要：树比我们长寿，花费至少三分之二的生命来巩固自己的根基，还为生态系统提供不可或缺的稳定性。树的经济体建立在所有个体都竭尽全力为社区做出贡献，因此没有谁是"失业"的。当树不再继续长高时，它们转而关注强化现有的支持系统，提高社区面对挑战的韧性，让经济发展关注本地社区的需求。个别人仅仅为了获利，甚至不惜让全社会承担风险，忽视对其他个体可能的影响。这种投机欲望更加恶化了过度消费的现状。当一场经济危机因为上述原因来临时，我们并没有足够的资源来延续过度消费的生活方式。所以，我们需要从树的生长智慧中汲取灵感：在经济体成熟之后，不是继续上升而是强化根基，批判性地看待经济增长。

动手能力
Capacity to Implement

你准备好承担风险了吗？你知道如何评估风险吗？衡量风险的数学公式是风险发生的可能性乘以可能的损失。但是，你在承担风险的同时也有可能会得到好处，所以需要对比风险和潜在收益来做出选择。风险有很多种：金融风险意味着你可能会损失钱财，策略风险意味着失去对现有资源的控制，而灾害风险意味着风暴等突发事件可能造成的损失。尝试画一个树状图来分析好处和坏处、潜在损失和收益，从而得出结论。为了保证你做的决定不受从众心理或泡沫经济影响，可以和家人朋友分享你的分析，一起讨论这些问题。

教师与家长指南

故事灵感来自
This Fable Is Inspired by

约瑟夫·斯蒂格利茨
Joseph Stiglitz

约瑟夫·斯蒂格利茨在美国马塞诸塞州的阿默斯特学院获得本科学位，并于 1965 年成为富布赖特学者。1967 年，他获得麻省理工学院博士学位，担任助理教授，开始了他的学术生涯。1966 年至 1970 年间，他在英国剑桥大学做研究员。他曾在耶鲁大学、斯坦福大学、普林斯顿大学做过研究。2001 年至今，他在美国哥伦比亚大学商学院和国际公共事务学院担任教授，同时也在巴黎综合理工学院教书。斯蒂格利茨与他人合作了多篇论文，分析多种避险情况（比如个人储蓄选择和企业生产决策）下理论上会造成的后果。斯蒂格利茨运用理性预期均衡假设模型来更加现实地分析资本主义。他的结论是资本主义偏离了以社会主义措施来纠偏的模型。

图书在版编目(CIP)数据

冈特生态童书.第八辑:全36册:汉英对照/
(比)冈特·鲍利著;(哥伦)凯瑟琳娜·巴赫绘;
何家振等译. —上海:上海远东出版社,2021
ISBN 978-7-5476-1773-1

Ⅰ.①冈… Ⅱ.①冈…②凯…③何… Ⅲ.①生态环
境–环境保护–儿童读物—汉、英 Ⅳ.①X171.1-49

中国版本图书馆CIP数据核字(2021)第249940号

策　　划	张　蓉
责任编辑	程云琦
封面设计	魏　来　李　廉

冈特生态童书
更高更强
[比]冈特·鲍利　著
[哥伦]凯瑟琳娜·巴赫　绘
周徽音　译

记得要和身边的小朋友分享环保知识哦!
八喜冰淇淋祝你成为环保小使者!

Education 284

心态至上

Mind over Matter

Gunter Pauli

[比] 冈特·鲍利 著
[哥伦] 凯瑟琳娜·巴赫 绘
何家振 译

上海远东出版社

丛书编委会

主　任：贾　峰
副主任：何家振　闫世东　林　玉
委　员：李原原　祝真旭　牛玲娟　梁雅丽　任泽林
　　　　王　岢　陈　卫　郑循如　吴建民　彭　勇
　　　　王梦雨　戴　虹　翟致信　靳增江　孟　蝶

特别感谢以下热心人士对童书工作的支持：

匡志强　宋小华　解　东　厉　云　李　婧　陈　果
刘　丹　熊彩虹　罗淑怡　旷　婉　杨　荣　刘学振
何圣霖　廖清州　谭燕宁　韦小宏　李　杰　欧　亮
陈强林　王　征　张林霞　寿颖慧　罗　佳　傅　俊
胡海朋　白永喆　冯家宝

目录

心态至上	4
你知道吗?	22
想一想	26
自己动手!	27
学科知识	28
情感智慧	29
艺术	29
思维拓展	30
动手能力	30
故事灵感来自	31

Contents

Mind over Matter	4
Did You Know?	22
Think about It	26
Do It Yourself!	27
Academic Knowledge	28
Emotional Intelligence	29
The Arts	29
Systems: Making the Connections	30
Capacity to Implement	30
This Fable Is Inspired by	31

帝企鹅用身体抵挡猛烈的暴风雪袭击，掩护着他的幼企鹅。一只木蛙注意到企鹅的英勇行为，她知道如果企鹅没有保护幼鸟的坚强决心，他们父子都会死去。

　　木蛙问道："你们企鹅如何顶住连续数日时速超过100千米的暴风雪？更别说零下40摄氏度的最低气温了……"

An emperor penguin is shielding his chick against a severe blizzard. A wood frog notices his heroic action and realises that without the penguin's strong determination to protect his chick, neither father nor son will survive.

"How do you penguins manage to survive winds of over a hundred kilometres per hour, that blow for days on end?" Frog asks. "Not to even mention minimum temperatures of minus forty degrees…"

帝企鹅掩护着他的幼企鹅……

An emperor penguin is shielding his chick…

……你妻子到海里调养身体去了……

… your wife goes to the sea to nourish herself …

"哦，有志者事竟成，我们就是活生生的例子。"

"我看到你妻子生完蛋后就到海里调养身体去了。你给了她很大的帮助。"

"我尽了最大努力，我们已经下定决心了，我们三个都要活下来。"

"噢，说起来容易做起来难。"

"Well, we are living proof that it is possible, if one is determined to do so."

"I see your wife goes to the sea to nourish herself after the arrival of the egg. You are a great help to her."

"I do my very best, but we made up our minds that all three will survive."

"Well, that is easier said than done."

"人们必须做的是，专心想积极结果，让大脑没有机会想负面东西。"

"我想你需要设想最坏的情况，比如想想自己濒临死亡，如此才能让你现在的一切努力看上去行得通。"

"那么，如果你想象你的整个身体都着火了，而且你正遭受着强烈的疼痛……"

"而事实上这个时候你正经受天寒地冻……"

"What one has to do, is focus so intently on thinking about the positive outcome, that there is no room to even think of the negative."

"And I thought you needed to imagine the very worst, like thinking of nearly dying, to make whatever you are going through appear doable after all."

"So, if you imagine that your whole body is on fire, and you are suffering that intense pain…"

"When you are, in fact, freezing…"

……事实上你正经受天寒地冻……

... you are, in fact, freezing ...

……你会感到疼痛似乎缓解了……

... you will feel relieved ...

"那么，这时天寒地冻意味着什么呢？"

"显然，你会感到疼痛似乎缓解了。"木蛙回答道。

"那么，如果一个人真的觉得很冷，却想象自己正在遭受灼伤的痛苦，会发生什么呢？"

"他会感到寒冷似乎缓解了，不是吗？"

"What will the freezing cold mean to you then?"

"Obviously, you will feel relieved," Frog replies.

"So what would happen if someone is really feeling very cold, but imagines he is suffering the pain of being burnt?"

"He will feel relieved, won't he?"

"就像一个被烧伤的人，想象自己在冰天雪地里玩雪球，这可以帮他忘记疼痛。"

"你觉得有那么容易吗？"木蛙问道。

"当然不是。但是如果你总是关注消极的一面……"

"Like someone suffering from burns, imagining that he is out in the freezing cold playing with snowballs, will help him forget his pain."

"Is it that easy, you think?" Frog asks.

"Of course it isn't. But if you are always focusing on the negative…"

……想象自己在玩雪球……

… imagining he is out playing with snowballs …

……心灵的力量。

... the power of the mind.

"听着,当我们面对巨大的危险时,我们有两个选择:战斗,或者逃跑。"

"勇于战斗才更有可能战胜困难。如果逃跑并放弃,我会被冻死。我的小企鹅也会死。"

"嗯,我很佩服你,"木蛙说,"但我仍然不明白为什么有那么多人不相信心灵的力量。"

"Look, when we face great danger, we have two options: to fight, or to flee."

"My fighting spirit is more likely to overcome hardship. If I flee and give up, I will freeze to death. And so will my chick."

"Well, I admire you for that," Frog says. "But I still wonder why there are so many people who do not believe in the power of the mind."

"我就是鲜活的例证,我不明白为什么人们会说意念不管用!"

"我知道保持积极的心态对我们大有益处……"

"我知道负面的念头会伤害我们。"企鹅补充道。

"如果我们相信自己能做到,我们就能成功!"

"And I wonder about people claiming that these thoughts do not work, when I am living proof of it!"

"I know for a fact that keeping our thinking positive, can help us…"

"And I know for a fact that negative thinking can harm us," Penguin adds.

"And if we believe we can do things, we will be able to do them!"

……如果我们相信自己能做到，我们就……

… if we believe we can do things, we will …

……正确对待不如意的事情……

... to deal with disappointment ...

"但是，我们也必须学会如何正确对待不如意的事情，因为在生活中，并不是每件事都像我们期望的那样。"企鹅叹息道。

"说得太对了。"木蛙回答道。"你知道吗？那些相信自己长寿的人，确实活得更久。"

"But then, we also have to learn how to deal with disappointment as, in life, not everything goes exactly the way we expect," Penguin sighs.

"Too true." Frog replies. "But did you know that those who believe they will have a long life, do live longer?"

"心态至上，真的！就像你一样，有信心和决心熬过每一个寒冷的冬季——而你确实还活着，每年都看到春天的阳光把冰雪融化……"

……这仅仅是开始！……

"Mind over matter, indeed! Like you, who have faith and the determination to survive each freezing winter – and you do live to see the sun thawing the ice every spring…"

... AND IT HAS ONLY JUST BEGUN!...

……这仅仅是开始！……

… AND IT HAS ONLY JUST BEGUN! …

Did You Know?
你知道吗？

When burn patients in excruciating pain play games imagining throwing snowballs at each other in freezing cold weather, their pain levels are lower. Winning the game leaves them with no time to think about pain.

当疼痛难忍的烧伤患者想象自己在寒冷的天气里互相扔雪球时，其疼痛程度会降低。在游戏中获胜的欲望让他们无暇感受痛苦。

The emperor penguin is able to swim at a depth of up to 500 metres, and can remain underwater for 20 minutes. It is able to shut down non-essential organs when needed to conserve energy to protect its offspring.

帝企鹅可以在水下500米的深度游泳，在水下能待20分钟。当需要保存能量以保护后代时，帝企鹅可以让非必要的器官停止工作。

The emperor penguin is the only penguin species to breed during the winter in Antarctica. Thousands of penguins trek up to 120 kilometres to reach their breeding grounds. The female lays only one egg per season.

帝企鹅是唯一在南极冬季繁殖的企鹅。成千上万只企鹅跋涉 120 千米到达繁殖地。雌企鹅每季度只产一枚蛋。

Penguin's feathers remain flat in the water, insulating it against the cold. When on land, the feathers are held up by tiny muscles, allowing for a layer of air next to the skin to reduce heat loss.

企鹅在水里时羽毛是平贴在身上的,这样能阻挡寒冷。上岸后,企鹅的羽毛由细小的肌肉支撑起来,在皮肤周边隔离出一层空气防止热量流失。

An Emperor penguin pair will travel up to 500 kilometres back and fro to collect food for their single chick, showing a clear purpose and strong dedication, overcoming obstacles, to ensure survival of the species.

一对帝企鹅会往返 500 千米，为自己唯一的幼仔寻找食物，表现出清晰的目标和很强的奉献精神，克服重重困难以确保物种的延续。

Most animals that hibernate live off stored fat, with thick coats of fur to protect them from the cold. The wood frog of Alaska and Canada survives the cold by "freezing to death".

大多数冬眠动物靠储存的脂肪生存，它们有厚厚的皮毛御寒。加拿大和美国阿拉斯加的木蛙以"冻死"的方式克服寒冷求得生存。

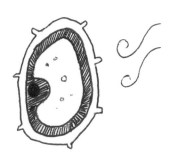

Wood frogs retain water in their cells when their tissues freeze. As a result, these amphibians can freeze and thaw many times every winter, without any ill effects.

木蛙肌体结冰时，其细胞内保留着水分。因此，每年冬天这种两栖动物可以多次冻僵再解冻，没有任何不良影响。

Excessive focus on everything that could possibly go wrong is harmful. Research shows the more a patient is told about possible negative side-effects of a treatment, the more likely it is those will be experienced.

过度关注每件可能出错的事是有害的。研究表明，病人被告知的潜在医疗不良反应越多，就越有可能出现这些不良反应。

Think about It

Would you still be feeling pain when you think of something funny?

当你想有趣的事情时，你还会感到疼痛吗？

When you make up your mind to do something, is it more easy or more difficult?

当你下定决心做某事后，做起来是更容易，还是更困难了？

Is the Emperor penguin a hero?

帝企鹅是英雄吗？

The more you know, the more you worry?

知道得越多，烦恼也越多吗？

Do It Yourself! 自己动手!

What is the best way for us to harness our mind's energy to overcome moments of fear or pain? Think about a moment in your life that you really did not like. Now think of two or three things that always make you smile, or make you feel strong and courageous. Can you feel the difference? Make a list of all the positive things that can act as a bright light in your mind, dispelling the darkness of negative moments. When the going gets tough, you can focus on these thoughts to give you courage and help you face adversity. Now encourage your friends and family members to do the same, and to support each other through trying times.

用心灵力量克服恐惧或痛苦的最好方法是什么？想一想你生命中最难受的时刻，然后再想想两三件总是让你微笑或者让你变得坚强勇敢的事。你感觉到不同了吗？列出所有可以充当驱散艰难时刻黑暗的明灯的积极事物。在困难时刻，多想想那些给你勇气的想法，这能帮你面对逆境。现在鼓励你的朋友和家人也这样做，并在困难时刻互相支持。

TEACHER AND PARENT GUIDE

学科知识
Academic Knowledge

生物学	木蛙可产下3 000枚卵；当身体暴露在极端低温环境下，会患低体温症；世界上有18种企鹅；帝企鹅的血红蛋白结构使其能够在低氧环境下生存；帝企鹅的腿上长有羽毛，以阻止热量流失。
化　学	木蛙产生的糖脂是一种特殊的防冻物质，可以阻止细胞内的水结冰。
物　理	身体在水中失温比在空气中快25倍；气压伤：由于内部压力造成的结构损伤；企鹅依靠独特的声音辨别彼此；企鹅有皮下脂肪保温；冬季过后，雄帝企鹅体重会减少一半。
工程学	木蛙细胞之间会结冰，但这不会破坏其细胞。
经济学	经济活力有赖于人民、政府和投资者对未来的信心；制定一个明确的目标，然后专注于这个目标。
伦理学	用恐惧驱使人们，而不是绘制一个宏大的共同愿景和承诺。
历　史	3 700万年前，南极洲生活着巨型企鹅（高2米，重100千克）。
地　理	暴风雪：风雪交加而成的冬季风暴，导致能见度很低；木蛙是唯一生活在北极圈以北的蛙类；极地涡旋：在大气层的中上层循环、旋转的冷空气。
数　学	卫星图像，包括冰层表面帝企鹅粪便形成的斑迹，可用于评估一个聚落中的帝企鹅数量。
生活方式	生活在乐观的氛围中让人更长寿；如果一对帝企鹅的蛋没有了，它们之间的关系就解除了，它们都会回到大海。
社会学	与做过相似决定的人交谈；听取做过同样事的人的建议。
心理学	"战斗或逃跑"的压力；坚定的决心让我们能够在困难面前坚持下去；畏惧不能推动决策；不要纠结于决策，要专注于执行；想象一下，现在的决定在未来会产生什么结果；没有完美无缺的决策，障碍和阻力总是随时发生，对此，我们要勇于接受；心态至上：控制对疼痛的感知；安慰剂效应和反安慰剂效应。
系统论	帝企鹅能够通过改变身体和生存方式适应极端恶劣的环境，即使在不利的条件下也能顽强地成长；人们希望在有限条件下，可以认识自己并做好自己。

教师与家长指南

情感智慧
Emotional Intelligence

木蛙　　当木蛙遇到另一种能够在寒冷的冬天生存下来的生物时,她被深深地吸引了。她很想知道帝企鹅是怎么做到的。她认为帝企鹅的行为是英勇的,并钦佩他的生存决心。她意识到心灵的力量,并分享了她对此的想法。她质疑,用决心克服困难是否那么容易。她钦佩帝企鹅的战斗精神,她不明白为什么人们会怀疑心灵力量。她深信积极想法的治愈能力,相信彼此有能力克服逆境。她同意每个人都需要学会应对生活中的不如意,并指出相信自己长寿的人实际上比不相信的人活得更长。

帝企鹅　　帝企鹅很清楚自己在寒冷中生存下来的决心,在自己和家人的生命中起着怎样的作用。他乐于与人分享他的方法,即专注于思考好的事物,不让负面的想法占据他的头脑。帝企鹅承认这不容易做到,并且同意要避免把注意力集中在负面的事物上。当木蛙提到战斗或逃跑的选择时,帝企鹅明白是他的战斗精神和决心确保了他自己和他的孩子的生存。他想知道,在他这样的案例面前,人们怎么能声称意念力量不起作用。他承认生活中并不是每件事都能如愿,我们需要学会应对不如意的事。谈话结束时,他称赞了木蛙的信心和决心。

艺术
The Arts

木蛙没有帝企鹅那么迷人,但木蛙的眼睛很有趣。研究木蛙眼睛的照片,它的眼睛在身体其他部位冻结的时候仍然完好无损。用简单的黑白铅笔画描绘木蛙,抓住木蛙看帝企鹅时入迷的神情。在你向别人解释意志力的作用时,把你的画展示给别人看。你的画可以帮助他们积极思考。

TEACHER AND PARENT GUIDE

思维拓展
Systems: Making the Connections

　　生活不是一帆风顺的。有些人和动植物的生存环境非常恶劣。在自然界中，木蛙和帝企鹅都生活在充满挑战的环境中。木蛙的冬眠方式类似于冻死，而帝企鹅已经适应了最恶劣的环境。对于独居的木蛙来说，它依靠能抵御严寒的身体机能来求生，而帝企鹅则通过与配偶、群落成员合作来求生，因为它不能独自生存。在聚落的安全区域内，企鹅蛋一旦产下就会被雄企鹅接管照顾，这样雌企鹅就可以返回觅食地补充能量储备。雄企鹅不仅照顾企鹅蛋，还与群落中其他雄企鹅聚集在一起形成一个肉体屏障，它们每几个小时轮流抵挡暴风雪的冲击，以保护身后的其他企鹅。由于无法吃东西，雄企鹅的体重在冬季会下降50%。每只帝企鹅都清楚地知道自己需要做什么，并以极大的决心完成任务，从而在速度超过100千米/时的寒风中生存下来。这种行为是为了群落的生存，也是为了整个物种的生存。我们一定会从这些动物为改善生活突破万难的奉献精神中学到很多东西。为了克服困难和障碍，我们要有一个明确的目标，并利用我们的意志力。对于那些常常需要应对危机的社区来说，这种精神力量是一种重要的支撑。通过适应与合作，帝企鹅抵御了那些可能让其他动物灭绝的恶劣环境，其种群实现了长久繁荣。人类也可以向它们学习。如果有正确的领导、意愿和决心，就可以取得超越客观条件限制的成就。这里的核心信息是：我们需要的是一个共同的目标，一个共同信仰的目标，一个以改善生活和保障我们子孙后代的未来为原则的目标。如果有了这样的目标，我们就很有可能战胜一切困难取得成功。

动手能力
Capacity to Implement

　　从生态、地区或国家的角度来看一下全球。找出那些环境得到保护、新产业得到发展的国家和地区，以及那些似乎资源很少或没有资源的地方，由于它们的规模和实力，进入世界市场的机会非常有限。在研究这些案例时，要特别注意少数有决心让他们的城市、地区或国家排除万难走向繁荣的人。与你的朋友和家人讨论，谁是这类人中的当代典范，谁又是过去的典范？

教师与家长指南

故事灵感来自
This Fable Is Inspired by

乔·马钱特
Jo Marchant

  乔·马钱特毕业于伦敦帝国理工学院，获得科学传播学理学硕士学位，之后继续深造，并获得英国伦敦圣巴塞洛缪医学院遗传学和医学微生物学博士学位。她担任《新科学家》和《自然》杂志的高级编辑。她曾在《纽约时报》《卫报》和《史密森尼》杂志上发表文章，吸引了世界各地的读者。她的书《自愈力的真相》于2016年出版。乔写过关于身心关系以及用头脑改善健康（包括虚拟现实、无药物治疗技术等领域）的著作。

图书在版编目(CIP)数据

冈特生态童书.第八辑:全36册:汉英对照/
(比)冈特·鲍利著;(哥伦)凯瑟琳娜·巴赫绘;
何家振等译.—上海:上海远东出版社,2021
ISBN 978-7-5476-1773-1

Ⅰ.①冈… Ⅱ.①冈…②凯…③何… Ⅲ.①生态环境–环境保护–儿童读物—汉、英 Ⅳ.①X171.1-49

中国版本图书馆CIP数据核字(2021)第249940号

策　　划　张　蓉
责任编辑　祁东城
封面设计　魏　来　李　廉

冈特生态童书
心态至上
[比]冈特·鲍利　著
[哥伦]凯瑟琳娜·巴赫　绘
何家振　译

记得要和身边的小朋友分享环保知识哦!
八喜冰淇淋祝你成为环保小使者!

Education 285

第六感
The Sixth Sense

Gunter Pauli

[比] 冈特·鲍利 著
[哥伦] 凯瑟琳娜·巴赫 绘
何家振 译

上海远东出版社

丛书编委会

主　任：贾　峰
副主任：何家振　闫世东　林　玉
委　员：李原原　祝真旭　牛玲娟　梁雅丽　任泽林
　　　　王　岢　陈　卫　郑循如　吴建民　彭　勇
　　　　王梦雨　戴　虹　翟致信　靳增江　孟　蝶

特别感谢以下热心人士对童书工作的支持：

匡志强　宋小华　解　东　厉　云　李　婧　陈　果
刘　丹　熊彩虹　罗淑怡　旷　婉　杨　荣　刘学振
何圣霖　廖清州　谭燕宁　韦小宏　李　杰　欧　亮
陈强林　王　征　张林霞　寿颖慧　罗　佳　傅　俊
胡海朋　白永喆　冯家宝

目录

第六感	4
你知道吗？	22
想一想	26
自己动手！	27
学科知识	28
情感智慧	29
艺术	29
思维拓展	30
动手能力	30
故事灵感来自	31

Contents

The Sixth Sense	4
Did You Know?	22
Think about It	26
Do It Yourself!	27
Academic Knowledge	28
Emotional Intelligence	29
The Arts	29
Systems: Making the Connections	30
Capacity to Implement	30
This Fable Is Inspired by	31

一条气象鱼（一种泥鳅）活跃起来，兴奋地在欧洲最长的河流——伏尔加河里游着。一只第一次从远方来到这里的鸭嘴兽说道：

　　"我注意到渔民们很紧张，他们一直在谈论即将来临的坏天气。"

A weatherfish is becoming very active, excitedly swimming around in the longest river in Europe, the Volga. A platypus, visiting for the first time from far away, remarks,
"I've noticed how nervous the fishermen are getting, talk about nothing else but the bad weather on the horizon."

一条气象鱼活跃起来……

A weatherfish is becoming very active ...

……看到我上下游动……

... see me swimming up and down ...

"当然了,他们一看到我上下游动,就知道暴风雨要来了。"

"你觉得暴风雨就要来了吗?"

"是的,这就是为什么我叫这名字了。我可不是无缘无故就被叫作气象鱼的。"

"Of course, they know that there is a storm coming, as soon as they see me swimming up and down."

"Do you feel that there is a storm coming?"

"Yes, that is how I have earned my name. I am not called the weatherfish for nothing."

"哦，你在浑浊的水中游泳，几乎没有氧气可供呼吸，就这样你还能预测天气？"

"是的，我有第六感。"

"不可能。只有五种感觉。"

"不对！我确实有第六感，能预测天气。"

"So swimming around in the murky water, with hardly any oxygen to breathe, you are able to predict the weather?"
"Yes, I have a sixth sense."
"Not possible. There are only five senses."
"Wrong! I do have a sixth sense, one for forecasting the weather."

……能预测天气。

... one for forecasting the weather.

我感受到气压变化。

I sense the change in air pressure.

"你是尝到了、闻到了,还是听到了?毕竟在这浑浊的水里,你什么也看不见。"

"我感受到气压变化。"

"哦,即使在水中,你也能感受到气压变化吗?"

"你看,我有能碰触东西的手指吗?"

"Do you taste it, smell it, or hear it? Because you can't possibly see anything in this murky water."

"I sense the change in air pressure."

"So, you feel the changing pressure even in the water?"

"Look, do I have fingers to touch anything with?"

"没有,但我想也许你的触须……"

"我的触须是用来找食物的。如果气压变化了,我全身都能感觉到。下来吧,看看你能不能感受我感知到的东西。"

"我可以尝试体验难以解释的东西。"

"No, but I thought that perhaps your barbs…"
"My barbs are for finding food. When air pressure changes, I feel it all over my body. Get in, and see if you can feel what I pick up."
"I could try to experience something that is difficult to explain."

我的触须是用来找食物的。

My barbs are for finding food.

……可以听到心跳……

... can pick up the heartbeat ...

"嗯,你自己就是那么不同寻常的动物,所以也许你能感觉到我正在解释的现象。"

"你知道吗?我也能用我的喙感知东西。"鸭嘴兽说。"我可以不接触猎物,从远处听到猎物的心跳。"

"你是说你的'鼻子'能听到它的心跳?"

"Well, you are such an unusual animal yourself, so perhaps you will feel what it is I am trying to explain."
"Did you know I can also sense things, but with my bill?" Platypus asks. "I can pick up the heartbeat of my prey, from a distance without touching it."
"That means your nose can hear its heart?"

"这还不是全部!我甚至能听到它的神经颤动。"

"听到神经?这真的令人毛骨悚然……我一点都不知道你能做到这些。"气象鱼回答道。

"我可以不用耳朵、眼睛和鼻子,只用我所说的'第六感'捕捉猎物。"

"That is not all! I can even hear its nerves."
"Hearing its nerves? Now this is really getting spooky... I had no idea you could do all of that," Weatherfish replies.
"I can hunt with my ears, eyes and nose closed, just by using what I call my 'sixth sense'."

我甚至能听到它的神经颤动。

I can even hear its nerves.

……感觉到蚯蚓的电磁场……

... feel the electromagnetic field of an earthworm ...

"我知道你不同寻常,因为你是一种下蛋的哺乳动物,但你还能先屏蔽其他所有的感官,然后使用不为人知的感官?好神奇!"

"如果需要,我可以使用我的'第六感'。我甚至能感觉到蚯蚓的电磁场。"鸭嘴兽自豪地说。"所以,你可以想象,我几乎从没有错失过一餐。"

"I knew you are unusual, being a mammal that lays eggs, but shutting down all your other senses, and then using one that no one knows much about? That is extraordinary!"

"When I need to, I can rely on my 'sixth sense'. I can even feel the electromagnetic field of an earthworm," Platypus boasts. "So, as you can imagine, I hardly ever miss a meal."

"哇,你有那么惊人的第六感!我想知道谁有第七感?"

……这仅仅是开始!……

"Wow, you have such an astounding sixth sense! I wonder who has a seventh?"

... AND IT HAS ONLY JUST BEGUN!...

......这仅仅是开始!......

... AND IT HAS ONLY JUST BEGUN! ...

Did You Know?
你知道吗？

After deciphering the duck-billed platypus's entire DNA, scientists determined that it shares genes with reptiles, birds, and mammals. It is a mammal without a stomach, something it has in common with certain fish.

在破译长着鸭喙的鸭嘴兽的全部DNA之后，科学家们确定它同时具有爬行动物、鸟类和哺乳动物的基因。它是一种没有胃的哺乳动物，这是它与某些鱼类的共同之处。

A platypus's bill comprises of thousands of cells that give it the ability to perceive the electromagnetic fields generated by very small living creatures, like earthworms and crickets.

鸭嘴兽的喙由许多细胞组成，这些细胞使鸭嘴兽能够感知蚯蚓和蟋蟀等非常小的生物产生的电磁场。

The platypus is one of the very few venomous mammals, which is one of its reptilian characteristics. Unlike snakes, a platypus's venom is not in its fangs but in the hollow spur, which the males have on each hind leg.

鸭嘴兽是一种罕见的有毒哺乳动物，这是它的爬行动物特征之一。与蛇不同，鸭嘴兽的毒液不是在毒牙里，而是在雄性后腿上的空心毒刺里。

When the first platypus specimen was sent back to England from Australia in the late 18th century, the scientists who examined it thought that someone was playing a trick on them.

18世纪末，当第一个鸭嘴兽标本从澳大利亚送回英国时，研究它的科学家们认为有人在捉弄他们。

The weatherfish (*Misgurnus fossilis*) can breathe atmospheric oxygen, using its intestines as an accessory respiratory organ. This enables it to live in water with low oxygen levels, where it survives long droughts and frosts.

气象鱼（纵带泥鳅）可以用它的肠道作为辅助呼吸器官，吸入大气中的氧气。这使它能够生活在含氧量低的水环境中，在那里它需要挺过长期的干旱和霜冻。

The Volga River is Europe's longest river, with a total length of more than 3,500 kilometres. It drains a catchment of over 150,000 streams and rivers, before flowing into the Caspian Sea.

伏尔加河是欧洲最长的河流，全长 3 500 多千米。在流入里海之前，有 15 万条小溪和河流汇入伏尔加河。

Aristotle, the Greek philosopher and a student of Plato, wrote extensively about physics, biology and zoology, and was the first to describe the five senses.

亚里士多德是一位希腊哲学家,还是柏拉图的学生。他写了大量有关物理学、生物学和动物学的著作,他是第一个描述五种感觉的人。

Animals rely on scent, and scent is highly dependent on atmospheric pressure. When pressure is low, scent will travel further and higher in the atmosphere. On high-pressure days, scent will dissipate quickly.

动物依赖气味,而气味很大程度取决于大气压强。当气压低时,气味会在大气中传播得更远更高。在高气压的日子里,气味会很快消散。

Think about It 想一想

Can you sense when it will rain soon? Perhaps your grandmother can feel it in her joints?

你能感觉到快要下雨了吗？也许，你奶奶能通过她的关节感觉到？

Is intuition a sixth sense?

直觉是第六感吗？

A mammal that lays eggs? Is that not surreal?

有一种哺乳动物会生蛋，这难道不离奇吗？

Can you "hear" with your heart?

你能用心脏"听"吗？

Do you have a sixth sense? Think about the following situations: someone enters the room and the atmosphere changes; the telephone rings and you sense who it is calling before you answer; your dog goes to the door, knowing that someone is about to knock. Talk about such experiences with your friends and family members to find out more. Ask if and how they have experienced anything they can ascribe to having a sixth sense.

你有第六感吗？想想以下情况：有人进入房间，气氛发生了变化；电话铃响了，你在接电话前就感觉到是谁打来的；你的狗走到门口，知道有人要敲门。和你的朋友、家人谈谈此类经历以发现更多情况。问一问他们是否有过可以归因于第六感的经历，以及怎样经历的。

TEACHER AND PARENT GUIDE

学科知识
Academic Knowledge

生物学	气象鱼（纵带泥鳅）是一种以水底物质为食的杂食性动物，它可以呼吸大气中的氧气，通过肠道中一个复杂的血管系统提取氧气，然后从肛门排出废气；鸭嘴兽和针鼹是仅有的卵生哺乳动物；鸭嘴兽通过探测猎物肌肉收缩产生的电磁场来定位猎物；鸭嘴兽的触觉和电感觉之间有着密切的联系；鸭嘴兽有双锥形眼睛。
化　学	化学天气预报：基于大气中化学物质迁移模式的天气预报；由火山喷发、闪电和日冕太阳粒子抛射引起的地球大气层的变化；大气化学包括对酸雨、臭氧损耗、光化学烟雾和温室气体的研究。
物　理	气压变化会对动物产生影响，因为许多动物依赖气味，而气味与气压密切相关。
工程学	用气象气球、卫星和多普勒雷达等预报天气；法国生物感应公司在水下天气预报机器人中安装了15个传感器；磁感受、回声定位、电感受、湿度感受、红外感应。
经济学	从旅游业到农业，天气预报在地方经济中发挥着关键作用，能够产生收入、降低成本，并帮助人们应对飓风、洪水和海啸等自然灾害。
伦理学	尊重那些明显有用，但用当前的科学无法解释的感知方式。
历　史	亚里士多德的《灵魂论》第2卷写于公元前4世纪，该书第7—11章论述视觉、听觉、嗅觉、味觉和触觉；1799年，欧洲博物学家第一次遇到鸭嘴兽，当时认为这是一场骗局。
地　理	高气压通常意味着明亮、蓝色、晴朗的白天或夜晚，而正常气压下，则有散乱的云层；死水潭是河道改变后留下的一个孤立的池塘。
数　学	提前12—24小时预报天气的关键手段是测量气压，准确率达70%。
生活方式	根据天气预报决定如何穿衣服，使用何种交通工具，以及一天的生活方式。
社会学	一些民族把鸭嘴兽当作图腾。
心理学	气压变化导致耳痛、头晕或头痛。
系统论	河道疏浚和水生杂草清理对气象鱼种群构成威胁。

教师与家长指南

情感智慧
Emotional Intelligence

鸭嘴兽

鸭嘴兽观察到渔民变得紧张不安，就自信地直接问气象鱼是否感觉到暴风雨即将来临。由于她不能肯定是否该相信气象鱼的话，于是她重复了自己的问题。她清楚自己有第六感，但她怀疑气象鱼是否也有第六感。在气象鱼自信地回答后，她要求气象鱼进一步解释。然后她描述了自己的第六感，这种第六感依靠和人类手指一样敏感的喙。她有信心描述难以解释的东西——如何用喙感知猎物的心跳。她毫无保留地介绍她感知蚯蚓电磁场的能力，这让气象鱼感到震惊。

气象鱼

气象鱼很自信，并指出他的名字已经说明了一切。当鸭嘴兽要求他作更多解释时，他只是简单地说他有第六感。当鸭嘴兽表示不相信后，他直截了当地说他使用这种感觉来预测天气，这是鸭嘴兽不能不接受的事实。当鸭嘴兽继续追问并迫使他提供信息时，他很恼火，于是非常自信地邀请鸭嘴兽到水中来。由于鸭嘴兽也是一种具有神奇能力的动物，气象鱼相信鸭嘴兽能够理解他一直试图解释的关于第六感的问题。他进一步了解鸭嘴兽的特征，发现这比他自己身上的特征更复杂。他敬畏鸭嘴兽的第六感，这使他怀疑是否还存在第七感，并好奇第七感会是什么样的。

艺术
The Arts

如何以艺术方式表现科学家难以描述的第六感呢？和你的朋友讨论什么是第六感，并在不同的表述中达成一致。写一两个小品，并指定不同的人作为演员，通过表演来展示第六感的存在。准备好上台了吗？

TEACHER AND PARENT GUIDE

思维拓展
Systems: Making the Connections

我们有时感知到了某些东西，却不知道自己是如何感知的，这通常被当作第六感的例子。这也被称为超感官知觉，因为已知的五种感官中没有一种能感知此类信息。动物也有特殊的感知能力，科学家们只是到现在才有更多发现。我们知道动物有人类所没有的感知方式，就像这篇童话展示的那样。科学家们已经在研究各种人类没有或失去的感知能力，比如回声定位、磁感受、电感受、湿度感受和红外感应，可能还有更多。动物能看到我们看不到的东西，而且它们对颜色的感知与我们不同。这些是我们已经理解的一些感知方式，但还有很多感知方式有待探索。套用传统科学方法的科学家很难合理地解释不符合简单的因果规律的现象，然而观察结果清楚地表明，值得我们关注的还有更多。人们对气象鱼的行为提出了一种可能的解释，那就是它会对气压变化作出反应。另一方面，鸭嘴兽代表了一种更复杂的情况，它不能像气象鱼那样用因果逻辑轻易解释。越来越多的对动物感知环境方式的研究，激起了人类对自身超感官感知能力的兴趣。这就是创新研究的力量，随着对动物世界案例的研究，越来越多的问题被提出。这是一项有影响力的、崭新的、探索未知世界的研究。因此，需要穿透表象、突破思维定式的创造性思维，探究尚不能解释的事物。是时候从大自然获取灵感了，研究动植物如何感知未知事物，以及如何常态化地使用我们的直觉。通过研究能够感知但尚不知如何感知的现象，找到与我们周围千变万化的客观存在保持联系并进行监测的方式与手段。不能被感知的，就无法被测量；不能被测量的，就无法应对。

动手能力
Capacity to Implement

狗可能是最常见的使用第六感的动物。首先，狗有高度发达的嗅觉。它们还会对着人看不见的东西吠叫，可能是因为它能听到人类听不到的声波频率。它们的视野也比人类宽广。除此之外，它们还有感知即将来临的自然灾害的能力。总结那些狗具有而人类没有的感知能力，并思考我们如何能从它们的独特能力中获益。现在与你的朋友、家人分享你的发现和想法。

教师与家长指南

故事灵感来自
This Fable Is Inspired by

鲁珀特·谢德瑞克
Rupert Sheldrake

鲁珀特·谢德瑞克毕业于英国剑桥大学卡莱尔学院，主修博物学、化学和生理学，并于1963年毕业。他在美国哈佛大学研究生院继续他的学术学习，在那里他学习哲学和科学史，然后在剑桥大学生物化学系攻读博士学位，并在1967年毕业。1968—1969年，他在马来亚大学继续他的研究，后来又在印度海得拉巴的国际热带半干旱地区作物研究所从事研究工作。自1989年以来，他一直是美国加利福尼亚州旧金山的思维科学研究所的研究员。1999年，他出版了《狗狗知道你要回家？——探索不可思议的动物感知能力》一书（由伦敦哈钦森出版社出版）。接着，他在2003年出版了《被盯着的感觉，以及扩展思维的其他方面》。

图书在版编目（CIP）数据

冈特生态童书.第八辑:全36册:汉英对照／
(比)冈特·鲍利著；(哥伦)凯瑟琳娜·巴赫绘；
何家振等译.—上海:上海远东出版社,2021
ISBN 978-7-5476-1773-1

Ⅰ.①冈… Ⅱ.①冈… ②凯… ③何… Ⅲ.①生态环境－环境保护—儿童读物—汉、英 Ⅳ.①X171.1-49

中国版本图书馆CIP数据核字(2021)第249940号

策　　划	张　蓉
责任编辑	祁东城
封面设计	魏　来　李　廉

冈特生态童书

第六感

[比] 冈特·鲍利　著
[哥伦] 凯瑟琳娜·巴赫　绘
何家振　译

记得要和身边的小朋友分享环保知识哦！
八喜冰淇淋祝你成为环保小使者！

Education 286

夜莺的歌
The Nightingale's Song

Gunter Pauli

［比］冈特·鲍利 著
［哥伦］凯瑟琳娜·巴赫 绘
何家振 译

上海远东出版社

丛书编委会

主　任：贾　峰
副主任：何家振　闫世东　林　玉
委　员：李原原　祝真旭　牛玲娟　梁雅丽　任泽林
　　　　王　岢　陈　卫　郑循如　吴建民　彭　勇
　　　　王梦雨　戴　虹　翟致信　靳增江　孟　蝶

特别感谢以下热心人士对童书工作的支持：

匡志强　宋小华　解　东　厉　云　李　婧　陈　果
刘　丹　熊彩虹　罗淑怡　旷　婉　杨　荣　刘学振
何圣霖　廖清州　谭燕宁　韦小宏　李　杰　欧　亮
陈强林　王　征　张林霞　寿颖慧　罗　佳　傅　俊
胡海朋　白永喆　冯家宝

目录

夜莺的歌	4
你知道吗？	22
想一想	26
自己动手！	27
学科知识	28
情感智慧	29
艺术	29
思维拓展	30
动手能力	30
故事灵感来自	31

Contents

The Nightingale's Song	4
Did You Know?	22
Think about It	26
Do It Yourself!	27
Academic Knowledge	28
Emotional Intelligence	29
The Arts	29
Systems: Making the Connections	30
Capacity to Implement	30
This Fable Is Inspired by	31

一只夜莺看到一只云雀被一组镜子里反射的早晨阳光迷住了,正在朝镜子走去。

"小心!这是陷阱。"夜莺警告道。

"陷阱?怎么是陷阱?我看不出有任何危险。"云雀环顾四周回答道。

A nightingale watches a skylark approaching a set of small mirrors, intrigued by their play of light in the morning sun.

"Watch out! It is a trap," the nightingale warns.

"A trap? How? I see no danger," the skylark replies, as he looks around.

……正在朝镜子走去……

... approaching a set of small mirrors ...

……人们喜欢圈养我们，因为我们的歌声……

... people do like keeping us, for our singing ...

"好吧,这是你的生活。但我告诉你,我曾见过有人在地上撒网,一次能捉到至少1000只你们的小伙伴。"

"是的,人们确实喜欢圈养我们,因为我们的歌声。"

"不要自欺欺人了,大多数被抓住的鸟最终不会被关进笼子,而是被放在盘子里吃掉。"

"Well, it is your life. But I tell you, I have seen people casting nets over a field, catching at least a thousand of your buddies in one go."

"Yes, people do like keeping us, for our singing."

"Don't fool yourself, most of those caught don't end up in a cage, but on a plate – to be eaten."

"嗯，听起来很有戏剧性，我得承认你很可能是对的……但是人们也会把我们关起来，为的是能欣赏我们优美的歌声。"

"你确实是很棒的歌星。不过在唱歌方面，我也是把好手。"

"我是在黎明第一个歌唱的鸟。我也曾听到你唱出自己的心声，甚至在大家都在熟睡的夜晚。"

"Well, that sounds dramatic, and I must admit that you are most likely right… But people also cage us so they can enjoy our beautiful birdsong."
"You are a great entertainer, granted. And I do my part too, when it comes to singing."
"I am the first to sing at dawn. And I've heard you sing your heart out too, even at night when everyone is asleep."

……人们也会把我们关起来……

... people also cage us ...

月光下为心爱的姑娘唱小夜曲……

Serenading ladies by moonlight ...

"人们可能睡着了,但我想吸引的夜莺姑娘是不会睡的。"

"好浪漫啊!月光下为心爱的姑娘唱小夜曲……"

"成功来之不易。我们需要一直练习拿手的歌曲。这是一个人做到最好所必须的。"

"People may be asleep, but not the nightingale ladies I am trying to attract."

"That is so romantic! Serenading ladies by moonlight…"

"Success does not come easy. We need to practice our great songs all the time. That is what one needs to do to be the best."

"那么你告诉我,夜莺,每年冬天,你向南迁徙到温暖的非洲,这时你都做了什么?"

"我坚持每天练习。我们夜莺必须掌握上千种发音和音节,组合在几百个短语里。"

"你一点也没夸大吧?发出1 000种不同的声音?这比人类的发音多多了。"

"And tell me, Nightingale, what do you every winter, when you migrate south, to the warmth of Africa?"

"I keep on practicing every day. We nightingales have to master more than one thousand sounds and syllables, combined in hundreds of phrases."

"Are you not exaggerating a little? Making a thousand different sounds? That is a lot more than people do."

……到温暖的非洲？

... to the warmth of Africa?

你应该钦佩和羡慕我……

You should be admiring and respecting me ...

"你不必如此惊讶！你应该钦佩和羡慕我能做到这一点。我们学歌唱就像人类学会说、读、写一样困难。"

"但是人的脑容量大。你的脑容量很小。"

"You should not be so surprised! You should be admiring and respecting me for being able to do this. It is as difficult for us to learn to do this as it is for a person to learn to speak, read and write."

"But people have big brains. You have a tiny one."

"你是说我是笨蛋吗？脑容量的大小无关紧要；关键是如何使用它。我的大脑大部分都用来发出完美的声音——别无他事。"

"那么，为什么人们想抓住我，把我关在笼子里？既然你们这么厉害，人们为什么不多抓你们呢？"

"Are you calling me a birdbrain? Size does not matter; it is how you use it. Most of my brain is used to make the perfect sound – and nothing else."

"So, why do people want to catch me and keep me in a cage? Why don't they catch more of you if you are such a marvel?"

大小无关紧要……

Size does not matter ...

……撞向笼子的铁栏杆,直到死去。

... against the bars of our cages, until we die.

"当我们被抓住的时候,我们会一直试图逃跑。在逃跑时,我们会把自己撞向笼子的铁栏杆,直到死去。"

"这太可悲了,人们已经为你花了钱,所以他们不会放你们自由。"云雀评论道。

"When we are caught, we keep on trying to escape. In doing so, we will smash ourselves against the bars of our cages, until we die."

"It is so sad that because people have paid for you, they will not set you free," Skylark comments.

"听着,每个人都有悲伤的故事。我会继续唱歌来寻找伴侣,保卫我的领地,发出警报信号。但我只在自由的时候下才会唱歌。"

"但愿人们能明白我们和他们一样喜爱声音的魔力,我们会给他们带来快乐,但不是在囚禁中。"

……这仅仅是开始!……

"Look, everyone has a sad story to tell. I will keep on singing to find a mate, to defend my territory, or to signal the alarm. But I only sing when I am free."

"If people could only understand that we enjoy the magic of sound as much as they do, and we give them joy, but not in captivity."

... AND IT HAS ONLY JUST BEGUN!...

……这仅仅是开始!……

... AND IT HAS ONLY JUST BEGUN! ...

Did You Know?
你知道吗？

Nightingales have a rich sound repertoire, producing over 1000 different sounds, compared with 340 by skylarks, and 100 by blackbirds. The part of their brain responsible for sound is bigger than in most birds.

夜莺发音的种类非常丰富，能发出1 000多种不同的声音。与之相比，云雀能发出340种声音，黑鹂的发音有100种。这些鸟类的大脑中负责声音的部分比大多数鸟类都要大。

The nightingale males that produce the best songs and have the largest repertoires will end up securing females. These arrive at breeding sites earlier, have longer wings, are heavier, and in a better condition.

能唱出最好的歌并且声音百转千回的雄性夜莺最终会得到雌性的青睐。这些雄性夜莺能更早到达繁殖地，翅膀更长，体重更重，身体状况更好。

Ludwig van Beethoven included, in Symphony 6, flute imitations of the nightingale's song; Frans Liszt featured a nightingale song in Mephisto Waltzes; Pyotr Tchaikovsky composed the song The Nightingale.

贝多芬在《第六交响曲》中用长笛模仿夜莺的歌声；李斯特在《梅菲斯特圆舞曲》中有一首夜莺之歌；柴可夫斯基创作了歌曲《夜莺》。

In classical Persian, Urdu and Turkish poetry, the love of the nightingale for the rose is a metaphor for the poet's love for his beloved, and a worshipper's love for God.

在古典波斯、乌尔都和土耳其诗歌中，夜莺对玫瑰的爱隐喻诗人对情人的爱，以及礼拜者对上帝的爱。

In birds, noise leads to sleep disturbances, triggers chronic stress, causes cells to age faster, and disrupts their immune system functioning. Air pollution has proven to reduce the size of songbirds' repertoires.

对于鸟类，噪声会导致睡眠紊乱，引发慢性压力，加速细胞衰老，并破坏其免疫系统功能。空气污染已被证明会减少鸣禽的发音丰富程度。

The average length of a skylark flight song is just over two minutes, but 30-minute performances have been recorded. Most skylarks sing in the sky from around an altitude of 50 metres, but some do so at 200 metres.

飞行中云雀鸣叫的平均时长只有 2 分钟多一点，但有记录显示云雀鸣叫可达 30 分钟之久。大多数云雀在 50 米高空鸣叫，但也有一些在 200 米高空鸣叫。

Skylarks have long been regarded as a delicacy, and are caught in nets dragged across fields at night that trap hundreds of birds at a time. The birds are lured there by ingenious lark mirrors.

长期以来，云雀一直被人们视为美味佳肴。人们会在夜晚把网拖过田野，这样一次能捕获几百只云雀。人们精心设计的诱镜会把云雀引诱到那里。

Europe's smallest skylark population (just ten pairs) is found on the Faeroe Islands, while Poland has an estimated nine million pairs. One Cape Verde Island is home to 100 critically endangered Raso larks.

欧洲最小的云雀种群（只有10对）生活在法罗群岛，而波兰估计有900万对。佛得角群岛的一个岛屿是100只极度濒危的拉索云雀的家园。

Think about It

Birds learn to sing like we learn to speak?

鸟学鸣唱就像人类学说话一样吗?

Would you like to be put in a cage to sing for your master, and food?

你愿意被关进笼子里为你的主人和食物唱歌吗?

Birds, with their small brains, know 1,000 songs?

鸟儿的脑容量很小,它们会唱1 000首歌曲吗?

How do you feel when you hear a birdsong at dawn?

听到黎明鸟鸣时,你有什么感觉?

Do It Yourself!  自己动手！

There are about four thousand birds species in the world. How many different birdcalls can you identify? Which birds can you identify by the first few notes of their songs? List at least six. Start with the birds common to your area. It may be the canary, wren, skylark or nightingale. Now find at least two more. Pay attention to the birds around you, until you can identify them by the first notes you hear them make.

世界上大约有 4 000 种鸟类。你能听出多少种不同的鸟鸣声？你能通过鸟鸣的前几个发音分辨出是什么鸟吗？从你所在地区常见的鸟类开始，列出至少 6 个。比如，金丝雀、鹪鹩、云雀或夜莺。现在至少再找两种。留意你周围的鸟儿，直到你能通过它们发出的前几个音符就能识别它们。

TEACHER AND PARENT GUIDE

学科知识
Academic Knowledge

生物学	夜莺是一种从欧洲迁徙来的食虫鸟类，在撒哈拉以南的非洲过冬；夜莺喜欢在榛树上筑巢；黄褐色猫头鹰是夜莺的主要天敌；人发声用喉咙，鸟则用鸣管发声；云雀是雀形目鸟类。
化 学	金属污染影响了鸟类的鸣叫行为，在污染最严重的地区，雄性鸣禽发音种类显著减少，在黎明合鸣中的鸣叫总量也显著减少。
物 理	节奏是指在鸟类持续啼鸣中单个音符的计时单位；鸟类可以同时发出两种声音，并自己唱二重唱，这是人类声音在物理上不可能做到的。
工程学	鸟语属于最复杂的动物交流系统。
经济学	在10 000种鸟类中，有4 000种是鸣禽；歌曲和音乐的存在使人们在工作中更快乐、更有效率。
伦理学	鸟儿唱歌如此之好，以至于得到诗人、音乐家和艺术家的赞赏；把会唱歌的鸟儿关在笼子里，让它们的主人享受的愿望；圈养的夜莺不会唱歌，也不能生存。
历 史	鸣禽进化于5 000万年前；夜莺经常出现在荷马、维吉尔、奥维德、莎士比亚、艾略特、约翰·济慈的作品中；伊戈尔·斯特拉文斯基的第一部歌剧《夜莺》（1914）以安徒生的同名故事（1884）为蓝本，歌颂夜莺歌声的美丽。
地 理	夜莺喜欢灌丛（通常靠近水源）、开阔或者低矮林地；夜莺是乌克兰和伊朗的国鸟；韩国宫廷舞蹈《夜莺舞》的灵感来自夜莺；西欧和小亚细亚的夜莺迁徙到撒哈拉以南的非洲，而高加索、波斯和中亚的夜莺往东非迁徙。
数 学	通过模式识别或其他数学算法来检测结构规则，以帮助理解夜莺鸣声；歌曲节奏具有强烈的多重分形特征，表现出可预测和不可预测模式之间的波动；多重分形分析是一种用于测试信号在不同时间尺度上的方差波动的方法。
生活方式	养宠物的文化，尤指养鸣禽的文化；鸣禽学习歌唱的方式就像人类学习说话，幼鸟模仿成年鸟的鸣叫，我们对孩子唱得多，孩子们也会学得多。
社会学	普通的夜莺被诗人们当作一种象征，一种灵感；夜莺被认为"能激发诗人灵感的超级艺术大师"，这与它能发出1 000多种声音有关；找到配偶后，雄夜莺会减少哨声，并在夜晚停止鸣叫，直到雌夜莺下蛋。
心理学	伦巴第效应：人们会根据周围环境的音量来调整自己声音的大小，如果周围的噪声增加，人们会不由自主地加大声音。
系统论	1995年至2008年间，英国灌木栖息地的丧失导致夜莺数量减少了50%；30多年来，云雀的数量下降了90%。

教师与家长指南

情感智慧
Emotional Intelligence

夜莺

夜莺很为云雀担心。当报警理由消失后,夜莺轻松了一些。当云雀对真实的危险浑然不觉时,夜莺直言不讳地说不应该自欺欺人。夜莺认识到云雀是一个很棒的歌星,但不是唯一的歌星。夜莺承认成功来之不易,需要大量的培训、训练和实践。夜莺甚至在非洲越冬时继续训练。夜莺说自己可以发出1 000多种声音,而这并没有任何夸张。当受到质疑时,夜莺表达了愤怒。当被问及脑容量时,夜莺否认了所谓的脑容量不足。他透露,夜莺永远不会停止逃跑,直到死去。

云雀

云雀不理睬夜莺的警告,相信人类是友好的。当夜莺强烈反对时,云雀承认夜莺多半是对的。尽管如此,云雀还是觉得自己的歌声受到了人们的喜爱,并且给了人们一定程度的安慰,即使这并不是真正的安慰。云雀改变了声调和语言,谈论起夜晚浪漫的歌声。云雀对夜莺练习唱歌很好奇,并提出质疑。云雀质疑夜莺大脑容量太小。云雀就想弄明白,如果夜莺唱得更好,为什么人们会猎取云雀而不搭理夜莺呢?云雀表达了一个愿望,希望人们能更好地理解和对待鸟类,因为鸟类能保持声音的魔力,但只有当不被囚禁时才能做到最好。

艺术
The Arts

你会吹口哨吗?如果不会,或许是时候学了,看看鸟类是如何发出美妙声音的。可能需要一些练习(以及周围人对你的容忍),一旦你学会如何发出基本的口哨音,就可以开始下一步了。你现在可以开始吹不同的声音,并把它们串在一起创作一首歌曲。在你创作之前,你可以和鸟儿们一起唱美妙的黎明大合唱了。

TEACHER AND PARENT GUIDE

思维拓展
Systems: Making the Connections

　　长期以来，人类认为自己拥有独一无二的使用语言的特权。随着科学的进步，我们研究动物（哺乳动物和鸟类）、植物甚至细菌之间的交流方式，我们越来越了解它们的能力，这远远超出了我们最初的认知。鸟类能做人类所不能做的事：同时发出两个音符，甚至自己唱二重唱。云雀和夜莺是声音大师，或许是无知让捕猎它们成为一项运动。这些鸟不仅受到大规模诱捕计划的危害，而且通常还受两种类型的污染影响。空气中的毒素和农业中广泛使用的化学物质危害了它们的栖息地，并导致它们鸣叫时间缩短。在城市，鸣禽也受到噪声污染的影响。随着噪声水平的上升，对鸟鸣的影响是明确的、可测量的。当周围噪声很多时，人们往往会更大声地说话，而鸟类往往会变得更寂静。鸟鸣沉寂影响了鸟类和人类的生活质量。鸣禽在音乐、戏剧、诗歌和寓言中存在了几千年。它们因对人类幸福的影响而闻名。只要一想到鸟鸣，就会让人感到某种快乐。鸟类是生态系统不可或缺且令人敬畏的一部分，但也非常脆弱。因此，人们希望科学的进步能让人更多地了解鸟类鸣叫的神奇，这种声音由脑容量如此小的生物发出，却完全奉献于促进地球生命的福祉。

动手能力
Capacity to Implement

　　有趣的鸣禽并非只有云雀和夜莺。在南美洲的亚马孙地区，我们发现了一种非常有音乐天赋的鹩鹩，叫作管风琴鹩鹩或音乐家鹩鹩。在互联网上查找管风琴鹩鹩的鸣声，欣赏它们独特的发音，包括丰富的、长笛的音符混合着啾啾声和格格声。在巴西，这种鸟被称为乌拉普鲁。播放这种鸟的鸣音，让大家有机会讨论这些声音只是鸟鸣，还是真正的音乐。你很可能会成为鸟类创作音乐而不仅仅是鸣叫这一理论的支持者。

教师与家长指南

故事灵感来自
This Fable Is Inspired by

克里斯蒂娜·罗斯克
Christina Roeske

克里斯蒂娜·罗斯克以研究语言学开始她的学术生涯。2000年，她作为交换生在巴黎第七大学学习了一年的语音和语法。她成功地通过了德国国家考试，成为一名中学教师，主讲生物、德语和文学（1996—2004）。她继续从事研究，并在柏林自由大学获得神经生物学博士学位（2004—2008）。2008年至2015年，她在美国纽约市立大学的市立学院和亨特学院担任科学家。她对动物交流，鸟类鸣声学习，音乐与鸟鸣的节奏和旋律，以及鸟类学习鸣叫时的鸟类社会环境，鸟鸣的神经机制等方面有专门研究。

图书在版编目(CIP)数据

冈特生态童书.第八辑:全36册:汉英对照 /
(比)冈特·鲍利著;(哥伦)凯瑟琳娜·巴赫绘;
何家振等译.—上海:上海远东出版社,2021
ISBN 978-7-5476-1773-1

Ⅰ.①冈… Ⅱ.①冈…②凯…③何… Ⅲ.①生态环境-环境保护-儿童读物—汉、英 Ⅳ.①X171.1-49

中国版本图书馆CIP数据核字(2021)第249940号

策　　划	张　蓉
责任编辑	程云琦
封面设计	魏　来　李　廉

冈特生态童书
夜莺的歌
[比]冈特·鲍利　著
[哥伦]凯瑟琳娜·巴赫　绘
何家振　译

记得要和身边的小朋友分享环保知识哦!
八喜冰淇淋祝你成为环保小使者!

Education 287

红皮绿血

Red Skin and Green Blood

Gunter Pauli

[比]冈特·鲍利 著
[哥伦]凯瑟琳娜·巴赫 绘
何家振 译

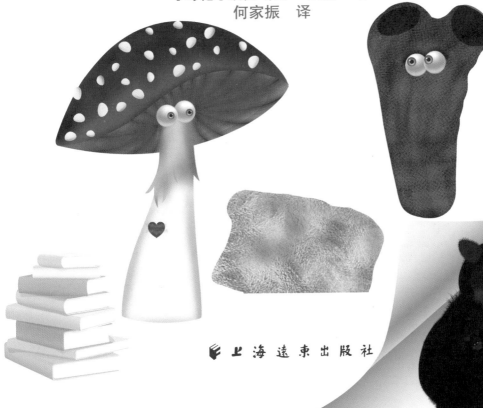

上海远东出版社

丛书编委会

主　任：贾　峰
副主任：何家振　闫世东　林　玉
委　员：李原原　祝真旭　牛玲娟　梁雅丽　任泽林
　　　　王　岢　陈　卫　郑循如　吴建民　彭　勇
　　　　王梦雨　戴　虹　翟致信　靳增江　孟　蝶

特别感谢以下热心人士对童书工作的支持：

匡志强　宋小华　解　东　厉　云　李　婧　陈　果
刘　丹　熊彩虹　罗淑怡　旷　婉　杨　荣　刘学振
何圣霖　廖清州　谭燕宁　韦小宏　李　杰　欧　亮
陈强林　王　征　张林霞　寿颖慧　罗　佳　傅　俊
胡海朋　白永喆　冯家宝

目录

红皮绿血	4
你知道吗？	22
想一想	26
自己动手！	27
学科知识	28
情感智慧	29
艺术	29
思维拓展	30
动手能力	30
故事灵感来自	31

Contents

Red Skin and Green Blood	4
Did You Know?	22
Think about It	26
Do It Yourself!	27
Academic Knowledge	28
Emotional Intelligence	29
The Arts	29
Systems: Making the Connections	30
Capacity to Implement	30
This Fable Is Inspired by	31

一只非常孤独的毒蝇伞感到被人嫌弃,想找个人说说话。她遇到一只海鞘。海鞘乐于聊天,这让毒蝇伞很高兴。

"你很难找到朋友,这并不奇怪。我们都知道你毒性非常大。人们从儿时听过的故事和看过的电影里就知道了。"

A very lonely amanita mushroom is feeling rejected, and looking for someone to talk to. She comes across a sea squirt and is pleased that he is open to having a chat.

"It is no surprise you have difficulty finding friends. We all know that you are very, very toxic. People have learnt this from the stories they've heard and the movies they've seen from an early age."

……孤独的毒蝇伞……

... lonely amanita mushroom ...

一定是因为我鲜红的皮肤……

It must be my bright red skin ...

"一定是因为我鲜红的皮肤,上面有很多白点。只要一看到我,所有人都会后退或绕道走。"

"是的,你红色皮肤上的白色斑点那么扎眼。"

"扎眼……什么?"

"It must be my bright red skin, with all these white dots. The mere sight of me makes everyone step back or walk around me."

"Yes, those white spots on your red skin make you so conspicuous."

"Con... what?"

"对其他人来说，一眼就看出你是麻烦！"
"我？是麻烦吗？就因为我长这样？"
"下雨时会发生什么？"海鞘问道。

"To others, it is immediately evident that you mean trouble!"
"Me? Trouble? Just because of my appearance?"
"What happens when it rains?" Sea Squirt asks.

下雨时会发生什么?

What happens when it rains?

……有很多蘑菇比我的毒性大得多。

… other mushrooms a lot more toxic than me.

"我的白斑消失了，"毒蝇伞回答说。

"于是，人们会误把你当成另一种蘑菇，吃掉你，然后死去。"

"有很多蘑菇比我的毒性大得多。为什么拿我说事，而且只说我？"

"I lose my white spots," Amanita Mushroom replies.

"Then people mistake you for another kind of mushroom, eat you and die."

"There are many other mushrooms a lot more toxic than me. Why pick on me, and me alone?"

"你的毒性比其他蘑菇小,但仍然有毒。不那么坏仍然是坏,知道吗?"

"我想不到一点活在这个世上的好理由。那你有什么用呢?你为什么会活在这个世界上?"

"Being less toxic than other mushrooms, you are still toxic. Less bad is still bad, remember?"

"I cannot imagine any good reason for me to exist. And what good are you? Why are you here in this world?"

……你有什么用呢?

... what good are you?

……那你的采矿设备在哪呢……

... where is your mining equipment, ...

"我们海鞘是海洋的矿工。"

"那你的采矿设备在哪呢?我看到的只是一条皮革管,有一个进水口和一个出水口。"

"我是一个很棒的采矿工。我在寻找金属的过程中不会破坏任何东西……"

"We sea squirts are the miners of the sea."
"So where is your mining equipment, then? All I see is a leathery tube, with an inlet and an outlet."
"I'm a great miner. But I don't destroy anything while looking for metals…"

"那不叫采矿！我觉得挖矿要在地表用炸药爆破，工人们在恶劣的条件下挖掘，矿石加工需要大量的热量，污染远超你想象。"

"好了，好了，好了！让我给你介绍一下我的滤食性摄食。我提取最纯净的稀有金属，然后储存在我的皮肤里。"

"Then don't call it mining! I was imagining dynamite blowing holes in the face of the Earth, workers excavating under harsh conditions, ore processed with a lot of heat and beat – polluting like you cannot imagine."

"Well, well, well! Let me introduce you to my way of filter feeding. I extract very rare metals, in their purest form. And then store them in my skin."

我觉得要用炸药爆破……

I was imagining dynamite ...

你听说过钒吗?

Have you heard of vanadium?

"你从海洋中收获最多的是什么？"

"你听说过钒吗？它让我的血液变绿了。"

"绿色血液！听起来不太好。但我知道钒能使钢变得非常坚固。"

"And what is it that you mine most from the ocean?"

"Have you heard of vanadium? It is what makes my blood green."

"Green blood! That doesn't sound good. But I do know that vanadium makes steel very strong."

"是的，所有看到我的血的人都以为我病了。不过，你对钒和它用途的了解让我很钦佩。"

"至少有你佩服我有知识——而且不害怕我有毒。谢谢，最近没什么人夸过我。"

……这仅仅是开始！……

"Yes, everyone who sees my blood thinks I must be sick. But I am very impressed that you know about vanadium and its uses."

"At least you are impressed by my knowledge – and not scared of me being toxic. Thank you, I haven't had many compliments lately."

... AND IT HAS ONLY JUST BEGUN!...

……这仅仅是开始！……

... AND IT HAS ONLY JUST BEGUN! ...

Did You Know?

你知道吗?

The tunicate may be a lifesaver. Several anti-tumour compounds have been found its tissue. One of these was the first marine-derived compound to enter Phase I and II clinical trials to fight cancer.

被囊动物有可能救人性命。在其组织中发现了几种抗肿瘤化合物,其中一种是首个进入I期和II期抗癌临床试验的海洋衍生化合物。

Tunicates have two main openings. They suck in seawater through an inhalant siphon, filtering out food particles and oxygen using a branchial basket, before expelling the water through the exhalant siphon.

被囊动物有两个主要开口。它们用进水管吸入海水,用鳃篮过滤掉食物颗粒和氧气,然后通过出水管排出海水。

Tunicate larvae look more like vertebrate animals than adults do. They are tadpole-like, and have a notochord, which is believed to be the earliest form of vertebra.

被囊动物的幼虫看起来比成虫更像脊椎动物。它们长得像蝌蚪，有一个脊索，被认为是脊椎的最早形态。

Several species of tunicates are edible, and can be eaten raw, cooked, dried or pickled. In Chile, local edible tunicates are known as *piure*.

有几种被囊类动物是可食用的，可以生吃、熟吃、干吃或腌吃。在智利，当地的可食用被囊动物被称为 *piure*。

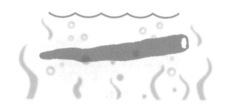

One group of tunicates, called pyrosomes, forms free-floating colonies that take the shape of a huge sock, one that can reach 20 metres in length.

有一组叫作鳞海鞘的被囊动物，能形成状如巨大袜子的自由漂浮的群落，长度可达20米。

Vanadium was discovered by Swedish chemist Nils Gabriel Sefström. Sefström named vanadium after the Norwegian goddess Vanadis, also called Freya. She is associated with beauty and fertility.

钒是由瑞典化学家尼尔斯·加布里埃尔·塞夫斯特姆发现的。塞夫斯特姆以挪威女神凡娜迪丝的名字为钒命名。她被认为美丽而多产。

Vanadium is a trace element that lowers blood sugar and cholesterol levels. In a 1994 study, it was found that goats with a diet deficient in vanadium gave birth to kids with skeletal deficiencies.

钒是一种可以降低血糖和胆固醇水平的微量元素。在1994年的一项研究中，人们发现钒元素摄入不足的山羊生的羊羔有骨骼缺陷。

Historically, the Amanita mushroom was used in Siberia and Finland, and also as a hallucinogen by the Vikings. It was considered toxic to people but no fatalities have been reported. It was also used to kill insects.

历史上，毒蝇伞曾在西伯利亚和芬兰使用，也曾被维京人用作致幻剂。它被认为对人体有毒，但没有死亡报告。它也被用来杀灭昆虫。

Think about It

想一想

Are you bad because a movie made you look bad?

你坏是因为某部电影让你看起来很坏吗?

What if no one wants to talk to you?

如果没人想跟你说话怎么办?

What would you think when you see green blood?

当你看到绿色的血液时,你会怎么想?

Mining without any equipment and without causing damage. Is this possible?

在没有任何设备和不造成破坏的情况下采矿,这是可能的吗?

Ask people if they know of the Amanita mushroom. They have most likely not heard of the name. Now show them a picture of the Amanita, with its red skin covered in white dots. What is their reaction to seeing what it looks like? Many people won't know the name, but will recognise the mushroom from the dancing mushroom scene in the Disney film Fantasia. Make sure you learn the difference between edible mushrooms that look like the Amanita, and the Amanita mushroom.

问问人们知不知道毒蝇伞（毒鹅膏菌）。他们很可能没听说过这个名字。现在给他们看一张毒蝇伞的照片，其红色皮肤上覆盖着白点。人们看到毒蝇伞的样子有什么反应？许多人可能不知道它的名字，但会在迪士尼电影《幻想曲》中看到舞动的蘑菇场景。一定要知道长得像毒蝇伞的可食用蘑菇和毒蝇伞的区别。

TEACHER AND PARENT GUIDE

学科知识
Academic Knowledge

生物学	被囊动物是一种海洋无脊椎动物；被囊类是固着的（不移动的），但有一个像蝌蚪的可移动的幼虫阶段；全球海洋中有3 000种被囊动物，主要生活在浅水区；最大的被囊动物是长1米的海郁金香；被囊动物有心脏和循环系统；被囊动物用一种从细菌"借来"的酶制造纤维素；毒蝇伞生长在桦树、松树和云杉树下，是一种菌根真菌。
化 学	毒蝇伞中的鹅膏蕈氨酸和毒蝇蕈醇是杀蝇活性药剂；由于有硫酸，被囊类体内的钒储存于pH值低于2的环境中；被囊动物将氨扩散到全身以清除含氮废物；被囊动物产生的纤维素，在动物中是绝无仅有的；被囊动物能产生有效的抗癌和抗病毒药物——膜海鞘素；北美驯鹿和驯鹿可以吃下伞形毒菌而不受毒素影响。
物 理	海鞘受到打扰时会收缩身体喷出水；被囊类动物富集钒的浓度是周围海水的1 000万倍；被囊动物是悬浮物捕食者。
工程学	通过滤食性摄食将锂、铌、钽等进行生物浓缩。
经济学	被囊类动物作为生物燃料的来源，其纤维素可转化为乙醇。其他部分可用作鱼饲料；被囊动物作为模式生物用于研究新药开发等生物过程；毒蝇伞历来是有机农场的杀虫剂。
伦理学	由于被虚构的电影和故事误导，主观地认为某件事或某个人是坏的，而不考虑每个物种或每个人所发挥的好的作用。
历 史	在寒武纪早期的化石中就有被囊动物的踪影。
地 理	最古老的被囊类化石在中国昆明附近发现；日本和韩国人吃海菠萝（被囊类动物）；毒蝇伞原产于北半球温带和寒带森林；西伯利亚萨满教的宗教仪式上使用的毒蝇伞。
数 学	用微分方程数学模型描述水生入侵物种（如被囊动物）的数量。
生活方式	通常把毒蝇伞与牛奶混合，以吸引和杀死苍蝇。
社会学	尽管没有死亡记录，人们仍然认为毒蝇伞的毒性很大，不过它确实会让人生病；在超级马里奥兄弟的电子游戏、格林兄弟的故事《糖果屋历险记》和迪士尼电影《幻想曲》中都有毒蝇伞。
心理学	当受到错误指责时，会造成身份丧失、耻辱、身心健康损害、关系破裂、朋友和家人态度改变；对收入和就业的影响。
系统论	作为船体上的附着生物或幼虫时进入压舱水从而成为具有侵害性的被囊类动物；毒蝇伞与树木有共生关系。

教师与家长指南

情感智慧
Emotional Intelligence

海 鞘

海鞘对毒蝇伞采取强势。他指出毒蝇伞很容易被认出来。他的话直截了当,没有任何外交辞令。他有信心不接受她的说法,即毒蝇伞的毒性比其他蘑菇小。当被问及他在地球上(和海洋中)扮演的角色时,他快速而坦率地给出回答。他深知自己是一个不会造成任何损害的采矿者。他很乐意介绍滤食性摄食的知识,并予以详述。起初,他并不认为毒蝇伞聪明,但后来意识到她知识广博,并对她表示钦佩。

毒蝇伞

毒蝇伞感到孤独和被孤立。当她发现有人愿意和她说话时,她很高兴。她很清楚,她的外表就足以把人吓跑。她对海鞘告知她媒体报道中她就是麻烦感到震惊。她大方地问,她活在世上是否有一个好的理由。她询问了海鞘活在世上的理由和他的采矿活动,并惊喜地发现他并没有像她想象的那样破坏环境。当她听说他有绿色血液时,她很担心他的健康。她对他开采的是什么表现出了兴趣,并与他讨论钒,证明自己在这方面的知识渊博。当海鞘对此表示钦佩并恭维她时,她表达了对海鞘的感激之情。

艺术
The Arts

让我们画一些线条来练习"快"艺术。看看你能不能在1分钟内画出一朵毒蝇伞。当你能够在一张纸或黑板上画一幅令人信服的毒蝇伞后,再试着在20秒内完成。然后看看你能不能在10秒钟内做到。

思维拓展
Systems: Making the Connections

在许多童话故事中，毒蝇伞被描绘成致命的有毒物质，几乎每个人都相信它是一种致命的蘑菇。事实上，最近并没有任何致人死亡的记录，但它在人们心中的形象并没改变。毒蝇伞在生态系统中扮演着重要的角色，作为一种菌根真菌，它可以与从云杉到桦树等树木交换关键的营养物质。自古以来，它就被用作杀虫剂，不仅有机有效，而且简单好用。自然界的一切都有其存在的理由，并且在生态系统中扮演着重要的角色。大众媒介制造了负面印象，夸大了危险，却忽视了其正面作用的信息。大自然总是充满了惊奇，常规之下却有许多例外。没有一个适用于一切情况的永恒标准。在这里，好奇心的力量以及拥有开放和求索的心态是很重要的。我们需要保持好奇心，时刻准备着看得更深，寻找突破成见的东西，以此表示对其他生命形式的尊重。这样，我们就有机会发现生物多样性的价值。在这个世界里，我们将不断发现从未听说过的新事实和新事物，比如不会对环境造成任何破坏的开采活动。对新信息保持开放的力量，展开联想，随时准备发现更多可能对环境以及这个星球上众生万物有益的事物。

动手能力
Capacity to Implement

如何自制杀虫剂？要非常小心，而且只能在老师或家长的监督下制作。如果你很幸运住在有云杉或桦树的森林附近，那就去采摘一些毒蝇伞。不要吃它。现在，把毒蝇伞晾干。接下来将它浸泡在加水稀释的牛奶中。用这种液体作为杀虫剂。如果你找不到毒蝇伞或者不想用它，可以选择另一项任务：研究三种滤食性动物，并了解它们的滤食性摄食是如何进行的。利用研究得来的信息设计一种不需要炸药或化学物质，只依靠物理定律的采矿方法。你可以想象一个采矿新时代！与朋友和家人分享这些信息。

教师与家长指南

故事灵感来自

This Fable Is Inspired by

詹妮弗·迪杰斯特拉
Jennifer Dijkstra

詹妮弗·迪杰斯特拉在加拿大新不伦瑞克大学获得文学学士学位。她在德国不来梅大学获得理学硕士学位。2007年，她继续学术生涯，在美国新罕布什尔大学获得动物学和动物生物学博士学位。2008—2013年，她在美国威尔斯国家河口研究保护区从事博士后研究。她目前是新罕布什尔大学海洋科学和海洋工程学院（新罕布什尔大学的海岸和海洋测绘中心）的研究助理教授。她的主要研究方向包括生物多样性和生物地理学的模式和过程、栖息地结构、引进物种等。她领导了入侵性海鞘扩散的研究，有些人认为这是一个问题，另一些人则认为这是一种生物采矿形式。

图书在版编目(CIP)数据

冈特生态童书.第八辑:全36册:汉英对照/
(比)冈特·鲍利著;(哥伦)凯瑟琳娜·巴赫绘;
何家振等译.—上海:上海远东出版社,2021
ISBN 978-7-5476-1773-1

Ⅰ.①冈… Ⅱ.①冈…②凯…③何… Ⅲ.①生态环境-环境保护-儿童读物—汉、英 Ⅳ.①X171.1-49

中国版本图书馆CIP数据核字(2021)第249940号

策　　划　张　蓉
责任编辑　程云琦
封面设计　魏　来　李　廉

冈特生态童书
红皮绿血
[比]冈特·鲍利　著
[哥伦]凯瑟琳娜·巴赫　绘
何家振　译

记得要和身边的小朋友分享环保知识哦！
八喜冰淇淋祝你成为环保小使者！

Education 288

小不点儿
Teeny-Weeny

Gunter Pauli

[比] 冈特·鲍利 著
[哥伦] 凯瑟琳娜·巴赫 绘
何家振 译

上海远东出版社

丛书编委会

主　任：贾　峰

副主任：何家振　闫世东　林　玉

委　员：李原原　祝真旭　牛玲娟　梁雅丽　任泽林

　　　　王　岢　陈　卫　郑循如　吴建民　彭　勇

　　　　王梦雨　戴　虹　翟致信　靳增江　孟　蝶

特别感谢以下热心人士对童书工作的支持：

匡志强　宋小华　解　东　厉　云　李　婧　陈　果
刘　丹　熊彩虹　罗淑怡　旷　婉　杨　荣　刘学振
何圣霖　廖清州　谭燕宁　韦小宏　李　杰　欧　亮
陈强林　王　征　张林霞　寿颖慧　罗　佳　傅　俊
胡海朋　白永喆　冯家宝

目录

小不点儿	4
你知道吗？	22
想一想	26
自己动手！	27
学科知识	28
情感智慧	29
艺术	29
思维拓展	30
动手能力	30
故事灵感来自	31

Contents

Teeny-Weeny	4
Did You Know?	22
Think about It	26
Do It Yourself!	27
Academic Knowledge	28
Emotional Intelligence	29
The Arts	29
Systems: Making the Connections	30
Capacity to Implement	30
This Fable Is Inspired by	31

一只侏儒海马想知道她是不是大海马家族中唯一很小的成员。当她遇到一种比其他兰花小得多的矮种兰花时,她问道:

"你这么小是因为吃得不够吗?"

A pygmy seahorse wonders if she is the only very small member of the big seahorse family. When she meets a pygmy orchid, one much smaller than other flowers in the orchid family, she asks,

"Are you this small because you didn't get enough food?"

一只侏儒海马……

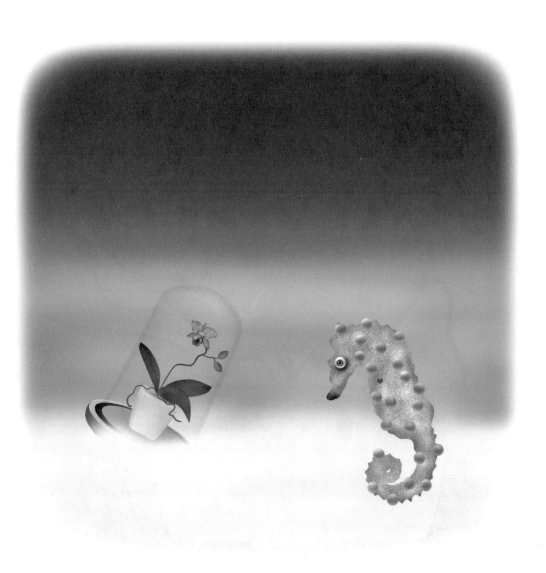

A pygmy seahorse ...

你觉得小就可爱吗?

you think that small is cute?

"哦，不！我有我需要的所有食物。我们家所有的兰花都这么小。我们很高兴自己这么小，这么可爱。"

"你觉得小就可爱吗？"

"嗯，瞧瞧你自己，你是一个小侏儒海马。所以毫无疑问，人们会觉得你很迷人。"

"Oh no! I had all the food I needed. All the orchids in my family are this small. We are happy to be so small, and so cute."

"You think that small is cute?"

"Well, look at yourself, you are a dwarf-sized seahorse. So there should be no doubt that people find you quite charming."

"我也见过非常小的人,还有马。看上去在自然界中,植物或动物有一些袖珍品种,这是普遍情况,而不是例外……"

"嗯,我非常满意自己只有其他海马品种的三分之一大小。"

"I have also seen people and horses that are very small. It seems that in Nature having a few miniature versions of a plants or animals is the rule, rather than the exception…"

"Well, I am completely content being about a third the size of other seahorse species."

……非常小的马。

... horses that are very small.

对，多样性……

Diversity, yes...

"无论我们体型是大是小,我们都可以尽自己最大的能力做出贡献。这就是一个群落成功的原因,有大大小小的成员,有年轻的和年老的……"

"对,多样性,我明白了!但我还是想知道,你和我是怎么变得这么小的?"

"Whatever our size, we can all contribute to the best of our ability. That is what makes a community so successful, having members that are big and small, young and old …"

"Diversity, yes, I get it! But I still wonder how did you and I turn out so teeny-weeny?"

"也许我妈妈在她还很年轻没长大的时候就生了我,所以我也很小吧?"

"要么是我们没有得到让肌肉和骨骼生长的合适食物。"

"Maybe my mother had me when she was still very young and still small, so I ended up very small as well?"

"Or, perhaps we did not get the right food to make our muscles and bones grow."

也许我妈妈……

Maybe my mother …

我见过剧院里用小矮人演出……

I have seen dwarfs used in theatres ...

"也许是因为我们的基因！"侏儒海马回答说。"不过，个子这么小，肯定也有好处。"

"有时候别人会因为我长得矮小而嘲笑戏弄我。我见过剧院里用小矮人演出——供人消遣！"

"Maybe it is in our genes!" The pygmy seahorse replies. "But to be this small must surely have some advantages."
"Sometimes others laugh at me and ridicule me because of my size. I have seen dwarfs used in theatres – for entertainment!"

"是的,一些小矮人扮演小丑,但我们这些侏儒生物却更受人喜爱。一旦我们被注意到,我们就会得到更多的关注。"

"恐怕我不能同意!看看人类多么喜欢巨型恐龙,再看看他们是如何恶劣地对待自己的同类俾格米人……"

"Yes, some teeny-weeny people are used as clowns, but we pygmy creatures are also loved more. Once we are noticed we do get more attention."

"I'm afraid I cannot agree! Look at how popular those huge dinosaurs are with people. And look at how badly people have treated the pygmies of their own species…"

……人类多么喜欢巨型恐龙……

... how popular those huge dinosaurs are ...

……你很不容易被注意到……

... easy for you to go unnoticed ...

"这么说,你认为作为一个极小的侏儒海马,优势之一是你很不容易被注意到,尤其是不容易被捕食者注意到?"矮种兰花问道。

"哦,是的,对我来说,要想过幸福的生活,最好不要引人注意。我们与环境融合得非常好,几乎不用担心被突然袭击。"

"So would you say that one of the advantages of being a teeny-weeny little pygmy seahorse is that it is easy for you to go unnoticed, especially by predators?" Pygmy Orchid asks.

"Oh yes, for me to lead a happy life, it is better if I go unnoticed. We blend so well into the environment that we hardly have to worry about surprise attacks."

"只要不是不被欣赏——仅仅因为我们特别,那么不被关注,特别是不被媒体关注就没有任何错!"

……这仅仅是开始!……

"Nothing wrong with going unnoticed especially by the media, as long as we do not go unappreciated – for our uniqueness!"

... AND IT HAS ONLY JUST BEGUN!...

……这仅仅是开始！……

... AND IT HAS ONLY JUST BEGUN! ...

你知道吗?

Pygmies are ethnic groups in the tropical forest areas of central Africa. Amongst those of pure blood, men have an average height of just 1.45 m and women of 1.33 m. Pygmies have been plundered and massacred on a large scale in history and are very close to extinction in the world.

俾格米人是非洲中部热带森林地区的民族。成年男性平均身高1.45米,成年女性平均身高1.33米。俾格米人在历史上遭到较大规模的劫掠和屠杀,在世界上濒临灭绝。

In the Congo, Pygmies make up 2% of the population. Many Pygmies were treated as slaves by the Bantu majority.

在刚果,俾格米人占人口的2%。许多俾格米人被占人口多数的班图人当作奴隶对待。

At the beginning of the 20th century, Pygmies were even put on display in zoos as a special species.

20世纪初，俾格米人甚至被当成一种特殊的物种，放在动物园里展览。

Pygmy seahorses are seven species of miniature seahorses that live in the Coral Triangle in the Pacific Ocean between Indonesia, Malaysia, Philippines, Papua New Guinea, Timor and Solomon Islands.

侏儒海马是生活在印度尼西亚、马来西亚、菲律宾、巴布亚新几内亚、东帝汶和所罗门群岛之间的太平洋珊瑚三角的七种微型海马。

Pygmy seahorses have a single gill opening on the back of the head (all other seahorses have a pair of gill openings) and the young are brooded within the male's trunk rather than a pouch on the tail.

侏儒海马在头部后面有一个单独的鳃孔（其他海马都有一对鳃孔），幼海马在雄海马的鼻子里而不是在尾巴上的育儿袋里孵化。

Pygmy seahorses are only found living on sea fans. The Bargibanti pygmy seahorse lives only on one genus of coral, Muricella spp, whilst the Walea pygmy seahorse is found living in association with soft corals.

侏儒海马只生活在海扇上。巴氏侏儒海马只生活在一个珊瑚属（类尖柳珊瑚），而瓦里侏儒海马被发现与软珊瑚生活在一起。

There are 50 varieties of pygmy orchids. They are endemic to Australia, New Zealand and New Caledonia, Cuba, and Jamaica. Some species can live more than 100 years.

矮种兰花有 50 种。它们是澳大利亚、新西兰、新喀里多尼亚、古巴和牙买加的特有种。其中有些品种可以活到 100 年以上。

An orchid seedpod can contain 3 million seeds. The seeds are so small that you can only see them under a microscope. Since they need a fungus to germinate and grow, there are no nutrients in the seed.

一株兰花的种子荚里有 300 万颗种子。这些种子很小，只有在显微镜下才能看到。它们需要真菌来发芽和生长，种子中没有营养。

Think about It

Are tiny ones more liked and better cared for than big ones?

小的比大的更受人们喜爱，能得到更好的照顾吗？

Is it better to have children young, or when you are older?

是岁数小的时候生孩子好，还是大一点的时候生孩子好？

Do you like big dinosaurs or a tiny seahorse?

你喜欢大恐龙还是小海马？

Can everyone, independent of size, be their best and contribute?

每个人无论高大或矮小，都能做到最好并有所贡献吗？

Do It Yourself! 自己动手！

How tall is everyone in your family? Is the tallest the happiest one, and the smallest the unhappy one? As we will see, happiness has nothing to see with size, but check for yourself to see if you come to the same conclusion. And when you have made your observations, then share them with friends and family.

你的家人都有多高？个子最高的是最幸福的，个子最小的是最不幸福的吗？正如我们将看到的，幸福与高矮无关，但你自己检查一下，看能否得出同样的结论。把你的观察分享给亲朋好友。

TEACHER AND PARENT GUIDE

学科知识
Academic Knowledge

生物学	世界上有3万至3.5万种兰花；兰花的大小、形状、颜色和香气随花粉传递者的不同而异；海马以微小的甲壳类动物为食，每天不敢冒险移动超过20—30厘米；侏儒症是一种疾病（成年人身高低于147厘米）；大多数侏儒症是由怀孕前卵子或精子细胞的基因变化引起的；至少有20个侏儒动物品种，包括变色龙、河马、大象、蓝鲸、猫头鹰等。
化 学	生长激素是一种多肽激素。
物 理	兰花的被膜就像海绵一样，可以从环境中快速吸收水分；兰花种子能经受长时间的冰冻和干旱而不丧失发芽能力；巴氏侏儒海马身体上的圆形结节，总是与它的宿主珊瑚的颜色和形状相匹配；侏儒海马没有眼睑，对光很敏感（请不要用闪光灯拍照）。
工程学	矮马是选择繁育的结果。
经济学	兰花具有观赏、药用、芳香剂、催情、礼节和饮食等用途，这创造了巨大的商业机会；全球兰花贸易价值至少60亿美元。
伦理学	袖珍物种并非异常，而是动植物生物多样性的一部分；不管体型大小，都要培养善良；由于身材矮小而受到灭绝人性的残酷对待；描述矮小者的正确术语。
历 史	迷你马最早于17世纪在欧洲培育，一个世纪后成为贵族的宠物；1939年上映的电影《绿野仙踪》中有124名侏儒演员。
地 理	兰花的最大种群分布在安第斯山脉，海拔1 000到3 000米之间；哥伦比亚有4 270种兰花，其中1 572种是哥伦比亚特有的；在哥伦比亚，兰花的种类比一个四季分明的国家的全部树木种类还要多；东京是唯一一个在其附近海域有侏儒海马的首都城市。
数 学	模拟两代之间间隔时间的"晚育"情景，直接受到女性生育年龄或育龄期出生人口的分布影响；侏儒症的遗传变化是随机发生的。
生活方式	侏儒应该受到尊重，而不是被嘲笑；足球运动员梅西个子很矮，但不妨碍他成为世界上最优秀的球员之一。
社会学	一个身材异常矮小的人，过去被称为侏儒（midget），这个词现在被认为带有贬义；残疾人面临的障碍问题，比如门把手太高够不着，门太重打不开。
心理学	将侏儒症患者称为侏儒（midget）会带来羞辱和孤立；一个人在小的时候受到关注会导致更强的个性发展；麦科洛视错觉效应。
系统论	兰花在生态系统运行中起着关键的生态作用，是生态保护状况的一个指标；珊瑚礁退化、栖息地丧失、海洋酸化与温度上升、爆破和刺网捕捞造成的栖息地破坏、污染以及沿海旅游开发，都威胁着侏儒海马的生存。

教师与家长指南

情感智慧
Emotional Intelligence

侏儒海马

侏儒海马好奇地想知道矮种兰花为什么这么小,是否和食物有关。她也想知道自己为什么这么小。她想知道自己的身份,想知道小是不是真的可爱。她认为拥有袖珍品种似乎是所有物种的通则。她以积极心态思考,相信每个人都能做出贡献,无论身材大小,无论老少。她在寻找自己这么小的成因,想知道基因、食物或怀孕的时间与其后代大小是否有关。在迸发了一系列令人激动的想法后,受到矮种兰花思维方式的启发,她看到了光明的一面,认可媒体和电影的影响。侏儒海马愿意承认小而不引人注意的优势。

矮种兰花

矮种兰花公开表示不同意侏儒海马的观点,并承认很高兴自己可爱而且小。她坚持认为,小就等于有魅力,虽然这意味着只有同类的三分之一大小。她还问了一个问题:在我们生命的早期阶段发生了什么,导致我们这么小?她考虑侏儒海马的思维方式,并要求人们看到生活的光明面。她认为,如果你个头小,就会受到更多的关注,这有助于头脑进化,变得更好、更强。然后,她同意了侏儒海马的观点:如果你想要快乐,那么最好不引人注目。

艺术
The Arts

如何融入环境,让自己隐形?可以通过一种独特的技术来实现,那就是视错觉。我们经常只看到明显的图像,而不能识别隐藏的图像。所以,试着运用艺术技巧,想象一张脸或一片森林,并包含只靠简单观察很难找到的昆虫或字符。

TEACHER AND PARENT GUIDE

思维拓展
Systems: Making the Connections

在自然界，有规则就有例外是正常的，这里所谓的规则就是多样性。这条规则不仅适用于生物多样性和共享这个星球的几百万物种，也适用于已知的单一物种内的多样性。据估计，在35 000种兰花中，大约有50种是矮种的，有些小到很难被认出是兰花科。说到海马，46种海马中有7种是侏儒海马。当我们研究所有的动植物时，我们经常发现袖珍物种。它们的矮小似乎主要是因为基因突变，而与营养或怀孕时间没有任何关系。然而，全世界对待侏儒的方式绝非可效法的，例如含有贬义的名字，如"矬子"。在过去半个世纪里，受新的职业机会、医疗进步、残疾人权利运动的影响，许多国家的侏儒获得了更多的肯定和认同。与此同时，我们也发现了人们对侏儒海马和矮种兰花的兴趣，仅提到其大小就对这种动物或植物产生了巨大的科学兴趣。这不是受探索新物种的刺激而进行的，而是对激动人心的生物多样性的承认。令人惊讶的是，变色龙、河马、大象、猫头鹰、响尾蛇等都有袖珍品种。这些微型物种的发现是非常及时的，不仅对科学研究，而且对现代社会中倡导侏儒的作用和重要性都有积极影响。正是由于数以千计的微型物种的发现，让我们能够一劳永逸地改变一种剥夺侏儒的文化，这种文化并不符合现代人类社会的基本伦理。

动手能力
Capacity to Implement

做一些研究，列出至少20种微型动植物品种。和亲朋好友分享你的清单，看看他们对此是否一无所知。现在讨论一下侏儒症的问题。当说到人类的侏儒时，帮助别人理解侏儒被认为是一种身体状况，而不是疾病或异常。讨论一下在谈论身材矮小的人时可以使用的词汇。

教师与家长指南

故事灵感来自
This Fable Is Inspired by

理查德·史密斯
Richard Smith

　　理查德·史密斯于 2002 年毕业于英国南安普顿大学，获得动物学学士学位。2005 年，他获得澳大利亚昆士兰大学海洋生态与进化硕士学位。理查德于 2007 年开始研究侏儒海马生物学与保护，并于 2011 年 4 月获得博士学位。这是在小型鱼类领域所做的第一项生物学研究，也是这个领域的第一个博士学位。理查德是伦敦皇家地理学会和林奈学会的会员。2016 年，理查德应邀加入 IUCN 海马、尖吻鱼和刺鱼专家组（SPSSG）。他最近还被任命为 iSeahorse.org 的"全球侏儒海马专家"，运用公民科学在世界各地进一步推动研究和保护海马。

图书在版编目(CIP)数据

冈特生态童书.第八辑:全36册:汉英对照/
(比)冈特·鲍利著;(哥伦)凯瑟琳娜·巴赫绘;
何家振等译.—上海:上海远东出版社,2021
ISBN 978-7-5476-1773-1

Ⅰ.①冈… Ⅱ.①冈… ②凯… ③何… Ⅲ.①生态环境–环境保护–儿童读物—汉、英 Ⅳ.①X171.1-49

中国版本图书馆CIP数据核字(2021)第249940号

策　　划	张　蓉
责任编辑	程云琦
封面设计	魏　来　李　廉

冈特生态童书

小不点儿

[比]冈特·鲍利　著
[哥伦]凯瑟琳娜·巴赫　绘
何家振　译

记得要和身边的小朋友分享环保知识哦!
八喜冰淇淋祝你成为环保小使者!

Water 253

苹果酒灭火

Cider Douses Fire

Gunter Pauli

［比］冈特·鲍利 著
［哥伦］凯瑟琳娜·巴赫 绘
闫世东 译

上海远东出版社

丛书编委会

主　任：贾　峰

副主任：何家振　闫世东　林　玉

委　员：李原原　祝真旭　牛玲娟　梁雅丽　任泽林
　　　　王　岢　陈　卫　郑循如　吴建民　彭　勇
　　　　王梦雨　戴　虹　翟致信　靳增江　孟　蝶

特别感谢以下热心人士对童书工作的支持：

匡志强　宋小华　解　东　厉　云　李　婧　陈　果
刘　丹　熊彩虹　罗淑怡　旷　婉　杨　荣　刘学振
何圣霖　廖清州　谭燕宁　韦小宏　李　杰　欧　亮
陈强林　王　征　张林霞　寿颖慧　罗　佳　傅　俊
胡海朋　白永喆　冯家宝

目录

苹果酒灭火	4
你知道吗？	22
想一想	26
自己动手！	27
学科知识	28
情感智慧	29
艺术	29
思维拓展	30
动手能力	30
故事灵感来自	31

Contents

Cider Douses Fire	4
Did You Know?	22
Think about It	26
Do It Yourself!	27
Academic Knowledge	28
Emotional Intelligence	29
The Arts	29
Systems: Making the Connections	30
Capacity to Implement	30
This Fable Is Inspired by	31

一棵苹果树和一棵梨树正在瑞典北部的一个果园中享受夏日午后的美好时光。

"今年我们的授粉情况非常好。你能想象到,仅凭我一棵树就能长出500个苹果吗?"

梨树回答道:"我们也仅在一棵树上就能长出多达100千克的梨。的确,我们都干得不错。"

An apple tree and a pear tree are enjoying a wonderful summer afternoon in an orchard in northern Sweden.

"We have been so well pollinated this year. Can you imagine me, just one tree, producing five hundred apples?"

"And we have up to a hundred kilograms of pears on just one tree. We are both doing very well, indeed," Pear Tree replies.

......果园夏日午后的美好时光......

... summer afternoon in an orchard ...

树木越多意味着每公顷土地产出的苹果越少

More trees mean less apples per hectare

"人们总想在同一个果园中种更多的树,给土壤浇更多的水,施更多的肥。但你知道这是有上限的,达到某个数量后树木越多意味着每公顷土地产出的苹果越少。"

"这些人——有的穿着白大褂,其他则穿西装打领带——总是在仔细计算每平方米果树数量的增加能否抵消每棵树苹果产量的减少。"

"People always want to plant more trees in the same orchard, adding more water and fertiliser to the soil. But there is a limit, you know, there is a point where more trees mean less apples per hectare."

"These people, some in white coats and others in suits with a tie, are always carefully calculating whether a drop in the number of apples per tree is compensated for by having more trees per square metre."

"人们为什么总是在寻找更多相同的东西?他们都对水果做了些什么?他们难到不知道吃当季的苹果和梨才是最好的吗?"

"没错!大自然充满了智慧,在人类一年中最需要营养的时候给予他们富含营养的水果。"

"What is it with them always searching for more of the same? What do people do with all the fruits? Don't they know it is best to eat the apples and pears when they are in season?"

"Exactly! Nature is so smart, providing people with the fruits that give them the nutrients in the time of the year when they need them most."

他们都对水果做了些什么?

What do people do with all the fruits?

榨出果汁和制成罐头……

Juicing and canning...

苹果树问道:"但是,如果每棵树上都结着500个苹果或梨,水果会不会太多了?"

"这就是人们加工处理水果的原因,这样他们就可以随时吃到水果了。"

"你的意思是榨出果汁和制成罐头,对吗?然而,吃新鲜水果或喝鲜榨果汁才是最理想的,或者至少使水果发酵。"

"使水果发酵?"

"But if every tree bears five hundred apples or pears, won't there be too much fruit?" Apple Tree asks.
"That is why they process fruits, so people can enjoy these all year round."
"You mean juicing and canning, don't you? Even when it is always best to eat our fruit or drink our juice fresh, or at least ferment it."
"Ferment it?"

"梨树，你不知道什么是发酵吗？"

"你听我说，学习和了解是需要时间的。我知道当我们的果实在树上成熟后，会有气味很特别的尘土散布在空气中。"

"那些不是尘土！那些是酵母菌，以果糖为食，使水果发酵。"

"Do you not know what fermentation is, Pear Tree?"
"Look, there is a time to learn and there is a time to know. What I do know is that when our fruit on the tree gets very ripe, then there is a lot of dust in the air that smells … different."
"That is not dust! That is yeast, feasting on the fruit sugars, fermenting it."

酵母菌，以果糖为食……

yeast, feasting on the fruit sugars …

一种美味的饮料，即苹果酒……

A delicious drink, called cider…

"发酵是不是意味着……破坏？"

苹果树回答道："不，发酵是改良。人类榨出果汁，让酵母菌把果汁变成一种美味的饮料，即苹果酒。"

"所以，这些漂浮在空气中的几乎看不见的小生物的工作就是发酵？"

"Fermenting means … destroying?"
"Fermenting means improving." Apple Tree replies. "People press the juice, and allow the yeast to turn the juice into a delicious drink, called cider."
"So, that is the job of those almost invisible little things floating in the air?"

"是的,酵母菌能出色地完成任务——把果汁变成随时都可以享用的瓶装饮料。"

"而且有些人喜欢在午餐和晚餐时享用。"

苹果树提醒道:"不管他们喝的是什么,只希望他们不要喝太多。"

"Yes, they do a remarkable job, turning juice into a bottled drink that can be enjoyed all year round."
"And that some people love to have it with their lunch and dinner."
"As long as they are careful to always drink in moderation, whatever they are drinking!" Apple Tree warns.

不要喝太多……

Careful to always drink in moderation ...

人们可以用苹果酒来灭火……

people can use cider to stop a fire …

"有一次，一个人把苹果酒洒到衬衫上了，然后当他回到工作中，他发现焊接时潮湿的衣服没有因为接触到火花而被点燃。"

"你是想告诉我人们可以用苹果酒来灭火吗？"

"没错！想象一下，还有多少大自然的奇迹有待人类去发现！仅仅是一种水果就可以用来灭火。"

"Once, someone spilled cider on his shirt, and when he went back to work he saw that while welding, the sparks didn't burn the wet fabric."

"Are you trying to tell me that people can use cider to stop a fire?"

"Yes, and imagine how many other marvels of Nature people have yet to discover! Just a fruit that can be used to douse a fire."

"我还不知道这些呢!只要人们愿意花时间近距离观察并不断探索,就会发现大自然充满了奇妙之处。"

……这仅仅是开始!……

"I did not know that! Nature is full of wonderful surprises – if only people would take the time to look closer and discover more."

... AND IT HAS ONLY JUST BEGUN!...

……这仅仅是开始！……

... AND IT HAS ONLY JUST BEGUN! ...

Did You Know?
你知道吗？

The wild apple is native to Central Asia, now Kazakhstan, where it has been growing for millions of years. The City of Almaty claims to be the birthplace of the apple. Alma-Ata means "The Father of Apples" in Kazakh.

野生苹果原产于中亚，在如今的哈萨克斯坦境内，那里苹果已经生长了数百万年。阿拉木图声称自己是苹果的发源地。阿拉木图在哈萨克语中的意思是"苹果之父"。

People have enjoyed apples since at least 6500 BC, and they are now the most widely farmed fruit in the world, with over 7,500 varieties. The crab-apple is the only native apple in North America.

早在公元前6500年，人们就开始享用苹果了。现在，苹果是世界上种植范围最广的水果，有超过7 500个品种。沙果是唯一的北美洲原生苹果。

The Romans discovered apples in Syria and introduced them throughout their Empire, practicing the skill of grafting apples (which belong to the rose family of plants).

罗马人在叙利亚发现了苹果，并将其推广到整个罗马帝国。他们不断实践并练习嫁接苹果（苹果属于蔷薇科植物）的技巧。

Apple trees can live for over 100 years. Apples blossom late in the spring, minimising frost damage, and can grow at much higher latitudes than other fruit. Apples account for 50% of the world's deciduous fruit trees.

苹果树可以活100多年。苹果树在晚春开花，减轻了霜冻的危害，而且相比其他水果可以在纬度更高的地区生长。苹果树占世界落叶果树总数的50%。

The pear is related to the apple, also from the rose family. Rich in fibre, it is traditionally used to treat constipation and intestinal inflammation, and to cure conditions such as cystitis and kidney stones.

梨是苹果的近亲，同样属于蔷薇科。它富含纤维，在传统医学中被用来治疗便秘、肠道炎症、膀胱炎和肾结石等疾病。

The book *Tsee Ming Yau Su*, written by Chia Shi-yi, in the sixth century describes the growing of pears. Pear trees were planted to ward off misfortune.

贾思勰于公元6世纪写的《齐民要术》中描述了梨的种植。人们种植梨树来辟邪。

In China, the pear is symbolic of justice, longevity, purity, wisdom, and benevolent administration. In Korea, the pear is a symbol of grace, nobility, and purity, whereas the pear tree symbolises comfort.

在中国，梨象征着正义、长寿、纯洁、智慧和仁政。在韩国，梨象征着优雅、高贵和纯洁，然而梨树则象征着舒适安逸。

The mixture of bases and acids from apples, pears and citrus fruit, absorb thermal energy, extinguishing flames and cooling the burning material. Carboxylic acids and inorganic alkalis produce a salt.

来自苹果、梨和柑橘类水果的碱和酸的混合物能吸收热能、熄灭火焰并使燃烧的物体降温。羧酸和无机碱反应会生成一种盐。

Think about It

Do you believe natural chemistry can be as good as petrochemistry?

你相信天然化学能和石油化学一样好吗?

Which do you prefer: an apple or a pear?

你更喜欢苹果还是梨?

Are fruits good medicine?

水果是良药吗?

Is Nature full of surprises?

大自然是否充满了惊喜?

Do It Yourself!

自己动手！

Study the Krebs cycle. It is more than a simple series of chemical reactions; it is chemistry at its best! The Krebs cycle lies at the core of all our bodily energy, and its logic allows us to imagine using fruit as a fire retardant. Not everyone quickly understands how it works, so enlist the help of friends and family members who can guide you, and together you will learn to understand the process. Once you have grasped it, put it in very simple language so that you can explain to even more people.

研究一下克雷布斯循环（三羧酸循环）。它可不只是一系列简单的化学反应，而是化学领域的最高峰。克雷布斯循环是我们体内能量循环的核心，它的逻辑使我们想到用水果作为阻燃剂。你可能难以迅速理解这一过程，因此你要想想有哪些朋友和家人可以帮你一起理解其中的原理。当你掌握了原理后，试着用简单的语言向所有人说明这一原理。

TEACHER AND PARENT GUIDE

学科知识
Academic Knowledge

生物学	三羧酸循环是细胞的主要能量来源，是有氧呼吸的一部分；蛋白质、脂肪和碳水化合物代谢的结合；种子发芽长成的苹果树不会结出像上一代那样的果实；果皮含有大量的营养成分；水果的发酵；酵母菌的作用。
化　学	受热膨胀物质往往是质子供体，包含不稳定的阳离子，这是一种不可燃气体，也是一种碳源；盐的比热容较低；甲酸盐；卤素，如氟、氯、溴；葡萄糖在葡萄糖代谢的过程中被氧化形成丙酮酸，然后继续被氧化成乙酰辅酶A并进入三羧酸循环；苹果含有甲醛。
物　理	以冷却（吸热）或阻燃的方式灭火；受热膨胀，体积增大，密度减小。
工程学	发芽和嫁接的区别。
经济学	苹果贸易额高达800亿美元；苹果可作为主粮，受经济变化影响小；种植密度更高、成本更高、每棵树产量更低但整体收入更高的经济模式；储存水果以便在非收获季享用。
伦理学	使用食品级阻燃剂可避免向家庭和自然界散播有害物质。
历　史	古希腊的荷马在3 000年前就写过"梨是神明的礼物"；老普林尼(公元23—79年)在一份有60多个版本的手稿中描述了几乎所有梨的品种；艾萨克·牛顿在17世纪通过观察苹果掉落而得出万有引力定律；汉斯·阿道夫·克雷布斯在1937年发现了三羧酸循环，被《自然》杂志拒绝后，该研究发表在《酶学》期刊上，于1953年获得诺贝尔奖。
地　理	苹果原产于哈萨克斯坦。
数　学	"梨形"的几何性质；一种特殊的钻石切割。
生活方式	即使很多营养成分都在果皮中，我们仍会削掉苹果和梨的皮；吃应季水果，能在我们最需要的时候提供矿物质和维生素。
社会学	分享美食和饮品的重要性，以及通过观察细节获得不同寻常的新见解的重要性。
心理学	学习新事物时惊喜的重要性；通过情感联想开展学习的重要性。
系统论	所有事物都是循环的一部分，我们仍在探索其中的联系。

教师与家长指南

情感智慧
Emotional Intelligence

苹果树

苹果树为自己的产量而自豪。然而他担心人们为了追求更高的产量会增加浇水量和施肥量,并在同一块地上种植更多的苹果树。他分享了他的看法:人们应该吃应季水果。他和梨树就如何处理全年的收成展开讨论。当梨树表示不知道何为发酵时,苹果树耐心地向其解释。他不断解释细节,确保梨树能正确理解事实。当梨树告诉他一些出人意料的事实时,他欣然接受并赞美大自然。

梨 树

梨树十分谦虚,并同样为自己的产量感到十分高兴。她对具体的经济核算很感兴趣,同时也赞美大自然的智慧。当她不知道何为发酵时,她并没有羞于提问,还分享了她对酵母菌的观察结果。她不在乎问一些答案显而易见的问题,因为她想要理解问题背后真正的原理。她会分享一些起初看似琐碎的观察结果,也就是浸上洒出来的果酒的衬衫不会被点燃。很快她就引起苹果树的注意,随后她在两件事间建立起联系并得出一个令人惊讶的结论。

艺术
The Arts

你会如何生动形象地描述苹果和梨的降温作用,也就是它们含有的天然化学物质能使物体的温度低于燃点?也许只用一张图很难描绘清楚,所以请试着用几幅连环画来说明原理。用一系列连环画能最有效地传达信息,赶紧把你想要展示的内容传达给你的读者吧。大多数人一开始可能会感到困惑,但在看了第二遍之后就会明白了。

TEACHER AND PARENT GUIDE

思维拓展
Systems: Making the Connections

　　现代社会中人们一直在尝试并寻找各种方法以应对他们面临的一些挑战。其中一个挑战就是火灾。人们一直在寻找能降低人身伤害风险的产品。人们在选择惰性建筑材料时引进被认为是安全材料的混凝土、水泥、钢铁和石棉。随着用石油和矿石生产的便利产品进入我们的生活，在产品中添加有毒化学物质以预防火灾和延缓火势成为了法律的强制规定。如今许多产品都添加了阻燃剂，毕竟没有防火证书就不能获得销售许可证，这意味着这些产品都含有这些有害化学物质。不幸的是，虽然这些化学物质具有耐火、阻燃的性质，但它们也很难降解。这些产品中的化学物质除了会在环境中富集外，还有害身体健康。结果，人们因过多使用这些产品导致其被禁止销售。而此时，普通人的观察结果激发了人们的灵感。一位细心的市民注意到洒上苹果酒的衣服无法燃烧，这促使人们发现了天然阻燃剂。与其说这个天然阻燃剂是一种有毒物质替代品，不如说是一种机遇，能使大量可用的天然产品的价值上升，从而创造一个新的行业。苹果和梨主要生长在北欧，得益于开花晚，它们能免受寒冬的侵袭。在短暂的收获季会有大量苹果被采集，这使得当地承包人不得不将多余的苹果酿成苹果酒。斯堪的纳维亚人生产了大量的苹果酒，以至苹果酒成为十分受大众欢迎的解渴饮料。由于苹果和梨是世界上种植最多的水果，并且分布在每一个大洲，这使得在局部建立一个小型可持续的化工产业成为可能。值得一提的是，虽然发现水果制品可用作阻燃剂相对简单，但这些产品背后的化学反应相当复杂，值得人们详细研究以掌握其中的原理。有时没有必要去全面了解和认识一件事物，知道它能发挥作用并且效果很好就足够了。

动手能力
Capacity to Implement

　　你能做些什么来降低火灾发生的风险呢？请从减少使用含有易燃物质的产品开始吧。有机棉、羊毛和乳胶制成的产品含有更多的氮元素和水分，需要更多的氧气才能燃烧，所以它们并不易燃。在考虑大自然中的阻燃剂之前，让我们先想想大自然中的哪些东西永远都不怕火。

教师与家长指南

故事灵感来自
This Fable Is Inspired by

马茨·尼尔森
Mats Nilsson

马茨·尼尔森在1969年于瑞典斯德哥尔摩大学获得数学学位后开启了他长期的学术生涯。他之后获得了工程物理学、电子与计算机技术学位。在斯德哥尔摩大学获得化学学位并毕业前，他获得了瑞典哥德堡查尔姆斯理工大学的专利技术学位。为了获得研究资金，马茨成为了数学、化学和物理教师。他参与设计了VSAT卫星系统，然后卖掉了他在该公司的股份。在大学期间，马茨就拿到了消防员证。他的一个客户，一家热能公司，希望他开发一种安全、无毒、环保的阻燃剂以替代现有的产品。2003年，马茨参加了瑞典的一项环保产品竞赛，他受自然启发研制的一款分子级噬热器使他成功挺进决赛。他还因此参加BBC的世界挑战，这是一项"真正改变世界"的创意竞赛。尼尔森不断改进并完成了产品设计，然后申请了专利。一个新的产业就此诞生。

图书在版编目（CIP）数据

冈特生态童书. 第八辑：全36册：汉英对照 /
（比）冈特·鲍利著；（哥伦）凯瑟琳娜·巴赫绘；
何家振等译. —上海：上海远东出版社，2021
ISBN 978-7-5476-1773-1

Ⅰ.①冈… Ⅱ.①冈… ②凯… ③何… Ⅲ.①生态环
境－环境保护－儿童读物—汉、英 Ⅳ.①X171.1-49

中国版本图书馆CIP数据核字（2021）第249940号

策　　划	张　蓉
责任编辑	祁东城
封面设计	魏　来　李　廉

冈特生态童书
苹果酒灭火
[比]冈特·鲍利　著
[哥伦]凯瑟琳娜·巴赫　绘
闫世东　译

记得要和身边的小朋友分享环保知识哦！
八喜冰淇淋祝你成为环保小使者！

Water 254

生物机器
Living Machines

Gunter Pauli

［比］冈特·鲍利 著

［哥伦］凯瑟琳娜·巴赫 绘

李原原 译

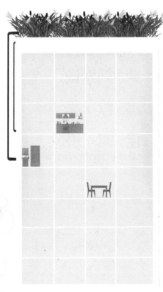

上海远东出版社

丛书编委会

主　任：贾　峰

副主任：何家振　闫世东　林　玉

委　员：李原原　祝真旭　牛玲娟　梁雅丽　任泽林
　　　　王　岢　陈　卫　郑循如　吴建民　彭　勇
　　　　王梦雨　戴　虹　翟致信　靳增江　孟　蝶

特别感谢以下热心人士对童书工作的支持：

匡志强　宋小华　解　东　厉　云　李　婧　陈　果
刘　丹　熊彩虹　罗淑怡　旷　婉　杨　荣　刘学振
何圣霖　廖清州　谭燕宁　韦小宏　李　杰　欧　亮
陈强林　王　征　张林霞　寿颖慧　罗　佳　傅　俊
胡海朋　白永喆　冯家宝

目录

生物机器	4
你知道吗？	22
想一想	26
自己动手！	27
学科知识	28
情感智慧	29
艺术	29
思维拓展	30
动手能力	30
故事灵感来自	31

Contents

Living Machines	4
Did You Know?	22
Think about It	26
Do It Yourself!	27
Academic Knowledge	28
Emotional Intelligence	29
The Arts	29
Systems: Making the Connections	30
Capacity to Implement	30
This Fable Is Inspired by	31

一只蜗牛和一根芦苇正享受着流过的营养丰富的水。这里的水温宜人,光线充足,周围的生物生机勃勃。芦苇看见自己的根部周围有一大群蜗牛,就说:

"你们是一个幸福的大家庭!喜欢那些蚊子卵吗?"

A snail and a reed are enjoying the nutrient-rich water flowing by. The water temperature is pleasant, there is abundant light, and all around them life is thriving. The reed sees the big family of snails around his roots and says,
"You are one happy family! Enjoying all those mosquito eggs?"

一只蜗牛和一根芦苇正享受着……

A snail and a reed are enjoying ...

我们不吃蚊子!

We do not eat mosquitos!

"我们不吃蚊子！鱼和蜘蛛才吃。我们吃植物，从而让自己长出坚固的壳。我们准备过冬，就像湿地一样。"

"但是在这个温室里没有冬天。任何生物都不需要冬眠。水，脏水，一年到头都在流。"

"脏水？谁说这是脏水？请称之为食物丰富的水。"

"We do not eat mosquitos! The fish and spiders do that. We eat plants to build a strong shell. We prepare for winter, just like the wetlands do."

"But here in this greenhouse there is no winter. No need for anyone to hibernate. Water, dirty water, keeps flowing through it all year round."

"Dirty water? Who calls this water dirty? Please, call this water rich in food."

"为什么人类会有这种奇怪的观念，认为他们用过的东西都是脏的，都是废物？对我们来说，这里面都是食物。"

"多亏了水中丰富的矿物质，看看这些植物长得多好啊！细菌是幸运的，它们可以最早吃到自己能吃的所有东西。"

"细菌、藻类、植物和动物能一起工作真是太棒了。"

"Why do people have this strange concept that whatever they've used is dirty, and should be considered waste? For us, it is simply full of food."

"Look at how well the plants grow thanks to abundant minerals in the water! Bacteria are the lucky ones, getting the first go at eating all they can."

"Wonderful that bacteria, algae, plants and animals can all work together."

……细菌、藻类、植物和动物……

...bacteria, algae, plants and animals...

……细菌和微小的藻类……

... the bacteria and the tiny algae ...

蜗牛说：“一旦细菌和微小的藻类享用了这顿丰盛的大餐，就该轮到我们进食和生长了。”

"这种循环永远不会停止。它被称为生物机器，因为人类会一直生产废水，而废水是我们的食物，所以我们可以不断生长。"

"And once the bacteria and the tiny algae have had their part of the sumptuous meal, it is our turn to eat and grow," Snail says.

"This cycle never stops. It's called a living machine, as people never stop producing wastewater, which is food for us, so we can grow and grow."

"好吧，多亏了这些丰富的资源，我们正逐步接管这片留给我们的空间。"

"别担心。鱼会来咬我的根，那些牙齿咬起来嘎吱嘎吱响的鱼会吃掉你们这些蜗牛以强化自己的骨骼。我也会一直长高，直到被人收获。"

"人们会把你编成漂亮的篮子吗？"

"Well, thanks to this abundance we are taking over the space given to us."

"Don't worry. The fish will come nibble on my roots, and those with crunchy teeth will feed on you snails, to strengthen their bones. I grow until I'm harvested."

"And do people turn you into pretty baskets?"

……鱼会来咬我的根……

...fish will come nibble on my roots...

……变成了蘑菇的基料。

...into feed for mushrooms.

"不！我们的数量非常多，我们的茎秆都变成了蘑菇的基料。我们是这台生物机器的一部分，所以生命一直在延续……"

"所以，来自城市的废水最终为人类提供了健康的食物。这才是正确的道路！我们应该用废水制造能源和食物，而不是处理它。"

"那么这些水都流到哪里去了？"芦苇问道。

"No! There is such a mass of us that our stalks are turned into feed for mushrooms. We are part of this living machine, so that life continues…"

"So, the wastewater from the city ultimately provides people with healthy food. That is the way to go! Instead of treating water, we make energy and food."

"But where does all the water flow to?" Reed asks.

"水被用来灌溉那些足球场大小的屋顶上的植物。"

"为什么人类要在大屋顶上种植物呢?"

"如果你在屋顶种了植物,然后给它们喷洒大量清洁的水,会发生什么?"

"The water is used to irrigate plants, on a roof the size of a football field."

"Now why would people plant plants on a big roof?"

"If you have plants on the roof and spray them with lots of clean water, what happens?"

……水被用来灌溉植物……

…water is used to irrigate plants…

……夏季植物会疯狂生长……

…in the summer the plants will grow wildly…

"嗯，夏季植物会在阳光下疯狂生长。它们沐浴在水中，然后水分就会蒸发。"

"那水分蒸发时会发生什么？"

"哦，我明白了！那能让建筑物内部保持凉爽，从而节省能源！"

"那冬天呢？"

"Mm, in the summer the plants will grow wildly, exposed to the sun. And they are bathed in water that will then evaporate."

"And what happens when the water evaporates?"

"Yes, I know! It will keep the inside of the building cool, saving on energy!"

"And what happens in the winter?"

"水会结冰。这样就会产生冰屋效应,也就是说热量会被锁在建筑物内部,让人们保持温暖。"

"这就是我说的生物机器!它节省了大量能源,生产食物,同时让一切都变得繁茂美丽!"

……这仅仅是开始!……

"The water will freeze. And you will have the igloo effect – meaning the heat will be trapped inside, keeping people warm."

"Now that's what I call a living machine! It saves lots of energy, and produces food while making everything lush and beautiful!"

... AND IT HAS ONLY JUST BEGUN!...

……这仅仅是开始!……

...AND IT HAS ONLY JUST BEGUN! ...

Did You Know?
你知道吗?

Living machines use bacteria, algae, plants and animals to purify water through sedimentation, filtration, absorption, nitrification and decomposition, with tedious processes performed by living organisms.

生物机器利用细菌、藻类、植物和动物,通过沉淀、过滤、吸收、硝化和分解来净化水,而这些繁琐的过程是由有机体来完成的。

The use of plants and animals gives it a unique aesthetic appearance. The facility is housed inside a greenhouse, creating a temperate climate that protects the process from freezing temperatures.

植物和动物的运用赋予生物机器独特的美学外观。该设施被安置在温室内,创造了一个温和的气候,使这个过程免受严寒的影响。

The first living machine for a factory was installed in 1992 in Malle, Belgium, treating the wastewater of a detergent factory. Warm wastewater was purified through reed beds before being pumped up to a grass roof.

1992年，人们在比利时的马莱安装了第一台生物机器，用于处理洗涤剂工厂产生的废水。温暖的废水经芦苇床净化后，再被泵到草坪屋顶。

The grass roof was irrigated summer and winter, bringing plant and bird life to an industrial zone. More than 30 families of birds were found nesting, safe from predators, pesticides and agricultural machinery.

人们在夏季和冬季灌溉草坪屋顶，草坪屋顶为工业区带来植物和鸟类。有30多个科的鸟类在此筑巢，从而远离捕食者、杀虫剂和农业机械。

The evaporation of the water on the roof kept the building cool. It eliminated the need for air-conditioning. Soap is an exothermic reaction releasing heat. The freezing of the water on the roof created a heat shield.

屋顶上水的蒸发使建筑物保持凉爽,因此它不需要空调。皂化反应是一种放热反应,能释放热量。屋顶上结的冰形成了保温层。

The first large-scale living machines for municipal wastewater was built in South Burlington, Vermont (USA), in 1995. It was also used as an educational centre for schools.

1995年,美国佛蒙特州的南伯灵顿市建造了第一台用于处理城市废水的大型生物机器。它也被用作学校的教育中心。

The living systems later added blending of shredded solid organic municipal waste into the wastewater treatment, in order to produce biogas. This reduces the waste to a landfill or incinerator by 50 percent.

后来，这个生物系统又将粉碎的城市有机固体垃圾与待处理的废水混合，以生产沼气。这使得50%的垃圾不用再被送到填埋场或焚化炉。

The combined treatment of the organic component of solid municipal waste with wastewater treatment generates revenues. The city doesn't pay for waste management, but gets paid for products and gas generated.

城市有机固体垃圾与废水的联合处理产生了收入。城市不仅不用花钱处理废物，还能通过生产的产品和天然气获得收入。

Think about It

想一想

Instead of paying taxes, the city gets paid for cleaning water?

市政府不仅不花纳税人的钱，反而还能靠处理废水来赚钱？

Would you have plants and flowers on the roof?

你会在屋顶上种鲜花或其他植物吗？

Can one turn dirty water into food and energy?

我们能把脏水变成食物和能源吗？

How can a frozen roof trap heat?

结冰的屋顶如何锁住热量？

Do It Yourself!

自己动手！

Think about all the wastewater that is flushed down the kitchen sink, bath and shower. Where does it go? Calculate how much water is flushed away every day. This is apart from the water flushed down the toilet, which goes elsewhere. Do you really consider the water from the kitchen and bath or shower dirty? Not if you use natural products without synthetic perfumes or colour agents. Study the water quality and think of ways it could be used to grow food. Think about what else you would need to use along with your wastewater to create a source of energy and food. Share your ideas with friends and family members.

所有从厨房水槽、浴缸和淋浴间被冲走的废水都去了哪儿？计算一下每天有多少水被冲走。从厕所冲下去的水则会流到其他地方。你真的认为从厨房、浴缸或淋浴间排出去的水是脏的吗（尤其当你不使用含合成香料或色素的产品时）？研究水质，想办法用这种水来种植食物。想想要用废水创造能源和食物，你还需要什么？与朋友和家人分享你的想法。

TEACHER AND PARENT GUIDE

学科知识
Academic Knowledge

生物学	生物机器基于湿地生态学原理；碳是湿地中被循环的主要养分。
化　学	湿地的水化学性质由pH值、盐度、养分、电导率、土壤组成、水硬度等因素决定；过量的氮会导致富营养化；用氯化水治疗疾病。
物　理	湿地水文研究涉及水域内的地表水和地下水的时空离散、流动及理化特征；水中氧气和二氧化碳的浓度取决于温度和压力。
工程学	人工湿地可以处理城市废水、工业废水和雨水；湿地提供生态系统服务，如蓄水、防洪、补充地下水、净化水、储存生物多样性、授粉；人工湿地可以去除45%的氮和60%的悬浮物。
经济学	花费纳税人的钱进行水处理与像重建湿地这样能为市政当局带来收入的公共服务之间的差异；水稻是种植在湿地上的主粮。
伦理学	模拟自然，处理废水并形成生态系统服务；付钱给别人来处理你的垃圾。
历　史	美索不达米亚人在公元前4000年左右发明了陶土下水道；公元前1650年，安纳托利亚的赫梯城开始使用陶土制成的下水道；奥克尼群岛上的斯卡拉布雷遗址拥有废水处理系统，可以追溯到公元前3500年；印度河城市洛塔尔（约公元前2350年）有污水处理系统；《拉姆萨尔公约》是重要的国际湿地公约，始于1971年。
地　理	湿地是一种生态系统，在这种生态系统中，洪水淹没产生的土壤以厌氧和好氧过程为主，迫使生物群落适应洪水；世界上最大的湿地是亚马孙河流域的沼泽森林和西伯利亚的泥炭地；湿地、沼池、沼泽与河岸带的差异；许多泥炭地是湿地；澳大利亚红树湿地具有抗菌、促伤口愈合、抗溃疡和抗氧化特性。
数　学	通过改变业务模式和集群活动将成本转化为收入。
生活方式	我们习惯于把水冲走，不考虑处理水的成本。
社会学	我们把责任交给地方政府，并准备为享受这些服务而纳税。
心理学	我们出钱让第三方为我们处理废物，这是一种傲慢的表现；家里干净的人比家里脏乱的人更健康；人体内有成千上万的完整的生物和神经化学系统，所有这些系统都是有组织的，细胞按照严格的时间表或昼夜节律运作；我们渴望家里整齐和清洁，这种渴望是人体内部运行规律的反映。
系统论	水处理系统是分散的，基于自然、美学、可持续性、安全卫生和健康标准，并可扩展到整个区域；湿地面临的危险来自人类开发产生的排水、过度放牧和采矿；河流通常是处理天然污水的一种原始形式。

教师与家长指南

情感智慧
Emotional Intelligence

芦苇

芦苇注意到蜗牛幸福的大家庭。他观察到，由于住在温室里，所有生物不受四季的影响。当蜗牛说人类不应该把动植物的食物来源归类为泥土或废物时，他马上表示赞同。他赞美细菌、藻类、植物和动物在一个持续的循环中共同从事的工作。芦苇告诉蜗牛，所有过剩的东西都会有后续的用途，所以不会有问题，这话让蜗牛安心。芦苇想知道被他们享用过的废水最后去了哪儿。当芦苇不明白为什么人类要在屋顶种植物时，他继续追问。直到明白了其中的逻辑关系，他才终于建立利用废水与造福所有物种之间的联系。

蜗牛

蜗牛表示，自己的首要任务是在冬季来临之前建造坚固的外壳。她坚持认为人类眼中的废水不仅不脏，而且有用。她知道这样一个事实：细菌获益最多，但最终所有物种都是赢家。当她得知芦苇被用作种蘑菇的基料，而蘑菇是人类的食物时，她赞赏这种将生物质转化为能源和食物的明智做法。蜗牛通过提问引导芦苇更好地理解生物机器的概念。最后她总结了生物机器带来的好处和对美好未来的展望。

艺术
The Arts

当看到大自然是如何利用我们的废物时，我们受到了启发，开始模仿大自然进行废水处理。想象一下，有一个过程可以把我们废弃的东西快速转化为其他生物需要的营养。使用你的艺术材料以一种吸引人的方式描绘这样一个过程。

TEACHER AND PARENT GUIDE

思维拓展
Systems: Making the Connections

　　大自然中的水可以通过一系列的生物、化学和物理过程实现自我净化,而我们对此一无所知。河流某一部分的水可能会被污染,如果不再受到额外的污染,这条河流就会继续向下游清澈流淌。然而,当水被合成肥皂、激素和抗生素污染时,天然的净化、处理和过滤系统就无法清除这些物质。然而,当我们明智地选择使用的材料,并开始关注可生物降解材料和天然材料时,我们给大自然带来的过度负担可以通过商业模式的根本转变来解决:我们不应该把废水看作必须处理和净化的水,而是设想如何将被污染的水用于种植或提供能源。目前工程学校还没有教授这样的课程,因为他们认为水需要通过沉降和过滤的方式来净化,并向水中泵入大量的空气以促进细菌的活动。或者,对废水进行厌氧处理,促使厌氧细菌产生甲烷气体。然而,这两种系统的缺点是都需要一定的费用。系统和持续地净化水需要大量的投资和维护费,这些费用导致市政税收的提高。如果商业模式从处理垃圾转变为生产食物和能源,那么就有可能将废水转化为推动当地经济的引擎。记住,在大自然里没有废物的概念。通过基础设施的巧妙设计保护系统不受四季变化的影响,并通过有效的工程技术让大自然的五界都发挥作用,我们就有可能模仿大自然的净水方法。这种系统的灵感来自湿地,一个动态的生态系统,可以处理迅速发生的洪水或缺水,提供一系列的生态系统服务。这种模式出现在近50年前,现在正迅速得到认可,原因很简单,它可以把目前的市政成本转化为收入。当我们能将使用过或被污染的水转化为收入、能源和食物的来源时,为什么还要继续为废水管理纳税?

动手能力
Capacity to Implement

　　让我们当一天的水处理工程师。收集一些早上洗澡后满是肥皂和洗发液的水。你要怎么处理这些灰色的水呢?首先想想什么东西会喜欢吃生长在这种水里的生物。一旦你找到了一个这样的物种,下一步要做什么?从大自然各界中找到更多的物种。仔细思考这个问题,运用你的想象力设想一个系统。你应该多观察大自然,毕竟大自然是这方面的大师。

教师与家长指南

故事灵感来自
This Fable Is Inspired by

约翰·托德
John Todd

约翰·托德在加拿大安大略湖的汉密尔顿湾附近长大，那里靠近被严重污染的沼泽和溪流。1961 年，他在加拿大蒙特利尔的麦吉尔大学获得农业理学学士学位，1963 年获得寄生虫学和热带医学理学硕士学位。他在美国密歇根大学获得海洋生物学博士学位，研究渔业和海洋学。1968—1970 年，他在圣地亚哥州立大学担任动物行为学助理教授，由此开始他的职业生涯。之后，他在伍兹霍尔海洋研究所担任助理科学家。他在马萨诸塞州科德角创立了新炼金术研究所，这是一个科研和反主流文化的机构。该机构提出了生物庇护所的概念，即包含生态系统的类似温室的建筑。1988 年，他在马萨诸塞州马里昂成立了生态工程协会，并开始建设污水处理厂。他继续在佛蒙特大学担任名誉教授和杰出讲师。

图书在版编目（CIP）数据

冈特生态童书.第八辑:全36册:汉英对照/
(比)冈特·鲍利著;(哥伦)凯瑟琳娜·巴赫绘;
何家振等译.—上海:上海远东出版社,2021
ISBN 978-7-5476-1773-1

Ⅰ.①冈… Ⅱ.①冈…②凯…③何… Ⅲ.①生态环
境–环境保护–儿童读物—汉、英 Ⅳ.①X171.1-49

中国版本图书馆CIP数据核字(2021)第249940号

策　　划　张　蓉
责任编辑　祁东城
封面设计　魏　来　李　廉

冈特生态童书
生物机器
[比]冈特·鲍利　著
[哥伦]凯瑟琳娜·巴赫　绘
李原原　译

记得要和身边的小朋友分享环保知识哦！
八喜冰淇淋祝你成为环保小使者！

Water 255

泥土、石头和木头

Earth, Stone and Wood

Gunter Pauli

[比] 冈特·鲍利　著
[哥伦] 凯瑟琳娜·巴赫　绘
李原原　译

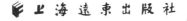

上海远东出版社

丛书编委会

主　任：贾　峰
副主任：何家振　闫世东　林　玉
委　员：李原原　祝真旭　牛玲娟　梁雅丽　任泽林
　　　　王　岢　陈　卫　郑循如　吴建民　彭　勇
　　　　王梦雨　戴　虹　翟致信　靳增江　孟　蝶

特别感谢以下热心人士对童书工作的支持：

匡志强　宋小华　解　东　厉　云　李　婧　陈　果
刘　丹　熊彩虹　罗淑怡　旷　婉　杨　荣　刘学振
何圣霖　廖清州　谭燕宁　韦小宏　李　杰　欧　亮
陈强林　王　征　张林霞　寿颖慧　罗　佳　傅　俊
胡海朋　白永喆　冯家宝

目录

泥土、石头和木头	4
你知道吗？	22
想一想	26
自己动手！	27
学科知识	28
情感智慧	29
艺术	29
思维拓展	30
动手能力	30
故事灵感来自	31

Contents

Earth, Stone and Wood	4
Did You Know?	22
Think about It	26
Do It Yourself!	27
Academic Knowledge	28
Emotional Intelligence	29
The Arts	29
Systems: Making the Connections	30
Capacity to Implement	30
This Fable Is Inspired by	31

一只老鼠正坐在一棵大橡树下。看着一只猫头鹰在树上筑巢,他说:

"听说这棵树要被砍了,我很难过。"

"不要为我难过。这里很安全,因为这棵树已经有500年历史,我相信人们不会允许它被人砍倒的。"

A mouse is sitting under a big oak tree. He sees an owl nesting in the tree, and says,
"I am so sorry to hear that your tree will soon be chopped down."
"No need to feel sorry for me. I am quite safe here, as this tree is five hundred years old, and I am sure people will not allow it to be cut down."

一只老鼠正坐在一棵大橡树下。

A mouse is sitting under a big oak tree.

……喜欢未雨绸缪的想法。

...like the idea of thinking ahead.

"是这样的,500年前建造这个大厅的时候,建筑师就种上了用于替换的橡树。现在这些树已经长成,正在等待砍伐。"

"太有远见了!虽然我第一次听说这件事,但我确实喜欢未雨绸缪的想法。"

"看来,即使像你这样聪明的人也能学到新东西。"老鼠笑着说。"你知道有些城市是用泥巴建起来的吗?"

"Well, when this hall was built five hundred years ago, the architect had replacement oaks planted. Those are now ready and waiting."

"Such farsightedness! News to me, but I do like the idea of thinking ahead."

"Even someone as wise as you can learn something new, it seems," Mouse laughs. "Did you know there are whole cities built out of mud?"

"我知道房子可以用泥土建造。但整个城市？下雨的时候，这些房子肯定都塌了。"

"嗯，那些建筑物的大屋顶能保护墙壁免受过多的水、风和阳光的侵蚀，因此完全有可能。"

"不过，这些房子能维持多久？总会有出现裂缝的时候，等到下次下雨时，整个建筑可能就被冲垮了！"

"I know houses can be made out of earth. But whole cities? Surely those won't last when it rains."

"Well, the large roofs of the buildings protect the walls from too much water, wind and sun, then it perfectly possible."

"Still, how long can they last? There always are some cracks, and with the next rain the entire building may simply wash away!"

我知道房子可以用泥土建造。

I know houses can be made out of earth.

这些建筑甚至能保持室内凉爽……

These buildings even keep the inside cool ...

"撒哈拉沙漠中有一片美妙的绿洲,多亏了人们的维护,那里所有用泥土建造的房屋在500年后仍然屹立不倒。"

"这可是很长一段承受阳光、酷热和暴雨的岁月。"猫头鹰说。

"是时候停止把泥土当作尘埃了。这些建筑甚至能在室外酷热难耐的情况下保持室内凉爽。"

"There is a wonderful oasis in the Sahara, where thanks to maintenance all the houses made from earth still stand five hundred years later."

"That is a very long time to withstand the sun, the heat and an occasional torrential rain," Owl remarks.

"It is time to stop thinking of earth as dirt. These buildings even keep the inside cool with the scorching heat outside."

"这么说，木制和土制的建筑甚至比用砂浆、砖、钢和水泥建造的现代建筑寿命更长。但里面的灰尘、蜘蛛网，还有老鼠，是不是很多？你们老鼠喜欢住在肮脏的地方吗？"

"你别去那儿……我们老鼠只去有食物的地方。"

"So, wooden and earthen buildings last longer than modern ones made from mortar, bricks, steel, and cement. But isn't there too much dust inside, and spider webs, and mice? Would you like to live there in the dirt?"

"Don't you go there… We mice only go where there is food."

······蜘蛛网······

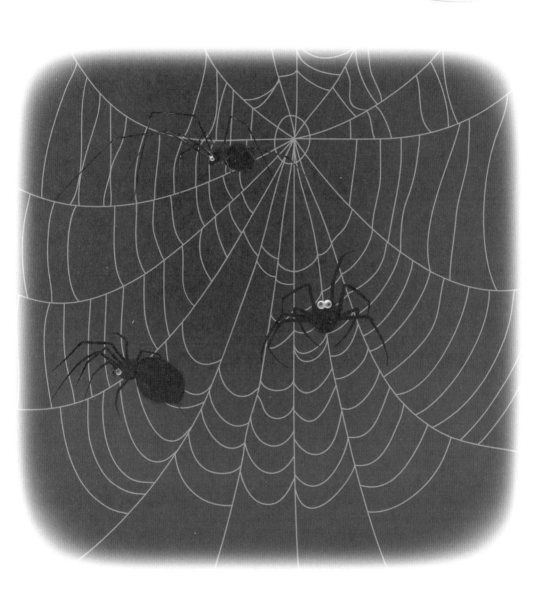

... and spider webs ...

······罗马的斗兽场······

...the Colosseum in Rome...

"那石头建筑呢?"

"就像土制建筑和木头建筑一样,石头建筑是永恒的,而且可持续。"

"中国的长城,希腊卫城的万神殿,罗马的斗兽场,这些都是石头建筑中的丰碑。糟糕的是,现在空气污染给它们造成的损害比战争还要严重。"

"And what about stone buildings?"
"Like earth and wood, stone buildings last forever, and are sustainable."
"The Great Wall of China, the Parthenon on the Acropolis in Greece, the Colosseum in Rome; all great achievements in building with stone. Too bad these now suffer more damage from air pollution than wars."

"嗯,你还可以加上金字塔、阿兹特克神庙和玛雅神庙。你知道,人们把石头、木头和泥土结合在一起,而不是用金属或钢筋混凝土。"

"我很好奇,这些材料是如何结合在一起的?"猫头鹰问。

"受传统文化启发的聪明设计和巧妙施工实现了这一点。最可持续的建筑方式是种植房屋和建筑。"

"Well, you can add the Pyramids, the Aztec and Mayan temples. You know, people join stone, wood and earth, without metals or reinforced concrete."

"I wonder how it all holds together?" Owl asks.

"Smart design, and clever engineering inspired by culture and tradition does it. The greatest sustainability would be to grow houses and buildings."

……你还可以加上金字塔……

...you can add the Pyramids...

……用草来建造……

...constructed from grass...

"一棵橡树需要500年才能长成,土和石头是惰性的,永远不会生长。那你是如何实现这个绿色梦想的呢?"

"近来,人们开始用草来建造建筑物。"

"哈哈!草?真的吗?任何草做的建筑都挺立不了一天。牛会把它吃了。"

"An oak takes five hundred years to grow to maturity, whereas earth and stone are inert and never grow. How could you realise this green dream?"

"Lately, buildings have been constructed from grass."

"Ha-ha! Grass? Really? No building made from grass will remain standing for more than a day. The cows will eat it."

"你凭什么说草不能用来盖房子?花时间去探索这些奇迹,再认识一下新生代的建筑师们——他们可能会为你设计一个美丽的鸟巢!"

……这仅仅是开始!……

"Why do you claim grass wouldn't work? Take the time to discover such marvels, and meet the next generation of architects – who may design a beautiful nest for you!"

... AND IT HAS ONLY JUST BEGUN!...

······这仅仅是开始!······

...AND IT HAS ONLY JUST BEGUN!...

Did You Know？

你知道吗？

There are 500-year old buildings made from earth, still standing in regions with low rainfall, like the Great Mosque of Djenné (Mali), the Bam Citadel (Iran), the village of the Siwa Oasis, and the Shibam City in Yemen.

在降雨量少的地区，仍有一些历史超过500年的土制建筑，比如马里宏伟的杰内大清真寺、伊朗的巴姆城堡、锡瓦绿洲村庄和也门的希巴姆古城。

The oldest and largest wooden buildings include the Todaiji Temple in Nara (Japan), the Old Government Building in Wellington (New Zealand), and the Kishi Pogost Church on Kishi Island in Russia.

最古老和最大的木制建筑包括日本奈良的东大寺、新西兰惠灵顿的旧政府大楼和俄罗斯基日岛的木结构教堂。

The Colosseum was built with travertine, a sedimentary rock from geothermal hot springs. The most famous natural travertine formations are Yellowstone National Park (USA), and Huanglong in Sichuan (China).

罗马斗兽场是用钙华建造的，钙华是一种来自地热温泉的沉积岩。最著名的天然钙华构造在美国的黄石国家公园和中国四川的黄龙景区。

The Romans used travertine for the construction of temples, bath complexes, and amphitheatres. The Sacré-Cœur Basilica of Paris and the Getty Centre in Los Angeles are modern buildings of the same type of rock.

罗马人用钙华建造庙宇、浴室和露天剧场。巴黎的圣心大教堂和洛杉矶的盖蒂中心都是用同一种岩石建造的现代建筑。

Bamboo is a grass. When cut, it will grow again. It is one of the fastest growing plants, and over a billion people in poverty-stricken areas use its poles as a structural building material.

竹子是一种草本植物。一旦被剪断，它还会再长出来。它是生长最快的植物之一，在贫困地区有超过 10 亿人把竹子的茎秆当作建筑结构材料。

When New College in Oxford received its Charter in 1379 AD, a grove of oak trees was planted, destined to replace the roof beams when these would become infested with beetles, as all oaks falls prey to beetles.

1379 年，牛津大学的新学院在获得办院的特许状的同时，在学院种植了一片橡树林，用来更换屋顶横梁，因为这些横梁受到甲虫的侵袭。

Siwa Oasis, on the edge of the Great Sand Sea, has walls built in Karshif stone particles bonded with silt mortar. Ceilings of palm wood and olive leaves strengthen the roof and protect against the rain.

锡瓦绿洲位于大沙海的边缘，城墙是用喀什夫石头颗粒与淤泥砂浆结合而成的。棕榈木和橄榄叶构成的天花板使屋顶更坚固，还能防雨。

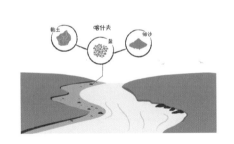

Karshif is a stone naturally formed at the shores of the salt lakes. Its particles consist mainly of salt bonded with clay and fine sand. Karshif stone has a high thermal insulation, outperforming modern insulation.

喀什夫是在盐湖岸边自然形成的石头。它的颗粒主要由盐、黏土和细沙结合而成。喀什夫具有很强的保温性能，优于现代保温材料。

Think about It

想一想

Who plans to build a house that will last for 500 years?

谁打算建造一座可以屹立500年的房子?

Are buildings that last 500 years sustainable, even when made from mined rock?

一栋建筑是不是只要能屹立500年就算可持续,即便它是用开采出来的岩石建造的?

Is bamboo only for the poor?

只有穷人才用竹子吗?

Can you grow a house?

你能种一栋房子吗?

Do It Yourself!

自己动手！

What is the house you live in built from? And the school buildings where you study? Make a detailed list all the materials that are used, and then verify if the building has the capacity to last for 500 years. If it cannot last that long, ask what could be changed so that the space remains useful, and would not need to be torn down and reconstructed using a lot more material, and therefore be a lot less sustainable. Share your findings with friends and family and ask for their opinion.

你住的房子和你就读的学校是用什么材料建造的？详细列出所有用到的材料，然后核实这些建筑能否屹立500年。如果不能坚持那么长时间，那就想想可以怎么改进，使得建筑超过最初设计寿命以后仍能安全使用，从而不需要拆除并使用更多的材料重建，否则建筑的可持续性就大大降低了。与朋友和家人分享你的发现，并听听他们的建议。

TEACHER AND PARENT GUIDE

学科知识
Academic Knowledge

生物学	6 000种树皮甲虫、天牛和象鼻虫破坏木材且不惧木材的化学处理；甲虫会破坏木材；橡树养育着数百种蝴蝶。
化 学	木材在生物和化学双重降解作用下通常会腐烂。
物 理	湿度高的地方更容易被甲虫侵扰；土墙比传统砖石墙更厚，因此可以储存更多热量；竹子较轻，抗拉强度与钢相当；夯土墙非常适合承重；钢、铁和铁锈影响混凝土结构耐久性；钢筋提高了强度，使悬臂结构和更薄、支撑更少的板成为可能，并加快了施工速度。
工程学	夯土中的防尘、抑尘技术；长城大部分结构是用夯土建造，或者以夯土为基础，用岩石筑成；夯土可以通过密封、涂漆、涂刷、抹灰、干衬或包覆等方式来防雨水。
经济学	黏土砖可以使用数百年，因此对生态有利，对消费主义不利；公认的预期是，一座新建筑至少能使用60年；当出售房屋或将其作为抵押物进行融资时，房屋的寿命至关重要；钢筋建筑的寿命最长可达100年，10年后开始老化，维修费用高达数十亿美元。
伦理学	选择可以快速施工但不能持久的建筑材料会增加使用的材料和碳排放。
历 史	西班牙南部格拉纳达的阿尔罕布拉宫大部分是用夯土建造的；罗马万神殿，世界上最大的无钢筋混凝土圆顶建筑，已有2 000年的历史；埃及人过去使用石膏砂浆，但这需要高温，进而导致森林砍伐。
地 理	针叶树和沼泽白栎能忍受潮湿的土壤，而猩红栎、栗树和白橡能忍受薄而干的土壤，清扩平栎和大果栎能忍受碱性土壤。
数 学	新鲜橡木在被切割并风干后，其水分损失导致径向收缩4%，切向收缩8%，而浸水橡木则径向收缩12%，切向收缩24%左右。
生活方式	提前计划，而不是每天疲于收拾昨天的烂摊子。
社会学	橡树是为后代而生的树，因为它能活200年以上，它的木材能使用500年以上；穷人只能住在竹屋里，进而竹屋成了贫穷的象征；持久的家庭或社会背景对社区创建的影响。
心理学	保持谦逊，认识到每个人，即使是最聪明的人，也总能学到新的东西；像竹屋这样美丽的建筑能使生活在其中的人内心平静。
系统论	树皮甲虫会攻击那些被烟雾、过于密集的种植和干旱削弱了树木韧性的木材；树皮甲虫与真菌有共生关系；有100多种脊椎动物吃橡果；水泥制造业产生了全球5%的碳排放和三分之一的需填埋的垃圾，混凝土通常是由海洋生物的骨骼与岩石磨碎混合而成，这需要数百万年的时间才能形成。

教师与家长指南

情感智慧
Emotional Intelligence

老 鼠

老鼠用一句简短的关心作为开场白,但脑子里已经准备好了一长串富有挑战性的对话。他向猫头鹰分享了一系列令人惊讶的事实。老鼠嘲笑猫头鹰,说猫头鹰这么聪明也得学习新知识。面对质疑,老鼠会提供事实和案例,而不是理论。他坦率地强调不应该把泥土当作尘埃。当猫头鹰提到老鼠喜欢住在灰尘多的地方时,老鼠警告猫头鹰不要提这个话题。为了给猫头鹰留下更深刻的印象,老鼠开始谈论石头建筑。老鼠将话题转移到自己的领域——用竹子建房。当猫头鹰嘲笑他用草建房时,老鼠很聪明地建议猫头鹰去探索更多新事物。

猫头鹰

起初,猫头鹰希望打消老鼠的顾虑,告诉对方自己待在树上是安全的。他对老鼠说的那些先贤的远见感到惊讶。猫头鹰对用夯土建造城市感到疑惑,也质疑老鼠提供的论据。他也在思考老鼠积极推广夯土建筑技术的原因。猫头鹰继续和老鼠交流,想了解老鼠对石头建筑的看法,以及石头、木头和泥土是如何结合在一起的。他提出更具体的事实来挑战老鼠,却落入了老鼠设置的圈套——用草建造房子。这种建筑技巧让猫头鹰发笑,直截了当地认为对方的这个建议不切实际,而老鼠的回答却让猫头鹰哑口无言。

艺术
The Arts

用石头创造艺术!出去散步时你会找石头吗?你最多可以叠几层石头而不倒塌?你应该选一块大石头放在底部,之后每次只添加一块。你必须缓慢而稳定地做这件事。这个练习将教会你如何控制力量并强化你的平衡感。

TEACHER AND PARENT GUIDE

思维拓展
Systems: Making the Connections

今天的建筑行业有两个明确的目标：快速建造和廉价建造。如果建造的时间被看得比建筑物的寿命更重要的话，那么就不可能以可持续的方式建造。建造可持续建筑的关键是选择可维持几个世纪的建筑材料。结构功能的持续时间似乎是过去的标准，而我们忽视了这样一个事实，即许多古老的建筑都经受住了岁月的考验，并且是由现成的材料如泥土、木材、竹子和岩石建造的。现代工业中，时间和劳动力被视为最宝贵的资源。大多数地区法律要求的建筑物最低使用年限不超过60年。建造持久的建筑不仅具有重大的生态优势，也确保了更好的财政基础。一栋还有很长使用寿命的房子可以作为新项目融资的担保。所以说，在过去的50年里，建筑业经历了如此大的转变，以至世界各地的建筑师在使用钢筋混凝土的过程中过分专注于降低成本，这是令人费解的。建筑的寿命，从生态和财政的角度来看，确实是第一个需要考虑的关键因素。第二是建筑的使用，主要涉及使用成本，而影响使用成本的一个关键因素是建筑的能源利用。对温度和湿度的有效管理是许多传统建筑技术的巨大优势。夯土、木材、竹子和石头无一例外地使建筑物能在建造和使用过程中实现零碳排放。我们迫切需要重新发现这些经受住了时间考验的建筑方法。当我们开始评估可供选择的建筑材料时，不要忘记或忽视过去几千年的丰富经验，很明显，我们确实有机会使用以前没有考虑过的材料。

动手能力
Capacity to Implement

建房子的开支只是费用的一部分。主要的费用是入住后的使用和管理成本。能源、废物处理和水是支出最多的项目。我们要如何建造一栋节约能源且能循环利用水的房子？首先要有清晰的思路，然后是坚持到底的决心。做不到这些，那就很难实现。研究所有已经实现上述目标的案例。报告你的发现，并记住，即使这听起来并不容易，也要有决心。

教师与家长指南

故事灵感来自
This Fable Is Inspired by

伊罗拉·哈迪
Elora Hardy

　　伊罗拉·哈迪，1980年出生，在印度尼西亚巴厘岛长大，是位加拿大设计师。她拥有艺术专业学位，曾为时尚品牌唐纳·卡兰工作。后来她回到巴厘岛，创办了Ibuku设计公司，有100多名员工。她的公司建造了现代史上最具代表性的竹结构建筑，包括巴厘岛的绿色学校。她带领一个充满活力的团队，包括技术精湛的匠人、建筑师、工程师和设计师，用竹子建造建筑和制作精致的家具。她与自己热爱的文化和景观重新建立了联系，并不断培养巴厘岛的匠人以及具有创新精神的设计师和建筑师。她的目标是使巴厘岛成为全球可持续设计领域的中心，并将这些设计传播到世界各地。

图书在版编目(CIP)数据

冈特生态童书.第八辑:全36册:汉英对照／
(比)冈特·鲍利著；(哥伦)凯瑟琳娜·巴赫绘；
何家振等译.—上海:上海远东出版社,2021
ISBN 978-7-5476-1773-1

Ⅰ.①冈… Ⅱ.①冈…②凯…③何… Ⅲ.①生态环境－环境保护－儿童读物—汉、英 Ⅳ.①X171.1-49

中国版本图书馆CIP数据核字(2021)第249940号

策　　划	张　蓉
责任编辑	祁东城
封面设计	魏　来　李　廉

冈特生态童书
泥土、石头和木头
[比]冈特·鲍利　著
[哥伦]凯瑟琳娜·巴赫　绘
李原原　译

记得要和身边的小朋友分享环保知识哦！
八喜冰淇淋祝你成为环保小使者！

Water 256

森林园丁
The Forest Gardener

Gunter Pauli

[比]冈特·鲍利 著
[哥伦]凯瑟琳娜·巴赫 绘
李原原 译

上海远东出版社

丛书编委会

主　任：贾　峰

副主任：何家振　闫世东　林　玉

委　员：李原原　祝真旭　牛玲娟　梁雅丽　任泽林
　　　　王　岢　陈　卫　郑循如　吴建民　彭　勇
　　　　王梦雨　戴　虹　翟致信　靳增江　孟　蝶

特别感谢以下热心人士对童书工作的支持：

匡志强　宋小华　解　东　厉　云　李　婧　陈　果
刘　丹　熊彩虹　罗淑怡　旷　婉　杨　荣　刘学振
何圣霖　廖清州　谭燕宁　韦小宏　李　杰　欧　亮
陈强林　王　征　张林霞　寿颖慧　罗　佳　傅　俊
胡海朋　白永喆　冯家宝

目录

森林园丁	4
你知道吗？	22
想一想	26
自己动手！	27
学科知识	28
情感智慧	29
艺术	29
思维拓展	30
动手能力	30
故事灵感来自	31

Contents

The Forest Gardener	4
Did You Know?	22
Think about It	26
Do It Yourself!	27
Academic Knowledge	28
Emotional Intelligence	29
The Arts	29
Systems: Making the Connections	30
Capacity to Implement	30
This Fable Is Inspired by	31

一只屎壳郎正在收集粪便准备产卵。她高兴地看到貘在他们的茅坑拉下很多粪便,她向其中一只道谢:

"你的便便对我们很有价值。谢谢你每天的供应。"

"嗯,在这片热带森林里,我吃了300多种不同的植物。无论你从我这里得到什么,都是有营养的。"

A dung beetle is collecting dung to lay her eggs in. She is happy to see how much dung the tapirs left at their latrine, and thanks one by saying,

"Your poop is very valuable to us. Thank you for the daily supply."

"Well, here in the tropical forest, I feast on more than three hundred different plants. Whatever you get from me is full of nutrients."

……屎壳郎正在收集粪便……

... dung beetle is collecting dung ...

......拉出最大便便的陆地哺乳动物......

... the largest pooping land mammal ...

"是的，你就是便便大亨！你拉下的每坨便便比一个人的脑袋还要大。难怪你被称为这片土地上拉出最大便便的陆地哺乳动物。"

"是的，我们尽最大努力在困难时期生存了下来。森林被大量砍伐，接着人类又烧毁土地！"

"Yes, you are a faecal fortune! Each of the piles you leave behind is bigger than a person's head. No wonder you are called the largest pooping land mammal on this continent."

"Well, we do the best we can to survive in these difficult times. There is so much deforestation – and then people burn the land!"

"我注意到你在被烧毁的森林里转悠。可那儿能吃的东西很少,你转悠什么呢?" 屎壳郎问道。

"我们通常生活在树冠下,在灌木丛中寻找食物。我们在开阔的地方晒太阳,因为阳光可以清洁我们的皮肤。"

"I've noticed you visiting the burnt down forests. Why, when there is so little to eat?" Dung Beetle asks.

"We normally live under the tree canopy, where we make our trails foraging through the bush. It is in open spaces where we get some sun, as the sun sanitises our skin."

我们通常生活在树冠下……

We normally live under the tree canopy ...

……在那些被烧毁的地区转悠了很长时间。

... spend a lot of time in those devastated regions.

"你似乎在那些被烧毁的地区转悠了很长时间。请给我仔细说说……"

"我们这些宽宏大量的排便者相信,通过在那里排便,我们有助于恢复因人类的无知而被破坏的森林。"

"没错,你就是一位伟大的森林园丁。有你在,周围的其他动物,从鹿到西猯,再到刺豚鼠,都得到了生命中新的机遇。"

"You seem to spend a lot of time in those devastated regions. Please tell me more…"

"We, the magnanimous defecators, believe that by pooping there, we help regenerate the forests that were destroyed through people's ignorance."

"Yes, you are a great forest gardener. When you are around, many other animals, from deer to peccaries and agoutis, get a new chance in life."

"西猯和刺豚鼠？以前从没听说过。这些是动物还是植物？"

"真丢脸！你不知道那些南美洲独特的有蹄猪和大型啮齿动物吗？"屎壳郎问道。

"你看，我只是一只貘。我只知道，自己四处走动时，要避开偷猎者，并留下自己的便便以造福其他动物。"

"Peccaries and agoutis? Never heard of those before. Are these animals or plants?"
"Shame on you! You don't know about the hoofed pigs and the big rodents of South America that are so unique?" Dung Beetle asks.
"Look I'm just a tapir. I know that when I keep moving around, I avoid poachers, and spreading my poop benefits to other animals as well."

西猯和刺豚鼠？

Peccaries and agoutis?

……从恐龙时代起就在这儿了。

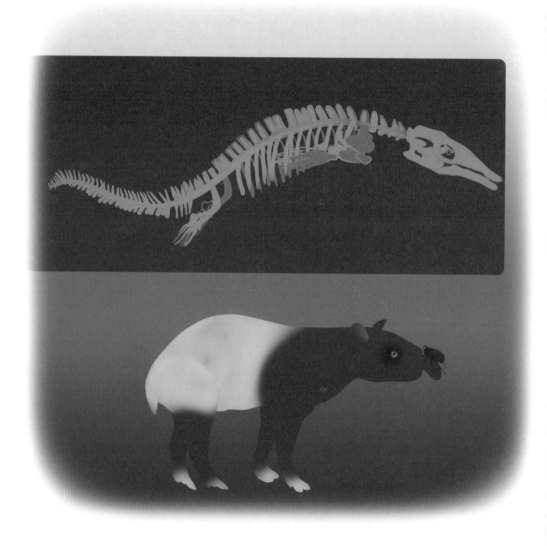

... around since the time of the dinosaurs.

"你扮演了如此重要的角色,有些人却认为你很愚蠢?"

"从恐龙时代起我就在这儿了,并没有多大变化。有些人可能会说貘愚蠢,但存活如此之久表明我们比其他许多现已灭绝的物种更聪明。"

"You play such a key role, and yet some people think you are stupid?"

"I have been around since the time of the dinosaurs, and haven't changed much. Some may call tapirs stupid, but having survived for so long shows we are smarter than many others – that are by now extinct."

"你是独一无二的!那么,谁是你最近的亲属呢?猪吗?"

"为什么是猪?我同马、犀牛有更多的共同点。顺便说一下,我最喜欢吃的东西是香蕉和海枣。"

"You are unique! So, who is your closest relative? The pig?"
"Why a pig? I have much more in common with horses and rhinos. And by the way, some of my favourite stuff to eat are bananas and palm dates."

……和犀牛。

... and rhinos.

海枣？

Palm dates?

"海枣？你是说那种有巨大种子的大水果吗？"

"你知道吗？我的肚子就像个种子银行。也许我是唯一能传播特大号种子的动物。只有鳄梨的种子会让我消化不良。我爱鳄梨，但不喜欢它们的种子。"

"Palm dates? You mean those big fruits with huge seeds?"
"Did you know my belly is like a seed bank? Maybe I am the only animal to spread outsized ones around. The only seeds that give me indigestion are the avocados. I love them, but not their seeds."

"你的便便一直很受欢迎。就像鲸的便便在北冰洋促进生命繁衍一样，你的便便帮助我们的森林再生。但我必须承认，你的超大便便有时对我来说确实太大了！"

"谁说你必须一人独自享用？也许共享更有意义！"

……这仅仅是开始！……

"Your poo is always welcome. Just like whales promote life in the Arctic with theirs, yours help regenerate our forests. But I have to confess, your huge droppings are sometimes just too much for me!"

"Who said you have to use it all? Maybe it makes sense to share!"

... AND IT HAS ONLY JUST BEGUN!...

……这仅仅是开始!……

… AND IT HAS ONLY JUST BEGUN! …

Did You Know? 你知道吗?

The tapir is considered a living fossil since it has been around since the Eocene Period. It has survived several waves of extinction. Tapirs are native to South and Central America, with one species found in Asia.

貘被认为是活化石,因为它出现于始新世。貘在几次物种灭绝浪潮中幸存下来。貘原产于南美洲和中美洲,在亚洲也发现了一种。

It is surprising that the tapir has survived, as it only bears one calf at the time, after a gestation period of 13 to 14 months. Should the tapir population decline further as it does today it's unlikely that it will recover.

令人惊讶的是,虽然雌貘的怀孕期为13—14个月,且每胎只产一只幼崽,这一物种还是存活下来了。如果貘的数量像现在这样持续减少,那么这个物种的重新繁荣就不太可能了。

A tapir weight is half of that of a horse. It has the strength to push trees over to get the fruit. Tapirs sleep most of the day. They eat enormous amounts of fruit, and distribute an enormous quantity of seeds.

貘的体重是马的一半。它有足够的力量推倒树木去摘水果。貘大部分时间都在睡觉。它们吃大量的水果，散播大量的种子。

At birth, a tapir calf is dark in colour, with yellow or white stripes and spots as camouflage. It resembles a watermelon with legs. The stripes fade after three months, and are completely gone by six months of age.

小貘刚出生时是深色的，带有黄色或白色的条纹和斑点作为伪装。它就像一个有腿的西瓜。3个月后条纹开始变淡，6个月后完全消失。

Tapirs have four toes on their front feet, and three on the back. They use their flexible noses as a snorkel when underwater. They will dive into water and feed on plants at the bottom of streams and lakes.

貘的前肢有四趾，后肢有三趾。它们在水下用灵活的鼻子充当通气管。它们会潜入水中，以溪流和湖泊底部的植物为食。

The Mountain Tapir of the Andes Mountains is one of the most endangered mammals in the world. The tapir uses its nose like a finger, plucking leaves from tree branches.

安第斯山脉的貘是世界上最濒危的哺乳动物之一。貘用它像手指一样的鼻子从树枝上采摘叶子。

The tapir makes a high-pitched whistling sound, like that of a car screeching to a halt. A snort, with feet in the soil, signals that the tapir is ready to defend itself. Tapirs, like dogs leave scent trails in the forest.

貘会发出尖锐的啸声，就像一辆汽车急刹车时发出的声音。一声鼻息，两只脚踩在土里，表示貘准备自卫。貘和狗一样，会在森林里留下气味痕迹。

In Chinese and Japanese folklore, tapirs are believed to eat people's nightmares. Tapirs avoid confrontation and favour running away, or submerging themselves in the water when feeling threatened.

在中国和日本的民间传说中，貘被描述成可以吃掉噩梦的神兽。貘避免正面冲突，喜欢逃跑，或在感觉到威胁时潜入水中。

Think about It

想一想

Tapirs have been around since the time of the dinosaurs?

貘从恐龙时代就存在了？

Can one animal play such a key role in regenerating forests?

一种动物能在森林再生中扮演如此关键的角色吗？

Why would the tapir be considered stupid?

为什么貘会被认为是愚蠢的？

A stomach as a seed bank?

胃可以被当作种子银行吗？

Do It Yourself!

自己动手!

How much land has been deforested in your region? How much of the land is still covered with forest? Governments around the world promote the regeneration of forests, but there seems to be more land lost than regenerated. Compile a list of the animals and plants that are key to the successful regeneration of a biodiverse forest. Is there one that stands out in your region? Identify a few candidates and discuss your list with friends and family to decide on which animal would be the most important at your local level.

你生活的地区有多少森林被毁掉了？还有多少土地仍有森林覆盖？世界各国政府都在促进森林的再生，但似乎失去的森林比再生的还要多。编制一份动植物名单，名单上是成功再造一片充满生物多样性的森林所需的关键物种。你生活的地区是否有这样的物种？找到几种可能的动物，并与朋友、家人讨论，以确定哪一种动物对你所在的地区最重要。

TEACHER AND PARENT GUIDE

学科知识
Academic Knowledge

生物学	2 000万年来，貘没有多大变化；裂唇嗅反应：长颈鹿、马、猫科动物和貘卷起上嘴唇，打开嘴里的一对细小的洞，将气味传递到一个叫作犁鼻器的特殊感觉器官，通过一种介于味觉和嗅觉之间的超强感觉官能来感知；上唇与鼻子融合；世界上有5 000种屎壳郎；西猯是长得像猪的有蹄哺乳动物；刺豚鼠是与豚鼠有亲缘关系的啮齿动物；貘的鼻子能够抓住东西；貘擅长游泳，还能在水下行走；貘和鹿身上的条纹和斑点，随着年龄的增长而褪色；奇蹄类的哺乳动物，如马和犀牛；长时间潜水会使某些寄生虫窒息；白貘的黑白色纹理在森林里起到了伪装的作用。
化　学	机械消化、化学消化和肠道消化；胃液刺激种子孵化；核果（有种子的水果，如棕榈果、咖啡果）的果肉被动物吃掉，这些种子在消化过程中经受化学处理，准备发芽。
物　理	受惊时，貘会躲在水下，用鼻子当通气管；貘在水中排便，以免它们的气味被发现。
工程学	在自然界，排泄物推动生命循环，而人类工程则消耗大量能源和财力去摧毁所有粪便。
经济学	人类认为处理废物需要成本，而在自然界，"废物"是新的生命周期的起点。
伦理学	为什么要浪费，甚至破坏有用和需要的东西呢？
历　史	貘被称为"活化石"，因为它们出现于5 500万年前（始新世早期），是有记录的历史最长的哺乳动物；玛雅人把西猯当作蛋白质来源。
地　理	马来西亚和苏门答腊的貘可以长到360千克，是南美洲以外唯一的貘；貘把种子传播到20千米外。
数　学	屎壳郎一晚上能在一个地下巢穴里埋下相当于自身重量250倍的东西，还能拖动超过自身体重1 000倍的物体（相当于一个人拉动6辆载满人的双层巴士）。
生活方式	屎壳郎不需要进食或饮水，因为粪便提供了所有必要的水分和营养。
社会学	对一个社会而言，传播种子、为种子提供营养的重要性，以及为下一代打下繁衍生命、生产食物的基础的重要性。
心理学	被认为愚蠢会导致情感创伤。
系统论	对貘而言，人类猎取貘肉是最大的威胁之一，同时还伴随人类对它们的栖息地的侵占；一份粪便样本包含100多颗植物种子，因此貘是关键的种子传播者，也是确保森林在火灾后再生的关键物种。

教师与家长指南

情感智慧
Emotional Intelligence

屎壳郎

屎壳郎很快乐，感激貘每天提供的粪便。他兴高采烈，对这么大量的食物感到心满意足。屎壳郎夸赞貘的存在和他拉下的大尺寸粪便。屎壳郎好奇貘为什么要去没有食物的地方。屎壳郎注意到貘在被烧毁的森林里待了很长时间，指出貘就是森林园丁。当屎壳郎发现貘不认识那些享受貘带来的好处的动物时，他直接表达了自己的惊讶和失望。貘能消化很大的种子，这点吸引了屎壳郎。屎壳郎深刻意识到貘在森林再生中的重要作用。

貘

貘非常清楚自己为其他动物提供了丰富的食物。貘意识到生存的艰难，并指出了人类造成的环境压力。貘不想给人留下深刻印象，所以把自己在森林转悠的理由，说成利用阳光清洁皮肤。貘不好意思谈论排便，也不愿意分享作为一个关键物种，自己在传播种子和森林再生方面起到的重要作用。貘坦率地承认，他并不知道周围所有动物的名字。当被问到貘是否愚蠢，貘并不生气，而是心平气和地分享貘从恐龙时代存活至今的事实，以此表明貘比其他许多现已灭绝的物种更聪明。貘讲述自己多么喜欢吃各种各样的水果，甚至包括那些种子又大又硬的水果，这使貘成为一个"种子银行"，在非常广泛的地区传播种子。

艺术
The Arts

貘是森林的园丁。所以，让我们创造一幅有趣的图画，展现貘使用它所有的特殊能力来重建森林。园丁种下种子并给土壤施肥，使种子发芽生长。用你的画来展示貘如何用它的鼻子采摘树叶和水果，以及它如何把粪便撒得到处都是，进而散播种子。还可以描绘貘藏在水里躲避捕食者的样子。

TEACHER AND PARENT GUIDE

思维拓展
Systems: Making the Connections

貘被认为是关键物种,因为它在自然景观的设计和架构以及生态系统的功能和结构中扮演着重要的角色,更因为貘是传播种子的"大师"。由于貘在水域和森林两栖,所以它对广泛的生态系统有贡献。然而,貘和它所在的生态系统都处境危险。必须制止偷猎和由于滥垦造成的貘栖息地的破坏。貘传播了大量高品质的种子。粪便中只有不到1%的种子在消化过程中受损,这意味着一旦它们沉积在森林的土壤中,几乎所有的种子都有机会发芽。貘喜欢在"茅坑"排便,研究表明,这些"茅坑"有益健康。这些"茅坑"为饥饿的屎壳郎提供种子,屎壳郎再将种子埋在地下,帮助种子发芽。此外,从植物的角度来看,种子储存在肥料中增加了存活的机会。森林是分阶段生长的。在森林大火等破坏性事件之后,喜欢阳光的先锋物种是第一批重新定居该地区的物种。生态系统需要耐阴生长的顶极群落物种。顶极群落物种在森林再生的后期往往占主导地位。貘粪便中含有的顶极物种种子大约是先锋物种的120倍,而且在粪便堆中发现的一些植物物种在该地区并不存在。貘的活动半径可达20千米。对貘的研究和思考让我们反思人类是如何处理产品和废物的。一些生物学家宣称,貘的消化系统效率低下,很多食物没有被消化,从而产生大量的粪便。然而,当我们研究这种"低效率"对大自然的影响时,就会发现它对恢复枯竭甚至被破坏地区起到了非常重要的作用。这是大自然给我们上的重要一课:大自然的目标不是只造福单一物种,而是让每个物种都能为整体作贡献。我们应该改变处理废物的逻辑,它们虽脏但贵。希望我们能够更多地学习和应用大自然的智慧,不再浪费众多有价值的"废物"资源。

动手能力
Capacity to Implement

当森林被烧毁,或者农民因土地枯竭而放弃土地时,我们能做些什么来恢复生态系统呢?设想一个受屎壳郎和貘启发的有创意的解决方案。制订一个使用当地资源和种子的计划,最好只种植当地的能结果的植物。设计出行动方案,分享和讨论,并就如何帮助大自然恢复力量达成共识。

教师与家长指南

故事灵感来自

This Fable Is Inspired by

帕特里夏·美第奇
Patrícia Medici

帕特里夏·美第奇于1995年获得圣保罗大学林业科学学士学位，并于2001年获得巴西米纳斯吉拉斯州联邦大学野生动物生态、保护和管理学硕士学位。2010年，她在英国肯特大学杜雷尔保护与生态研究所获得生物多样性管理博士学位。1996年，刚从大学毕业的帕特里夏就开始从事貘保护工作。她与生态研究所的一个团队一起，研究并推动了巴西貘的保护。2000年，帕特里夏成为世界自然保护联盟貘专家组的主席。2008年，她在巴西潘塔纳尔发起了"低地貘保护倡议"，在那之前潘塔纳尔没有开展过关于貘的研究。帕特里夏使我们对这种过去知之甚少的物种有了更多了解。她将貘称为"森林的园丁"。她的项目为巴西的貘保护工作确立了一个标准。

图书在版编目(CIP)数据

冈特生态童书.第八辑:全36册:汉英对照/
(比)冈特·鲍利著;(哥伦)凯瑟琳娜·巴赫绘;
何家振等译.—上海:上海远东出版社,2021
ISBN 978-7-5476-1773-1

Ⅰ.①冈… Ⅱ.①冈… ②凯… ③何… Ⅲ.①生态环
境-环境保护-儿童读物—汉、英 Ⅳ.①X171.1-49

中国版本图书馆CIP数据核字(2021)第249940号

策　　划　张　蓉
责任编辑　祁东城
封面设计　魏　来　李　廉

冈特生态童书
森林园丁
[比]冈特·鲍利　著
[哥伦]凯瑟琳娜·巴赫　绘
李原原　译

记得要和身边的小朋友分享环保知识哦!
八喜冰淇淋祝你成为环保小使者!

Water 257
从食物到纤维
From Food to Fibre

Gunter Pauli

[比]冈特·鲍利 著
[哥伦]凯瑟琳娜·巴赫 绘
靳维筠 译

上海远东出版社

丛书编委会

主　任：贾　峰
副主任：何家振　闫世东　林　玉
委　员：李原原　祝真旭　牛玲娟　梁雅丽　任泽林
　　　　王　岢　陈　卫　郑循如　吴建民　彭　勇
　　　　王梦雨　戴　虹　翟致信　靳增江　孟　蝶

特别感谢以下热心人士对童书工作的支持：

匡志强　宋小华　解　东　厉　云　李　婧　陈　果
刘　丹　熊彩虹　罗淑怡　旷　婉　杨　荣　刘学振
何圣霖　廖清州　谭燕宁　韦小宏　李　杰　欧　亮
陈强林　王　征　张林霞　寿颖慧　罗　佳　傅　俊
胡海朋　白永喆　冯家宝

目录

从食物到纤维	4
你知道吗?	22
想一想	26
自己动手!	27
学科知识	28
情感智慧	29
艺术	29
思维拓展	30
动手能力	30
故事灵感来自	31

Contents

From Food to Fibre	4
Did You Know?	22
Think about It	26
Do It Yourself!	27
Academic Knowledge	28
Emotional Intelligence	29
The Arts	29
Systems: Making the Connections	30
Capacity to Implement	30
This Fable Is Inspired by	31

一颗鳄梨和一颗芒果正在比谁的果核更大。这些巨大的果核占了整颗果实的大部分重量。

"听说人们叫你水果之王。"鳄梨说。

"被人称作王真是殊荣啊。可我宁愿被称作女王。我能理解人们误会的原因,因为我们芒果既有雄花又有雌花。"

An avocado and a mango are comparing the size of their seeds. These are huge, and make up most of the weight of their fruit.

"So, you are called the King of Fruit," the avocado says.

"What a privilege to be called the King. I'd rather be called the Queen, but I understand the confusion, as we mangoes have male and female flowers."

一颗鳄梨和一颗芒果……

An avocado and a mango...

……那又怎样？

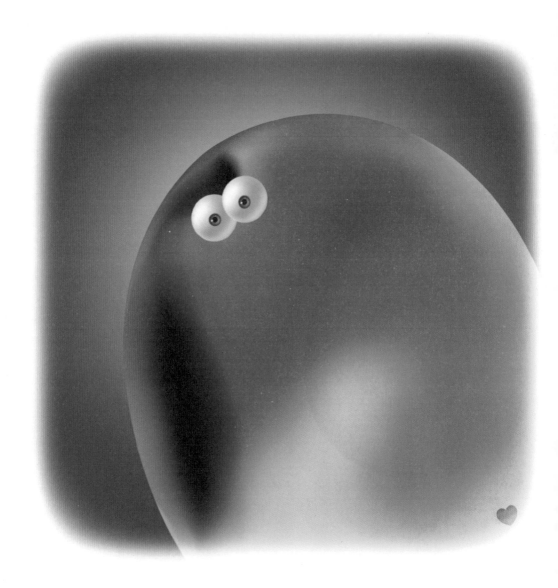

...what's the problem?

"你那浓厚的滋味让人们为你倾倒,而且你还是非常健康的食物,但是你的果核远远大过你的果肉!"

"所以呢,那又怎样?"芒果问。

"果核不能食用,它们会被扔掉。"

"You are celebrated for your rich taste, and as a healthy food, but you are almost more seed than flesh!"

"So, what's the problem?" Mango asks.

"Seeds are not eaten. They are thrown away."

"现在人们不吃果核,并不意味着果核永远不能被食用,也不代表果核不能被用于其他方面。我的果核有着丰富的营养,所以任何一个都不该被浪费。"

"你的果核硬得像石头,任何一个人咬的时候都会崩坏牙齿!"

"Because seeds are not eaten now doesn't mean they can never be eaten, or are no good for anything else. My seeds are so rich in nutrients that not a single one should go to waste."

"Anyone trying to chew your rock-hard seeds would break their teeth!"

我的果核有着丰富的营养……

My seeds are so rich in nutrients...

……古人的智慧……

... ancient wisdom ...

"你听我说,古人的智慧告诉人们,芒果树的任何一个部位,果实、花朵、树皮,甚至包括果核,都是可用的。"

"别一直和我说这些我已经知道的好处。请听我讲!人们为什么会牺牲自己的牙齿去咀嚼比皮革还难咬的东西呢?"

"Look here, ancient wisdom has taught people that every part of a mango tree can be used: the fruit, the flower, the bark and even the seed."

"Don't insist on telling me about all these good things we already know. Please listen! Why would anyone sink their teeth into something that is tougher than leather?"

"你知道吗？人们会用我的果实、果皮、果核来制作皮革。甚至也用你那坚硬的果核。"

"不会吧！芒果啊，我的果核虽然硬，但是它们可没有你的大。"

"果核、果皮，或者果肉，都可以用来制造皮革以及更多的东西。或者说你就像那些人一样，只知道利用我身上的一部分，比如果肉或果汁？"

"Did you know that people do make leather from my fruit, my skin and my seed? Even from your hard seeds too."

"Oh no, my seeds may be hard but they are not as big as yours, Mango."

"Seed, skin, or flesh: all can be used for leather, and more. Or are you like those people who only look at using one part of me, like my flesh or juice?"

……人们会用我的果实来制作皮革……

... people do make leather from my fruit ...

……对人类的头发和皮肤也是大有益处的。

...great for hair and skin.

"当然你是对的。我的果核对人类的头发和皮肤也是大有益处的。"

"我的也是呦。"芒果回答道。

"人们应该永远记住,每一个果核都有成长为一棵新树所需的一切。它具备助长下一代的力量。"

"You are right, of course. My seeds are also great for hair and skin."
"And so are mine," Mango replies.
"People should keep in mind that every seed has everything a plant needs to grow a new tree. It is equipped to feed the next generation."

"我的果核是很好的食物,同时也被用于制作窗户清洁剂、牙膏和婴儿长乳牙时用来磨牙的东西。"

"这下可就有趣了。我的果核的汁液可以用于制作永不褪色的墨水。它还可以用来制造能够杀虫、去霉,甚至杀菌的产品。"

"My seeds are great food, but also serve as a window cleaner, a toothpaste and something for babies to chew on when teething."

"Now that is interesting. The juice from my seed is used to make an ink that never fades. It can also be made into an product to kill insects, fungi, and even bacteria."

……永不褪色的墨水……

...ink that never fades...

……给棉花染上美妙的粉色……

... dye cotton a wonderful pinkish colour ...

"看起来我们有很多共同点哦。"

"你知道吗?人们会用我的果核和果皮给新羊毛或棉花染上美妙的粉色哦!"鳄梨问道。

"怎么用绿色的水果让一条裙子变粉啊?"

"Seems like we have a lot in common."
"Did you know that people use my seed and skin to dye virgin wool or cotton a wonderful pinkish colour?" Avocado asks.
"How does a green fruit make a dress turn pink?"

"我不清楚,但事实是我们的果核永远不应该被当作垃圾扔掉。它们有太多美好的用途了,不该被视为废物。"

"没错!人们至少应该把我们切碎,然后把粉末加到面包和墨西哥薄饼中,这样他们的孩子们都可以分享美食的一部分啦。大自然已经为我们准备好了。"

……这仅仅是开始!……

"I have no clue, but the truth is that our seeds should never end up in the garbage. They have too many good uses to be considered waste."

"Exactly! The least people can do is to shred us fine and add the powder to bread and tortillas, so that their children can also have a share of all the good things. Nature has intended for us."

... AND IT HAS ONLY JUST BEGUN!...

……这仅仅是开始！……

…AND IT HAS ONLY JUST BEGUN!…

Did You Know? 你知道吗?

There are over 1,000 varieties of mango in India alone. It has been estimated that the mango's biodiversity could be as much as 4,000 varieties. A mango tree can bear fruit for more than one hundred years.

仅在印度就有上千种芒果。据估计芒果的生物多样性可以达到4 000种。一棵芒果树结果的时间长达100多年。

The mango has been cultivated in India for over 400 years. Its tree, bark, flower and fruit have been widely described in Ayurvedic medicine.

芒果已经在印度被培育了长达400年之久。它的树、树皮、花朵、果实都在阿育吠陀医学中被广泛描述。

Mango flowers are pollinated by houseflies. The Central Institute for Subtropical Horticulture in Lucknow has 633 mango species. The State of Uttar Pradesh (India) is the world's largest producer of mangoes.

芒果花通过家蝇授粉。位于勒克瑙的中央亚热带园艺研究所拥有633种芒果。印度北方邦是世界上最大的芒果产地。

The mango originated in the Indo-Burmese region, and is family of the cashew. It was introduced to Brazil by Portuguese explorers in the 16th century, and to Florida in 1860.

芒果起源于印缅地区，与腰果属于同一科。它在16世纪被葡萄牙探险者带到巴西，在1860年被带到佛罗里达。

Avocado seeds produce a milky liquid that turns red when exposed to air. It has been used as a natural ink. Some documents from the 16th century, kept in Popayan, Colombia, are preserved to this day.

鳄梨的种子产出牛奶般的液体，并在接触空气后变为红色。它被当作天然的墨水。在哥伦比亚波帕扬，有一些16世纪的文件被保存至今。

A seed contains everything required for a new tree to grow. These are rich sources of trace minerals and natural oils. Yet, most fruit processing industries consider skin and seeds waste.

种子包含了一棵新树成长所需的一切。它富含微量矿物质和天然油。但是水果加工行业多数时候将果皮和果核视为废料。

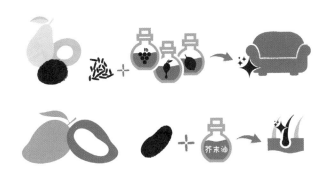

Finely shredded avocado seed, mixed with olive oil, vinegar and lemon juice, offer an ideal furniture polish for wood. Mango seeds mixed with mustard oil eliminate dandruff.

切得细碎的鳄梨核与橄榄油、醋、柠檬汁混合，就是一种理想的木制家具上光剂。芒果核与芥末油混合可以去除头皮屑。

California has 500 wild avocado varieties. Only 8 are farmed. Avocados thrive in a mineral rich soil, and need a lot of sun. The Mexican avocado relish called guacamole has gained great acceptance worldwide.

加利福尼亚有500种野生鳄梨。仅有8种被种植。鳄梨的生长需要富含矿物质的土壤，还需要充足的阳光。墨西哥鳄梨酱在全世界广受欢迎。

A seed carries all a new plant needs - and we throw it away?

一颗种子有着一株新生植物成长所需的一切，我们就这样将它们扔掉？

A mango tree has both male and female flowers?

一棵芒果树既有雄花也有雌花？

Would you like a pink avocado shirt or dress?

你想要一条粉红色的鳄梨衬衫或者裙子吗？

Shall we write a letter with avocado ink?

我们要不要用鳄梨墨水写一封信？

Do It Yourself!

自己动手！

Do you eat avocados at home? It is easy to grow your own avocado tree by germinating the seed in a glass of water. First wash the seed. Then use three toothpicks to suspend it broad-end down over a glass filled with water so that about a two centimetres of the seed is in the water. Put the glass in a warm place, out of direct sunlight, and replenish the water as needed. You should see the roots and a stem starting to sprout in about two to six weeks.

你在家吃鳄梨吗？自己种植鳄梨很简单，只需将果核放入有水的玻璃杯中使其发芽。首先洗干净果核。然后使用三根牙签将宽边控制在下方，架在有水的玻璃杯上，让果核浸入水下 2 厘米。将玻璃杯放在一个温暖的地方，避免阳光直射，必要时记得换水。2—6 周后，你可以看到根和茎开始萌发。

TEACHER AND PARENT GUIDE

学科知识
Academic Knowledge

生物学	鳄梨是一种单一种子浆果；核果是一般只有一个果核的水果。
化学	鳄梨果肉和果核含有鞣酸；鞣酸帮助山羊或绵羊这类动物反刍；鞣酸通过化学反应与铁结合；金属元素和鞣酸通过螯合作用形成配合物，这种配合物不被消化直接通过身体；芒果富含钾、镁、B族维生素、维生素A、维生素C、维生素E、维生素K；芒果可以增强食欲、促进消化，因为它含酯类、萜烯类和醛类；芒果含25种不同的类胡萝卜素。
物理	鳄梨墨水是绿色的；鳄梨核墨水是红色的；鞣酸浓缩物是产生红色的原因；很多水果经过消毒，并被冷却到很低的温度，因此不能发芽。
工程学	鳄梨核可以被转化为生物塑料；腐烂的芒果可以被转化为植鞣革。
经济学	所有鳄梨都是手工摘收，可以提供大量岗位；50%的鳄梨都被浪费了，是浪费率最高的食物；鳄梨被用来生产增甜剂、增稠剂、化妆品、精油、生物塑料、生物燃料。
伦理学	怎么可以将50%的水果都扔掉，它们有丰富的营养，甚至还能被转化为生物塑料呢！由腐烂芒果制成的植鞣革有助于动物福利。
历史	"鳄梨"（avocado）这个词来源于纳瓦特尔语 ā huacatl，意思是"形成"；鳄梨在树上成对生长，因为鳄梨是雄雌同树的。
地理	墨西哥是鳄梨的主要产地，占世界产量的32%，米却肯州是种植培育的中心。
数学	乘数效应：当50%的水果被浪费，那么产生的收入也只剩50%。
生活方式	芒果有丰富的维生素A和β-胡萝卜素，可以缓解年龄造成的视力衰退。
社会学	在印度，"芒果"这个词的意思是希望。
心理学	自尊，认为自己不应该被浪费；当我们发现与不认识的人有共同点时，我们受到鼓舞并更加自信。
系统论	世界上很多人因饥饿而受苦的同时，50%的芒果和鳄梨被浪费。

教师与家长指南

情感智慧
Emotional Intelligence

鳄梨

鳄梨引入话题,想确认芒果是不是被称为水果之王。看来鳄梨非常嫉妒芒果,而且也想要同样的称号。鳄梨质疑芒果的这个称号,因为芒果一半的果实(果核)都被扔掉了。鳄梨指出芒果的果核虽然有营养,但是太硬以至无法食用。鳄梨感觉到芒果并没有听他的话,于是改变语气。当芒果开始向鳄梨提问,鳄梨发现他们有很多共同之处。鳄梨的态度变得更积极,他不再质疑芒果,而是更多地谈论自己的特点,而且不得不承认自己无法解释为什么绿色的鳄梨可以产生红色和粉色。

芒果

芒果非常谦虚,觉得被称为王是一种莫大的殊荣。芒果对鳄梨的质疑没有太多回应。当鳄梨要求芒果倾听时,芒果改变了对话姿态,不再局限于自我辩护和解释,而是开始提问。芒果质问鳄梨:鳄梨是不是就像人类一样只关注金钱?芒果发现那些废弃的垃圾有多种用途,还发现自己和鳄梨有很多共同点。能做的事情非常多,这让他们很惊喜,而且当鳄梨无法解释自己如何由绿变粉时,芒果并没有趁势追问,而是提供了另一个简单的利用果核的方法,这在彼此之间建立了信任感。

艺术
The Arts

用鳄梨制作天然染料吧。你只需要准备果核。首先,请洗那些滑溜溜的果核并晾干。然后将它们放入锅中煮,锅中放入足量的水,能够淹没你想要染色的织物就行(大约每250克织物需要5个果核)。当果核在液体中滚动时,慢慢地敲打果核,果核的外壳会裂开并脱落,液体会变成明亮的红宝石色。浸泡过夜,第二天放入潮湿的布料,染成你想要的颜色。多试几次,但不用力求完美。

TEACHER AND PARENT GUIDE

思维拓展
Systems: Making the Connections

　　饥饿与营养不良持续影响着近10亿人，与饥饿和贫穷相关的世界问题也在持续恶化。我们不能奢望地球再赠予我们更多了，而应当利用现有资源做更多的事情。本书提到了一种解决方法：将原本被浪费的50%的鳄梨和芒果视作一个重要机遇。这个机遇被世界认可，并有大量研究作为支撑，可如何解释面对这样的机遇人们几乎什么都没有做？令人惊讶的现实是，这一切众所周知，却没什么人去做。我们不得不问问为什么。是缺少企业家精神？是不在乎？或者只是缺乏动力？一个人独自探索和实践确实非常困难。因此关键在于你要花时间去发现志同道合且有能力的人，这样你们就可以联手创建团队。就像大自然中的所有物种一样，我们有互补的能力，这让我们一起变得更强大。不仅因为我们的思考方式很相似，更因为我们确信自己有能力从他人身上看到我们没有亲眼看过的事物，体验我们没有亲身体验过的经历，而非仅仅把他人看作竞争对手。团队不必建立在所谓的强项基础上，甚至可以建立在他人眼中的弱项上。比如，芒果和鳄梨的弱项都是50%的果实被浪费，双方都意识到这样做是没有道理的，两者可以共同努力，使双方都受益匪浅。以色素为例，鳄梨的果实呈绿色，种子呈粉色甚至红色；芒果叶子和果皮呈棕色和灰色。如果这两种水果搭配在一起，那么就会产生更丰富的颜色组合。这样的合作不仅起到互补的作用，还能降低风险和成本，因为原先创造一种颜色所消耗的精力，现在可以用来创造四种颜色。大自然教导我们，只要共同合作，即使最初被视为问题的事物，也可以被转化为一种巨大的机会。不仅仅可以从以前被浪费的东西里获得收入，而且可以满足所有人的基本需求并为全社会作贡献。

动手能力
Capacity to Implement

　　市场上有一种新产品，称为植鞣革，这是用熟过头的水果制成的。我们也来试着用水果制作皮革吧！切碎果实并将它们煮烂，然后把水果汤铺洒在坚固的台面上。剩下的就是等液体成形。这样不能吃的水果就可以变为更有价值的东西了。制作时请小心热水。记录下整个过程并与他人分享。

教师与家长指南

故事灵感来自
This Fable Is Inspired by

路易斯·弗拉维奥·席勒·罗德里格兹
Luis Flavio Siller Rodriguez

弗拉维奥出生于墨西哥，2012年在墨西哥蒙特雷理工学院获得生物技术与生物加工工程学士学位。他在沙特阿拉伯的阿卜杜拉国王科技大学继续深造，并于2013年毕业，取得化学与生物工程学硕士学位。毕业后，弗拉维奥决定成为社会企业家。他注意到大量的食物被浪费，于是致力于开发针对食品产业问题的工程解决方案。他同恩里克·冈萨雷斯一起创建了天才食物公司，并开始解决鳄梨浪费问题，这是墨西哥最主要的废物来源之一。在过去的几年里，该公司凭借科学理论和工程技术方面的创新取得的成果向人们证明：改变是可能的。

图书在版编目（CIP）数据

冈特生态童书.第八辑:全36册:汉英对照／
(比)冈特·鲍利著；(哥伦)凯瑟琳娜·巴赫绘；
何家振等译.—上海：上海远东出版社,2021
ISBN 978-7-5476-1773-1

Ⅰ.①冈… Ⅱ.①冈…②凯…③何… Ⅲ.①生态环境-环境保护-儿童读物—汉、英 Ⅳ.①X171.1-49

中国版本图书馆CIP数据核字(2021)第249940号

策　　划	张　蓉
责任编辑	祁东城
封面设计	魏　来　李　廉

冈特生态童书

从食物到纤维
[比]冈特·鲍利　著
[哥伦]凯瑟琳娜·巴赫　绘
靳维筠　译

记得要和身边的小朋友分享环保知识哦！
八喜冰淇淋祝你成为环保小使者！

Food 258
从农场到餐桌
From Farm to Fork

Gunter Pauli

[比]冈特·鲍利 著
[哥伦]凯瑟琳娜·巴赫 绘
颜莹莹 译

上海远东出版社

丛书编委会

主　任：贾　峰

副主任：何家振　闫世东　林　玉

委　员：李原原　祝真旭　牛玲娟　梁雅丽　任泽林
　　　　王　岢　陈　卫　郑循如　吴建民　彭　勇
　　　　王梦雨　戴　虹　翟致信　靳增江　孟　蝶

特别感谢以下热心人士对童书工作的支持：

匡志强　宋小华　解　东　厉　云　李　婧　陈　果
刘　丹　熊彩虹　罗淑怡　旷　婉　杨　荣　刘学振
何圣霖　廖清州　谭燕宁　韦小宏　李　杰　欧　亮
陈强林　王　征　张林霞　寿颖慧　罗　佳　傅　俊
胡海朋　白永喆　冯家宝

目录

从农场到餐桌	4
你知道吗？	22
想一想	26
自己动手！	27
学科知识	28
情感智慧	29
艺术	29
思维拓展	30
动手能力	30
故事灵感来自	31

Contents

From Farm to Fork	4
Did You Know?	22
Think about It	26
Do It Yourself!	27
Academic Knowledge	28
Emotional Intelligence	29
The Arts	29
Systems: Making the Connections	30
Capacity to Implement	30
This Fable Is Inspired by	31

一只蚂蚁正在哥伦比亚桑坦德茂密的森林里游逛，它发现有人正准备抓住它。一根绿色的香蕉挂在树上，注视着抓捕蚂蚁的行动。

"嘿，你这只大蚂蚁！你最好跑开藏起来，否则你今晚就会成为大城市里的盘中餐。"

As an ant is running around the lush forest in Santander, Colombia, it spots some people ready to scoop it up. A green banana hanging in the tree watches the hunt for ants unfolding.

"Hey, you big ant! You'd better run and hide or you will end up on a plate in the big city tonight."

一只蚂蚁正在茂密的森林里游逛……

An ant is running around the lush forest …

……抓到我以获取新鲜食材……

... they can serve me fresh ...

"是的,我的屁股富含蛋白质,人们宁愿冒着被我咬的风险,也要抓到我以获取新鲜食材。"

"我知道蚂蚁屁股被运往伦敦和东京,那么你一定有些独特之处。"

"Yes, with my bottom so full of protein, people are willing to risk my bite to catch me, so that they can serve me fresh."
"I believe ant butts are shipped to London and Tokyo, so there must be something unique about you."

"哦，我们蚂蚁被当作食物已经有几千年历史了。我很高兴得知人们想要得到我最好的部分。"

"你的屁股？"

"Oh, we ants have been served as food for thousands of years. And it pleases me to know that people want the best part of me."

"But your bottom?"

你的屁股?

But your bottom?

……切片，捣烂或油炸……

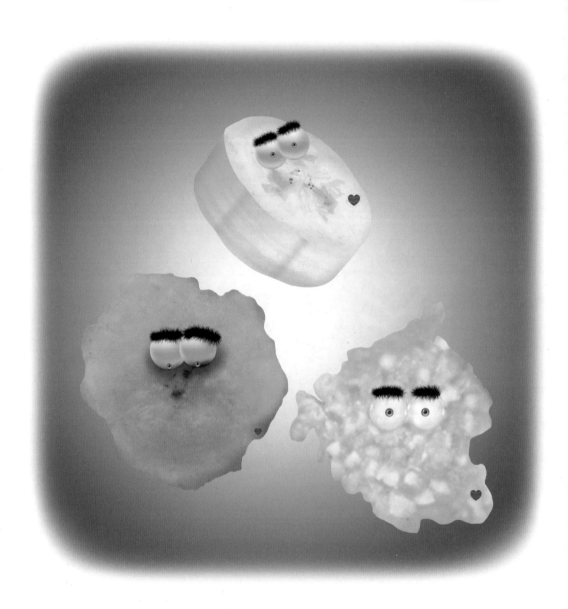

... sliced, mashed or fried ...

"听着，不管你喜不喜欢，食物无非就是谁吃谁的某一部分，不管是动物、植物、昆虫还是蘑菇。至少我不会像你一样被切片，捣烂或油炸。"

"我看到他们确实把你包在一些绿叶里，使味道更可口，"香蕉回答。

"Listen here, whether you like it or not, food is always about one being eating some part of another, be it animal, plant, insect, or mushroom. At least I am not being served sliced, mashed or fried, like you."

"I see they do wrap you in some green leaves to make their dish even tastier," Banana replies.

"绿叶？是的，有150多种植物都有可食用的叶子，它们也是美味。"

"嗯，真有意思……以前不被人注意的叶子现在成了美味佳肴。"

"人们需要利用现有资源。所得收益也应该流向供应这些资源的农民。"蚂蚁强调说。

"Leaves? Yes, there are more than a hundred and fifty different plants with edible leaves that can be harvested to add great taste."

"Hmm, interesting… that leaves that used to go unnoticed are now celebrated as delicacies."

"People need to use what is available. And the money should go to the farmers who supply it," Ant insists.

绿叶……

Leaves ...

……直接由农民供应给厨师……

... straight from the farmers to the cooks ...

"完全正确!我们在这里生长得很好,也为农民带来一些收入。但这还不够!这些收成应该直接由农民供应给厨师和厨师长,他们则将其变为美味大餐。"

"他们应该多加利用像我们这样美味的食物,而且为了健康,应该提倡使用最好的。就像在汉堡包这样的快餐出现之前一样!"

"Exactly! We grow well here, and we offer the farmer some income too. But that is not enough! The harvest should go straight from the farmers to the cooks and chefs, who turn it into great meals."

"They should make more use of tasty foods like us, and celebrate using the best – for the sake of their health. As it was before fast foods like hamburgers arrived!"

"的确。如果农民可以把庄稼直接卖给那些食用的人赚取足够的钱,他们就能够生存下去,也就不会考虑种植任何非法作物了。"

"正是如此!在农民和最终食用我们的消费者之间有太多人了。"

"人们如何确保农民既能赚足够的钱,又能照料好我们呢?"香蕉问道。

"Indeed. If farmers could make enough money selling their crops directly to the people who eat it, they would be able to make a living, and wouldn't even consider growing any of the illegal stuff."

"Exactly! There are too many people between the farmer and the consumer – the one who will ultimately eat us."

"How can people ensure that farmers make enough money, and take good care of us?" Banana asks.

……又能照料好我们呢?

... and take good care of us?

……每天提供近1亿个汉堡包……

... 100 million hamburgers per day ...

"坚持食物的多样性,正如大自然所赐。当然,还要美味,把营养变成盛宴和惊喜。"

"然而,他们的餐馆每天提供近1亿个汉堡包,却并没有消除饥饿。"

"生产大量食物不是解决问题的办法。为有需要的人提供有营养的食物,并因此获得丰厚的报酬,这才是秘诀。"

"By insisting that their food is as diverse as is offered by Nature. And tasty too, of course, by turning nutrition into a feast, and a surprise."

"And yet, they have restaurants serving nearly 100 million hamburgers per day – and that has not eradicated hunger."

"Producing tons of food is not the answer. Getting nutritious food on the table for those who need it – and getting paid well for it – that is the secret."

"我同意，人们的食物应该从森林、农场直接到餐桌。"

"他们的茶和咖啡也应该直接从作物到杯子，"蚂蚁补充道。

"是啊！这样钱才能进到那些照料土地的人的口袋里。"

……这仅仅是开始！……

"I agree, people need to get their food straight from the forest, and from the farm to the fork."

"And their tea and coffee from the crop to the cup," Ant adds.

"Yes! So they put money in the pockets of the people who take care of the soil."

... AND IT HAS ONLY JUST BEGUN!...

……这仅仅是开始！……

... AND IT HAS ONLY JUST BEGUN! ...

Did You Know?
你知道吗？

The farmer receives less than one-tenth of one percent of the price (around US $3) that the customer pays for a cup of coffee sold in a coffee shop.

顾客在咖啡店购买一杯咖啡约3美元，而农民得到的却不足其价格的千分之一。

Even when one pays fair trade prices, it is often not enough to offer the farmer sufficient income to make ends meet. Organic farming commands a higher price but has reduced yields, which reduces overall income.

即使向农民支付公平的贸易价格，他们的收入也往往不足以维持生计。有机种植的价格更高，但产量少，从而使得整体收入减少。

在农民和咖啡的最终消费者之间，有多达 100 个中间商，如采购商、经销商、运输商、烘焙商和包装商。当农民自己烘焙咖啡时，他的收入能提高 5 倍。

Between the farmer and the final consumer of a cup of coffee there are up to 100 intermediaries, such as buyers, handlers, transporters, roasters, and packagers. A farmer increases his revenue by a factor of five, when roasting his own coffee.

咖啡农平均拥有 1.5 公顷土地。为了提高咖啡产量，农民们减少了竹子的种植，甚至放弃了菜园。如果咖啡收入低，这些农民将遭受缺水和饥饿。

Coffee farmers own on average 1.5 hectares of land. In a drive towards more output, farmers eliminate bamboo and even give up a their vegetable gardens. If harvest income is low, such farmers will have no water and will go hungry.

高科技农业

Large corporations offer farmers seed, fertilisers, agrochemicals, time management, harvest planning and fixed prices. This reduces the farmer to someone who "rents" land, has a minimum income, and has his or her experience disregarded.

大公司为农民提供种子、化肥、农药、时间管理、收割方案和定价。这使得农民沦为"租种"土地的人，拿着最低收入，而且他们的体验被忽视。

Farmers are the custodians of land. They must ensure that the soil is continuously replenished with nutrients and carbon. If this is not done, then the land will be "mined" until it becomes completely infertile.

农民是土地的守护者。他们必须确保土壤一直保持充足的养分和碳。要不然，土地就会被"开采"到完全贫瘠。

A tropical forest generates 500 tons of biomass annually. An agricultural field, with all the genetics and chemistry applied, barely generates 10 tons. The farmer can generate more income from these 500 tons than from any form of monoculture.

一片热带森林每年产生500吨生物质。而一片农田，使用所有遗传学和化学的成果，也只能产生10吨生物质。农民可以从这500吨中获得比任何形式的单一种植更多的收入。

"Crop to cup" and "farm to fork" logic strengthen the local economy since money circulates in the local economy. As more value is generated and more goes from hand to hand, income grows due to a multiplier effect.

"从咖啡豆到咖啡"和"从农场到餐桌"的逻辑因货币只在当地流通而增强了当地经济。随着越来越多的价值产生和转手，收入也就因乘数效应而增长。

Think about It 想一想

Would you like to eat insects?

你愿意吃昆虫吗?

Should food be local and fresh, or preserved and shipped around the globe?

食物应该是本地新鲜的,还是保存起来并运往世界各地?

Is it possible for us to have high quality food and feed everyone in the world?

我们有可能拥有高质量的食物,并养活世界上的每个人吗?

Do hamburgers help fight hunger?

汉堡包有助于战胜饥饿吗?

Do It Yourself! 自己动手！

How many people have handled the tealeaves that you use to make your cup of tea? Do some research and find out who has handled the leaves, from the moment of harvest to the moment you pour hot water into your teacup. It may be as many as ten or twenty people, who each need to earn an income. Now, calculate which percentage of the price you pay for the product ultimately goes to the farmer, and how much goes to everyone else.

有多少人经手过你用来泡茶的茶叶？调查一下，从收获茶叶的那一刻直到你把热水倒进茶杯，都有谁接触过茶叶。可能有10人或20人之多，他们都需要从中赚取收入。现在，计算一下你为产品支付的价格中，有多少最终归农民所有，又有多少归其他人所有。

TEACHER AND PARENT GUIDE

学科知识
Academic Knowledge

生物学	蚂蚁、蟋蟀、蚱蜢和白蚁富含蛋白质；蚁后富含脂肪；蝎子的味道像螃蟹；为土壤补充养分；可食用树叶的生物多样性。
化 学	当受到威胁时，蚂蚁会分泌蚁酸作为防御，使它们有一种类似醋的味道；叶子富含维生素A、B、C、E和K，以及镁、铁、钙等矿物质，还含有类黄酮和植物素。
物 理	清洗树叶会使它们枯萎。
工程学	生产几亿个汉堡包并不能消除饥饿；将食品送到最需要的人手中的供应链管理。
经济学	世界上有20亿人常吃昆虫，把其作为蛋白质的主要来源；农民应该关注一体化收割，除了豆子，还有叶子，叶子有很好的食用价值和经济价值；从农场直接到餐厅厨房的一体化供应链，会产生更多的收入。
伦理学	对人们想要最好的感到高兴；当国际市场价格无法满足农民的生存需求，就会诱发种植非法产品。
历 史	200万年前，人们吃果实、树叶、树干和树皮；游猎采集部族的人认识几百种可食用的叶子、浆果和根茎。
地 理	亚洲热带地区的人吃荷叶，非洲中部地区的人吃香蕉叶，俄罗斯人吃橡树叶和腌菜，日本人和韩国人吃紫苏叶，中东和印度人吃无花果叶，安第斯人吃旱金莲叶子，地中海地区的人吃葡萄叶，东南亚地区的人吃竹叶。
数 学	农民根据市场价格及其波动来调整种植，然而其他人都赚取佣金，使得最终成本增加，但不会增加农民的收入。
生活方式	食用昆虫，如哥伦比亚的蚂蚁和非洲的蝗虫和毛虫；吃生鲜食物的新文化。
社会学	厨师的作用，是将不同的食材转化成美味健康的菜肴。
心理学	一种生物对于其他生物来说好吃又营养；愿意冒痛苦的风险来获得利益。
系统论	从农场到消费者的链条越短，食物就越新鲜，越有营养，收益也越多。

教师与家长指南

情感智慧
Emotional Intelligence

香蕉

香蕉警示蚂蚁危险迫在眉睫。他认为蚂蚁的独特之处在于蚂蚁是广受全球欢迎的美食。他感到惊讶,并打听更多有关蚂蚁屁股的事情,那是蚂蚁身上最好吃的部分,通常与另一道美味——叶子一起食用。当蚂蚁把话题从味道转到农民的未来,香蕉意识到农民卖水果却赚不到足够的收入。香蕉关心的是农民,他们需要更大的份额才能生存。他认为多样化才是解决世界饥饿问题的办法,而不是像生产汉堡包那样生产更多同类产品。他强调挣来的钱应该进入农民的口袋,因为农民是土地的守护者。

蚂 蚁

蚂蚁自知蚁类含有丰富的蛋白质,也意识到人们愿意冒着被咬的危险来获取美味。尽管她知道自己会被吃掉,但她为人们认识到可食用蚂蚁的高营养价值而感到骄傲。当香蕉质疑蚂蚁被吃掉这一事实时,她自信地揭开谜底,证实食物无非就是一个物种捕食另一个物种。她关心农民,认为他们需要提高收入,而消费者需要更美味的食物,而不是大量生产的汉堡包。蚂蚁表达了她坚定的观点:中间商太多了。她认为问题的关键在于,食物应该像大自然一样多样化,这样厨师才能为消费者提供新奇、非凡的美食,农民才能因供应食物得到丰厚的报酬。

艺术
The Arts

味道是一种艺术。让我们来了解不同种类、可以提味的叶子,以及它们是如何为我们日常饮食补充矿物质和维生素的。我们周围有很多可食用的植物,城市、森林、花园或校园都有它们的踪影。就近寻找至少10种知道名字的无毒、可食用的植物叶子,它们是你们本土或其他地域文化烹饪艺术的一部分。

TEACHER AND PARENT GUIDE

思维拓展
Systems: Making the Connections

　　多年来，农民一直被迫接受那种制造更多相同产品的生产模式：为动物生产饲料，饲养动物以获取生产汉堡包所需的肉类。虽然这增加了产量，但并没有增加农民的收入。农业作为一种职业变得越来越没有吸引力，农民的子女目睹了他们父母的艰辛，宁愿离开土地去从事其他职业。由于国际市场价格无法覆盖种植和收割成本，咖啡种植者数量减少了50%。为了以更低的成本生产更多的同类产品，从收割到加工并最终交到顾客手中，都由中间商来操作。这或许会让商品价格更便宜，却加重了农民的负担。因此有必要采取其他模式。其中之一就是接受多样性。蛋白质的来源有多种，包括昆虫。还有许多可供选择的健康添加剂和口味，我们没有必要限制自己只使用盐和胡椒。农民可以与餐厅厨师形成紧密联盟。有了如此丰富的生物多样性，农民可以充分发挥土地的潜力，关注土地的产出（包括蚂蚁和其他昆虫），从而增加收入。首先要知道大自然能提供什么，然后再去寻找并确定可持续产出的农产品的附加值。此外，让产品离消费者更近，就越有可能提高农民的收入水平。

动手能力
Capacity to Implement

　　你能直接联系到哪些生产食品的农场？你认识哪些可以直接购买农产品的农户？问问亲朋好友，有谁愿意加入你的"合作社"，这样就可以直接从农民那里购买产品而不需要任何中介。接下来就农产品的价格达成一致。它可能不会比你在超市里支付的低很多，但对农民来说，这可能是他平常收入的好几倍。这种做法可行吗？还是说这需要太多额外工作？虽然会有一些不便，但在现代包装和物流条件下进行直销，可能对农民未来的生存产生重大影响。

教师与家长指南

故事灵感来自
This Fable Is Inspired by

莱昂诺·埃斯皮诺萨
Leonor Espinosa

　　莱昂诺·埃斯皮诺萨在哥伦比亚卡塔赫纳大学学习经济学和美术。在迷上烹饪艺术之前,她曾从事营销和广告工作。经过多年对烹饪技术的研究和实践,她使用哥伦比亚传统食材,于2007年创建了自己的餐厅,名为LEO。在研究了本地非洲裔农民社区的日常饮食和营养结构后,莱昂诺建立了一个从当地居民的产地直接到她餐厅的供应链。她所创造的价值不仅是为农民提供了收入,而且还使他们不再为了不义之财而在森林中从事非法活动。她的目标是保护当地的饮食传统,同时践行可持续发展,促进当地经济。由于她的努力,哥伦比亚正成为拉丁美洲的美食热点,正如酸橘汁腌鱼之乡秘鲁以及玉米饼和鳄梨色拉酱之乡墨西哥一样。

图书在版编目(CIP)数据

冈特生态童书.第八辑:全36册:汉英对照/
(比)冈特·鲍利著;(哥伦)凯瑟琳娜·巴赫绘;
何家振等译.—上海:上海远东出版社,2021
ISBN 978-7-5476-1773-1

Ⅰ.①冈… Ⅱ.①冈…②凯…③何… Ⅲ.①生态环
境-环境保护-儿童读物—汉、英 Ⅳ.①X171.1-49

中国版本图书馆CIP数据核字(2021)第249940号

策　　划	张　蓉
责任编辑	程云琦
封面设计	魏　来　李　廉

冈特生态童书

从农场到餐桌

[比]冈特·鲍利　著
[哥伦]凯瑟琳娜·巴赫　绘
颜莹莹　译

记得要和身边的小朋友分享环保知识哦!
八喜冰淇淋祝你成为环保小使者!

Food 259
可以吃的咖啡
Eat Your Coffee

Gunter Pauli

[比]冈特·鲍利 著
[哥伦]凯瑟琳娜·巴赫 绘
颜莹莹 译

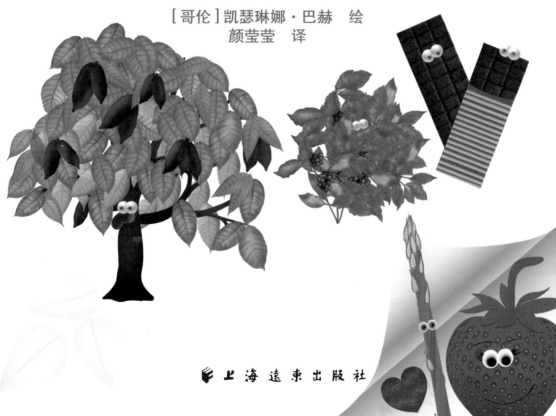

上海远东出版社

丛书编委会

主　任：贾　峰

副主任：何家振　闫世东　林　玉

委　员：李原原　祝真旭　牛玲娟　梁雅丽　任泽林
　　　　王　岢　陈　卫　郑循如　吴建民　彭　勇
　　　　王梦雨　戴　虹　翟致信　靳增江　孟　蝶

特别感谢以下热心人士对童书工作的支持：

匡志强　宋小华　解　东　厉　云　李　婧　陈　果
刘　丹　熊彩虹　罗淑怡　旷　婉　杨　荣　刘学振
何圣霖　廖清州　谭燕宁　韦小宏　李　杰　欧　亮
陈强林　王　征　张林霞　寿颖慧　罗　佳　傅　俊
胡海朋　白永喆　冯家宝

目录

可以吃的咖啡	4
你知道吗?	22
想一想	26
自己动手!	27
学科知识	28
情感智慧	29
艺术	29
思维拓展	30
动手能力	30
故事灵感来自	31

Contents

Eat Your Coffee	4
Did You Know?	22
Think about It	26
Do It Yourself!	27
Academic Knowledge	28
Emotional Intelligence	29
The Arts	29
Systems: Making the Connections	30
Capacity to Implement	30
This Fable Is Inspired by	31

一株咖啡树正在和一棵可可树聊天,他们都觉得人们种植咖啡和可可赚的钱不够养家。

"我们的农民没日没夜地工作,还总是忧心忡忡。他的孩子们经常挨饿。他觉得每公顷有两头奶牛就能让他脱离窘境。"

A coffee bush is chatting to a cacao tree, and they agree that the people growing them are not making enough money to care for their families.

"My farmer works day and night, and he worries all the time. His children often go hungry. He thinks that two cows on a hectare will save him from the worst."

一株咖啡树和一棵可可树聊天……

A coffee bush chats to a cacao tree ...

……两头奶牛的鬼话……

...nonsense about two cows...

"是谁在他脑子里灌输每公顷两头奶牛的鬼话的?"

"他确实欠银行很多钱,我想知道那是怎么回事。"

"一定是有人劝过他,说多种植和销售咖啡才是明智的,所以他才去砍倒了所有的竹子和果树。为此他甚至放弃了家里的菜园。"

"Who put that nonsense about two cows per hectare in his head?"

"He does owe the bank a lot of money, and I wonder how that happened."

"Someone must have convinced him it is wise to grow and sell more coffee, so he went and cut down all his bamboo and all his fruit trees. He even gave up the family vegetable garden to do so."

"那成功了吗？没有！因为大家生产得越多，价格就越低。多生产咖啡豆甚至无法弥补多种树的成本。如今他们既没有水，也没有水果和蔬菜。"

"经济并不公平！"可可树评论道。

"谁会说当前的全球贸易和市场经济是公平的？持续以较低成本生产更多相同的产品，这只会使农民和工人的收入越来越低。"

"And did it work? No! As everyone produced more, the price was lower. So producing more coffee beans did not even cover the cost of planting more trees. And now they don't have any water, or fruit and veggies."

"Economics just is not fair!" Cacao Tree remarks.

"Who would call the current global trade and the market economy fair? Always producing more of the same, and at lower cost, only means a lower income for the farmer and the workers."

……多生产咖啡豆……

... producing more coffee beans ...

……以为生产得越多……

... thought that by producing more ...

"农民以为生产得越多,他就会赚得越多。"

"只有当价格保持不变,甚至上升——因为更多的人在喝更多的咖啡——收入才会增加。"

"这真是一团糟!如果他不种咖啡而是选择养牛,那么他将失去所有——最终银行将控制一切。"

"The farmer thought that by producing more, he would earn more."

"Income will only rise when the price stays the same, or even goes up, since more people are drinking a lot more coffee."

"What a mess this is! If he choses keeping cows instead of growing coffee, then he will lose everything – and the banks will end up controlling it all."

"我们都将失去所有:你、人们和大自然。我们必须找到出路!"

"是的。但是告诉我,为什么你有这么明亮鲜艳的红色果实?"

"哦,我是一枚抗氧化炸弹,"咖啡树回答。

"We will all lose: you, the people, and Nature. We have to find a way out!"

"True. But tell me, why do you have such bright, colourful, and red cherries?"

"Oh, I am an anti-oxidant bomb," Coffee Bush replies.

我是一枚抗氧化炸弹……

I am an anti-oxidant bomb ...

磨成粉,做成巧克力……

Ground to powder to make chocolate ...

"我不明白。你是在和氧气对抗吗?"

"你没听说过抗氧化物吗?它们对人的健康有好处。"

"我只是一棵树。我的果实被晒干,磨成粉,做成巧克力。我还产生一种气味和味道独特的油脂。"

"I don't understand. Are you fighting against oxygen?"
"You've never heard of anti-oxidants? They are good for people's health."
"I am just a tree. My fruit is dried and ground to a powder to make chocolate. I also produce a butter, one that smells or tastes like nothing."

"这太有趣了!你就生长在我旁边,可我却从未发现你身上有油脂。"

"别把我和奶牛产的黄油弄混了。我生产一种植物油。"

"我喜欢植物油!那一定非常健康。让我们看看,如果把我的果皮跟你的油脂混合在一起,会发生什么……"

"How interesting! You grow near me, and yet I've never had a clue that you have butter."

"Don't confuse me with butter from a cow. I produce a plant butter."

"I love that! Must be super healthy. Let's see what happens if we mix the husks of my beans with your butter…"

……我的果皮跟你的油脂混合在一起。

... the husks of my beans with your butter.

我的果皮没什么问题……

Nothing wrong with my husks ...

"你想把你满是化学品的咖啡果皮,和没人想要的可可油混合在一起?"可可树问道。

"我的果皮没什么问题。我是有机的,看不到任何合成化学物质。人们会发现,我们俩混合起来不仅很好吃,还会帮他们集中注意力——而且还会给农民一大笔钱。"

"哦,我以前听说过这个承诺……"可可树回答。

"You want to mix your coffee husk, full of chemicals, with my butter, which no one wants?" Cacao Tree asks.

"Nothing wrong with my husks. I am organic, with not a chemical in sight. People will not only find a mixture of the two of us tasty, it will also help them focus – and it will offer the farmer a bundle of cash."

"Oh, I have heard that promise before…" Cacao Tree replies.

"让我们做一种可以吃的咖啡吧，帮助人们变得超级健康、超级专注！"

"你先是迷惑我，现在又来骗我！咖啡是用来喝的！"

"那可可只能吃吗？"

……这仅仅是开始！……

"Let's make a coffee that people can eat, to help them become super healthy and super focused!"

"First you confuse me, and now you try and trick me! Coffee is for drinking!"

"And is cacao only being eaten?"

... AND IT HAS ONLY JUST BEGUN!...

……这仅仅是开始!……

... AND IT HAS ONLY JUST BEGUN! ...

Did You Know?
你知道吗?

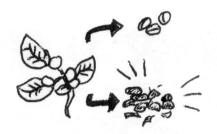

The cascara, which is the husk of the coffee bean, is the richest source of antioxidants of all plants, richer than goji berries. Due to the focus on caffeine, cascara has been discarded, or at best composted.

咖啡果皮即咖啡豆的外壳，是所有植物中抗氧化物含量最丰富的，比枸杞还要丰富。由于只关注咖啡因，咖啡果皮多被丢弃，最多被用来堆肥。

Cacao was first known in Europe as a drink. However, coffee and tea became more popular. Cacao was mixed with sugar and milk and modern day chocolate was born.

在欧洲，可可最初是作为一种饮料。然而，咖啡和茶更受欢迎。人们将可可与糖和牛奶混合，现代巧克力就这样诞生了。

King Leopold II of Belgium grew cacao. Due to over harvest he had chocolate made, served at the Palace alongside a cup of coffee, making solid cacao a popular sweet.

比利时国王利奥波德二世也种植可可。由于可可丰收，他让人做成巧克力，与咖啡一道在宫殿里享用，这使得固体可可成为一种受欢迎的甜点。

Cascara has long been considered unfit for human consumption. The rampant use of toxic chemicals turned this source of anti-oxidants into a waste product that would, most of the time, not even be composted.

人们一直认为咖啡果皮不适合食用。由于滥用有毒化学品，富含抗氧化物的咖啡果皮成为废弃物，在大多数情况下，甚至不会用来堆肥。

Cascara and coffee, roasted to eat, not to drink, and emulsified in cacao butter, provide a slow release source of energy, one that stimulates concentration and memory with the caffeine it contains.

咖啡果皮和咖啡经烘焙后食用而不是饮用,并且与可可脂一起乳化,能持续提供能量,其中含有的咖啡因可以提高注意力和记忆力。

The best coffee and cacao grows under a forest canopy. While the yield per hectare is lower than that of industrialised and mechanised farms, the entire 100% of the biomass offers the farmer more income.

最好的咖啡和可可生长在森林的树冠下。虽然每公顷的产量低于工业化和机械化农场,但100%的生物质能为农民提供更多的收入。

The entire harvest of coffee cherries is washed, then has the husks and skins removed, before only the best beans are sold as green beans. The amount of dried beans for sale is only 20% of the harvest.

只有最好的咖啡豆才能作为咖啡生豆出售，在此之前，所有收获的咖啡果实都要洗净，然后去壳去皮。被制干出售的咖啡豆只占收成的20%。

If the farmer processes whole cherries, and plants cacao under the tree canopy, and mixes coffee husks and cacao butter, then the harvest value increases five-fold. Solid coffee earns five times more than just selling the green beans.

如果农民处理全部的咖啡果实，在树冠下面种植可可，并将咖啡果皮与可可脂混合，所收获的价值会增加五倍。售卖固体咖啡的收入是单纯只出售咖啡生豆的五倍。

Think about It

Is coffee solid or liquid?　　　　　　　咖啡是固体还是液体？

When producing more of the same coffee beans – will income automatically rise?　　当生产越来越多同样的咖啡豆时，收入会自动上升吗？

Is economics fair?　　　　　　　　　　经济是公平的吗？

How easy is it to change habits?　　　　改变习惯有多容易？

Do you know about anti-oxidants? Do some research on why we need these. Now find out where we can find them: in our food, our gardens, in city parks, and in forests. You may find anti-oxidant rich foods growing right next to the road, or discarded as "waste" from fruit and vegetables. Is your daily anti-oxidant intake enough? Make a list of all your findings and share the information with friends and family members.

你了解抗氧化物吗？研究一下为什么我们需要抗氧化物。看看我们在哪里可以找到它们：我们的食物，花园，城市公园，还有森林。你或许会发现富含抗氧化物的食物就生长在路边，或者被当作水果和蔬菜的"废物"丢掉。你每天的抗氧化物摄入量够吗？列出你的发现，并与朋友和家人分享。

TEACHER AND PARENT GUIDE

学科知识
Academic Knowledge

生物学	富含抗氧化物的植物；大自然会产生黄烷醇，可以降血压、提高记忆力。
化 学	抗氧化剂的作用；植物脂肪和动物脂肪的区别；饱和脂肪和不饱和脂肪的区别；咖啡因作为杀虫剂；可可富含类黄酮；可可脂由棕榈酸、硬脂酸和油酸组成。
物 理	抗氧化物不耐热；可可脂的熔点低于体温，所以巧克力在我们手中融化；竹子能保持土壤中的水分。
工程学	保存热敏性营养素的加工工艺；抗氧化剂可以延长产品的保质期2至5年。
经济学	需求与供给的作用；"期望值"在未来价格设定中的作用；土地的生产力：用草喂牛，或种植单一作物，或利用现有的收成；单一种植占用了所有土地，即使土壤肥沃也会导致营养不良。
伦理学	鼓动农民生产更多同样的东西，这使他们债务累积，并可能因没有担保而失去所有。
历 史	咖啡是在也门被牧羊人发现的，他们观察到山羊摄入咖啡豆会表现得精力旺盛。
地 理	咖啡原产于埃塞俄比亚、中非和马达加斯加；可可源自拉丁美洲；美国要求产品含有100%的可可脂才称为巧克力，欧盟要求95%；象牙海岸和加纳是最大的可可生产地；位于南北回归线之间的咖啡和可可种植带。
数 学	价格是供求关系的结果。
生活方式	饮用或食用咖啡或可可，不同的生活方式为农民提供了不同的收入模式。
社会学	小农户咖啡种植者的减少导致棚户区贫困人口数量的增加。
心理学	在生活中怀有预期的作用；当人们怀有固执的想法时，不要和他们争论，而是问一些引导性的问题。
系统论	当银行取消抵押品赎回权时，整个系统就会瘫痪：体现在对农民及其家庭，以及对生态系统的影响上，停止生产会减少供给，有助于调整市场价格。

教师与家长指南

情感智慧
Emotional Intelligence

咖啡树

咖啡树对失去生计的农民表示担忧。他善于观察，注意到农民的孩子经常挨饿。他认为农民的解决方案根本不能解决问题。他花时间去掌握全面的信息，遵循种植更多同样作物的逻辑，却发现这只能以灾难告终。他认为当前的经济模式对农民是不公平的。他将失去收入的农民及其损失联系起来。他花时间向可可树解释什么是抗氧化物，并表示出同情心，因为他们面临同样的挑战。为了解决问题，他寻求协作共赢的办法。面对怀疑，咖啡树用提问却不给出答案的方式，让可可树自己思考。

可可树

可可树无法理解农民的思维逻辑。她发现农民的一切举动都与种植更多咖啡目标一致，最终却仍没有达到预期。她得出结论，发展并不是线性的：生产更多并不意味着收入更多。她意识到，当银行控制了农场，可可的未来也将不复存在。可可树开始通过对话寻找解决方案。当她不知道什么是抗氧化物时就勇敢去问，并不怕显得无知。她告诉咖啡树，她的植物油与黄油大不相同，几乎没人想要。当咖啡树提出一个可信的方案时，她并不相信它有用，并坚称咖啡只是一种饮料。

艺术
The Arts

当我们想到咖啡，会想到棕色，那是烘焙咖啡豆的颜色。然而，咖啡花是纯白色的，咖啡豆一开始是绿色的，然后变成黄色甚至鲜红色，最后变成深紫红色。用你喜欢的艺术材料来做一个壁挂，使用一系列大自然提供的色彩，从咖啡花的白色，一直到咖啡果实的深红色。

TEACHER AND PARENT GUIDE

思维拓展
Systems: Making the Connections

　　虽然农民一直期望利用灌溉、化学和基因技术提高产量，但土壤为植物提供营养的能力有限。由于当前的经济模式，农民已经转向单一种植，所有可用空间都只用来在各地种植大量相同的作物。虽然收成更多了，但更多农民做同样的事情，导致全球在售咖啡豆的数量增加。如果对咖啡的需求没有同样增长，就会出现供过于求。市场调节这一问题的唯一方法就是降价。这对农民来说是个坏消息，因为他只能用增加产量所获得的额外收入来支付种植的额外投资。当价格低到农民甚至无法收回最初的成本时，他面临着另一个挑战，那就是养活家人，并有足够的水源。唯一的办法便是改变商业经营模式。我们该关注的不是提高同一种作物的产量，而是利用现有的资源创造更多的价值。可可脂是可以替代动物脂肪的一种植物脂肪，而咖啡果皮富含抗氧化物。对于农民和咖啡豆收购商来说毫无价值的两种成分混合后，再经过创意设计，如今提供了一种新的盈利模式。我们需要抛弃传统逻辑。这意味着你现在要吃咖啡，而不是喝咖啡。我们只需提醒一下，可可最初是一种饮料，后来才变成固体食品。这是一种商业模式创新，它突破了供求逻辑以及随之而来的价格设定。

动手能力
Capacity to Implement

　　解决方案往往比我们想的要简单，只要我们能跳出思维定势。如果我们只把咖啡当作咖啡豆经烘焙研磨后冲泡出的饮料，就无法想出除了给予农民低价补贴之外的解决方案。然而，当我们看到可可和咖啡的混合不是作为饮料，而是作为固体食品，就能获得高收益。两种产品的结合为农场创造了更多工作机会，也为农户提供了一个可持续的未来。关键是"这个新产品会畅销吗？"问问大家是否愿意购买这种新型的能量产品——一种可以提高注意力和记忆力的美味食品。它应该不难卖！

教师与家长指南

故事灵感来自
This Fable Is Inspired by

莱维斯·维特昆斯
Raivis Vaitekuns

2001—2003 年，莱维斯·维特昆斯在拉脱维亚的里加理工大学学习平面设计和包装。在景观精品酒店艾娜瓦斯酒店工作多年，并学习了更多关于服务行业的知识后，他加入了惠普。2010 年，莱维斯在里加市中心创建了 Miit 咖啡店。他接受了咖啡师培训，并成为拉脱维亚特色咖啡行业的领军人物之一。之后，他在杰莫斯咖啡工作室担任咖啡师培训师，后来发明了一种固体咖啡，也就是 Pixels 咖啡。他和亲戚一起发明了可食用的咖啡条，这种咖啡制品不仅可以优化咖啡因的摄入，而且可以随时随地享用。这种方法帮助咖啡和可可种植者增加了 5 倍收入。

图书在版编目(CIP)数据

冈特生态童书.第八辑:全36册:汉英对照 /
(比)冈特·鲍利著;(哥伦)凯瑟琳娜·巴赫绘;
何家振等译.—上海:上海远东出版社,2021
ISBN 978-7-5476-1773-1

Ⅰ.①冈… Ⅱ.①冈… ②凯… ③何… Ⅲ.①生态环
境－环境保护－儿童读物—汉、英 Ⅳ.①X171.1-49

中国版本图书馆CIP数据核字(2021)第249940号

策　　划　张　蓉
责任编辑　程云琦
封面设计　魏　来　李　廉

冈特生态童书
可以吃的咖啡
[比]冈特·鲍利　著
[哥伦]凯瑟琳娜·巴赫　绘
颜莹莹　译

记得要和身边的小朋友分享环保知识哦！
八喜冰淇淋祝你成为环保小使者！

Food 260

六足农夫

Six-legged Farmers

Gunter Pauli

[比]冈特·鲍利 著
[哥伦]凯瑟琳娜·巴赫 绘
颜莹莹 译

上海远东出版社

丛书编委会

主　任：贾　峰
副主任：何家振　闫世东　林　玉
委　员：李原原　祝真旭　牛玲娟　梁雅丽　任泽林
　　　　王　岢　陈　卫　郑循如　吴建民　彭　勇
　　　　王梦雨　戴　虹　翟致信　靳增江　孟　蝶

特别感谢以下热心人士对童书工作的支持：

匡志强　宋小华　解　东　厉　云　李　婧　陈　果
刘　丹　熊彩虹　罗淑怡　旷　婉　杨　荣　刘学振
何圣霖　廖清州　谭燕宁　韦小宏　李　杰　欧　亮
陈强林　王　征　张林霞　寿颖慧　罗　佳　傅　俊
胡海朋　白永喆　冯家宝

目录

六足农夫	4
你知道吗？	22
想一想	26
自己动手！	27
学科知识	28
情感智慧	29
艺术	29
思维拓展	30
动手能力	30
故事灵感来自	31

Contents

Six-legged Farmers	4
Did You Know?	22
Think about It	26
Do It Yourself!	27
Academic Knowledge	28
Emotional Intelligence	29
The Arts	29
Systems: Making the Connections	30
Capacity to Implement	30
This Fable Is Inspired by	31

一只切叶蚁正在访问斐济。他是蘑菇种植专家，正满怀期待与他的表亲聊聊。他的表亲被称为有史以来首类种植庄稼的蚂蚁。

"从巴西远道而来，这对我来说真是一次漫长的旅行，"他说。

A leafcutter ant, an expert in mushroom farming, is visiting Fiji. He is looking forward to a chat with his cousin, who is known as the first type of ant to ever farm plants.

"This was such a long trip for me, all the way from Brazil," he says.

一只切叶蚁正在访问斐济。

A leafcutter ant is visiting Fiji.

……把一片叶子切成小碎屑……

... just to cut one leaf into little pieces ...

"对你们这些种蘑菇的来说,跑来跑去并不是什么新鲜事。你要走3千米的路才能把一片叶子切成小碎屑。"

"没错,我们花好几个小时在树和我们的蘑菇园之间走来走去。"

"你的耐力令我印象深刻,"种植庄稼的蚂蚁说。

"Running around is nothing new for you mushroom farmers. You walk up to three kilometres just to cut one leaf into little pieces of confetti."

"That is right, we spend hours walking back and forth between the tree and our mushroom farm."

"I am impressed with your stamina," Plant-farming Ant replies.

"但回报很丰厚,你要知道!无论发生什么,我们都不缺食物。"

"但是总吃同样的食物,你不觉得无聊吗?"

"嗯,我们的巢穴里住着500万到1000万只蚂蚁,几百万年来都吃同样的食物。它们如此美味,供应又这么稳定,我们为什么要改变呢?"

"But the rewards are great, I tell you! Whatever happens, we will always have food."

"But don't you get bored, always eating the same food?"

"Well, our nests house five to ten million ants, and we have all eaten the same kind of food, for millions of years. Why would we want to change, when it is so tasty, and our supply is so secure?"

……我们都不缺食物……

... we will always have food ...

……有一些细菌朋友……

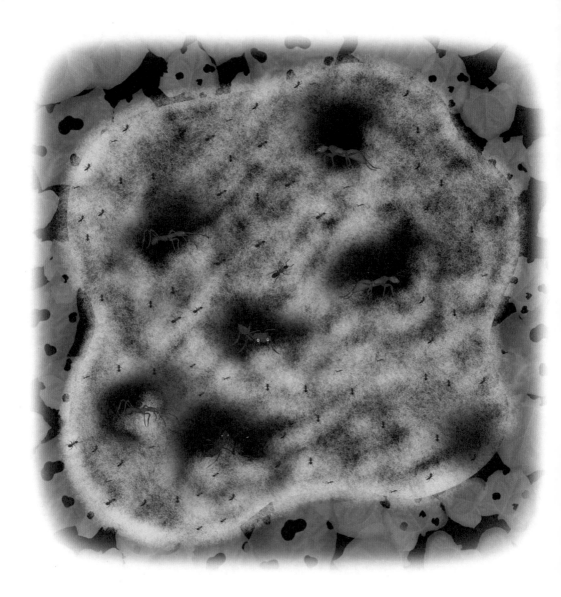

... We do have some friendly bacteria ...

"你用一些化学物质来防止其他真菌吃你新切好的叶子,这是真的吗?"

"嗯,是的。是有一些细菌朋友生活在我们身体的小口袋里,一旦发现有陌生的真菌想要抢我们的粮食,我们就放出这些细菌。他们保卫粮食时非常勇猛。"

"你们都是种植蘑菇的能手,就像人们养猪养鸡一样。我听说你们在世界各地都建立了家园。恭喜!"

"Is it true that you use some chemicals to control other fungi wanting to eat some of your freshly cut leaves?"

"Well, yes. We do have some friendly bacteria living in tiny pockets in our bodies, and as soon as we spot some strange fungi trying to rob us of our crops, we release the bacteria. They are ferocious in protecting our crops."

"You all farm mushrooms so well, like people rear pigs and chickens. And I hear you have established colonies all over the world. Congratulations!"

"我们到哪里都带着蘑菇孢子。我们的蚁后从一丁点真菌开始培育,很快就会长成一个大型的真菌园,养活数百万只蚂蚁。关于我就说这么多吧,我大老远赶来就是为了了解你们家种植植物的情况。"

"哦,是的,我们学会了在高高的树上种植植物。我们同住一个房子,给彼此需要的东西,这样我们就能一起成长。这棒极了,"种植庄稼的蚂蚁回答道。

"We take our mushroom spores with us wherever we go. Our queen starts with a tiny bit, and soon it grows into a huge farm, feeding millions. Enough about me, I travelled from afar to learn about your family growing plants."

"Oh yes, we've learnt to grow plants high up in a tree. We share a house, and give each other what we need, and so we thrive together. It's a great," Plant-farming Ant replies.

……我们学会了在高高的树上种植植物……

... we've learnt to grow plants high up ...

……这些植物长在树上的？

...those plants to grow in a tree?

"这么说，你并不是在地下筑巢或盖土堆来控制土壤的湿度和温度，而是把所有东西都带到空中？"

"是的。我们的植物喜欢晒太阳。他们身姿曼妙，并对我们所做的一切报以甜美的花朵。我们喜欢他们的糖果！"

"你是怎么让这些植物长在树上的？"

"So, you are not making underground nests or building mounds in the soil to control humidity and temperature, you take everything up into the sky?"

"Right. Our plants love to be in the sun. They grow beautifully, and give us their sweet flowers for all the work we do. We love their sugar!"

"How do you get those plants to grow in a tree?"

"很简单，我们在森林的地面上找到他们的种子，并把他们带到树上，小心翼翼地放在树皮下。在那儿，他们在我们的保护下成长。"

"那你住在哪儿？你不可能整天整夜在树上走来走去吧。"

"在我们的关爱照料下，我们的植物长得很快。他们慷慨地为我们开辟了专门的房间，而且这些房间非常舒适。"

"Simple, we find their seeds on the forest floor, and carry them up the tree, where we place them carefully under the bark. There they grow, under our protection."

"And where do you live? You can't possibly be walking around, up and down trees all day and night."

"Our plants grow quickly, with our love and care. And they generously grow special separate rooms for us. Very cosy ones, too."

……小心翼翼地放在树皮下。

...place them carefully under the bark.

我们有养育孩子的房间。

We have rooms where we raise our young.

"真是难以置信!"种蘑菇的蚂蚁惊呼。"你们在植物里面有居住单元,就好像你们住在同一栋房子里一样?"

"正是这样。我们有养育孩子的房间。但还不止这些。我们甚至有放置粪便的房间,还有存放垃圾的地方。"

"Unbelievable!" Mushroom Farming Ant exclaims. "You get living quarters – inside the plant – as if you are living in the same house?"

"Indeed. We have rooms where we raise our young. But that is not all. We even have rooms for our droppings, and a place to store our garbage."

"那么复杂？可别告诉我你们还有冲水马桶！"

"嗯，我们所有的废弃物都是我们植物的肥料和饲料。这是一个真正的社区，或者说单身公寓，但人们却只能梦想拥有它。"

……这仅仅是开始！……

"That sophisticated? Don't tell me you have flushing toilets as well!"

"Well, all our waste fertilises and feeds our plants. It is a real community – or the odd couple, but it is one people can only dream of."

... AND IT HAS ONLY JUST BEGUN!...

……这仅仅是开始！……

… AND IT HAS ONLY JUST BEGUN! …

Did You Know?

你知道吗？

People started farming 12,000 years ago. Leafcutter ants started sowing, tending to and farming mushrooms an estimated 15 millions years ago already.

人类于 12 000 年前开始耕种。切叶蚁在大约 1 500 万年前就开始培育和种植蘑菇了。

The leafcutter attine ants' fungus farms compare well with farming by people. Both kinds of farmers do things that may look unsustainable, such as growing single crops on a vast scale and applying insecticides.

切叶蚁族的真菌农场堪比人类的农业种植。二者都在做一些看起来不可持续的事情，比如大规模种植单一作物和使用杀虫剂。

Today, each nest of attine ants grows a single genetic strain of fungus, on an industrial scale. Their farm is a vast monoculture farm that can grow big enough to feed 7 to 10 million residents.

今天，每个切叶蚁群都在以工业规模种植单一菌种的真菌。它们的农场是一个巨大的单一种植农场，足以养活700万到1000万成员。

The ants don't eat the greens, but add their droppings to it, as a pre-treatment that encourages the fungi. The ants then deposit their little leaf bits on the fungus heap, and tend to the mycelium garden.

蚂蚁并不吃叶子，而是在其中加入它们的排泄物，作为真菌的培养基。然后蚂蚁把它们的小叶子放在真菌堆上，照料真菌园。

Ants have cultivated monocultures for millions of years. Some use pesticides to fight pests, killing invading foreign fungus with a toxin from Pseudonocardia bacteria, which thrive in a specialised ant pocket.

蚂蚁耕种单一作物已经有几百万年了。有些蚂蚁会使用杀虫剂，用一种毒素杀死入侵的外来菌类，这种毒素来自假诺卡氏菌，它可以在一种特殊的蚁腺中生存。

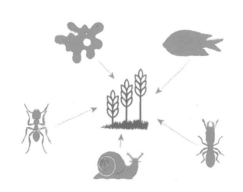

The ability to farm food is not limited to ants and termites. Amoebas, marsh snails, and damselfish encourage food to grow where and when they want it.

能耕种食物的不止蚂蚁和白蚁。变形虫、沼螺和小热带鱼都能随时随地种植食物。

Ambrosia beetles are classified as true farmers by scientists. The invading redbay ambrosia beetle (Xyleborus glabratus) raises a fungus that can destroy the inside of an avocado tree.

食菌小蠹被科学家们列为真正的种植者。入侵的石果材小蠹（Xyleborus glabratus）会产生一种能破坏鳄梨树内部的真菌。

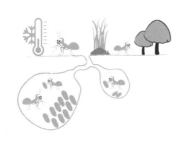

Using humidity and temperature control in their mounds, Macrotermitinae termites nurture fungi in cosy caves, feeding them dead plant material. These tiny farmers then feast on the fungi.

白蚁在它们舒适的巢穴中通过控制湿度和温度培育真菌，用死去的植物为它们施肥。这些小型农民就以真菌为食。

Think about It 想一想

Did you know that ants and termites have been farming for 15 million years?

你知道蚂蚁和白蚁已经耕种了1500万年吗?

What do you think about plants that create space for ants to live, and leave their waste as their fertiliser and food?

你是如何看待植物为蚂蚁创造生活空间,并把它们的排泄物当作肥料和食物的?

Ever heard of an ant city of 10 million?

你听说过拥有千万成员的蚂蚁之城吗?

Plants that grow without soil only on ant droppings?

你听说过不需要土壤,只在蚂蚁粪便上生长的植物吗?

Do It Yourself! 自己动手!

Take a look at the history books, looking for the first early people who went from a hunter-gatherer lifestyle to one of growing crops and leading a settled life in a village, building a community. The power of their new lifestyle was their self-sufficiency in food, all year around. Ask yourself for how long you would be able to supply your own food, in case of a crisis? Is it possible to redesign our ways of food production and consumption, so that we can feed ourselves in difficult times, all the time? Think out of the box, since we cannot all have access to land to plant seeds, harvest and process crops. Perhaps we need to start thinking like the ants do, and farm in the air?

看看历史书，查阅一下最早期人类的资料，他们如何由狩猎采集的生活方式转变为种植庄稼，建立社群，在村庄里定居。他们全年在食物上的自给自足促成了这种新的生活方式。如果遇到危机，你能自给自足多久？有没有可能重新规划我们的食物生产和消费方式，让我们在困难时期也始终能养活自己？我们不可能都到土地上去播种、收获和加工作物，所以要跳出思维定势。或许我们需要像蚂蚁一样，开始在空中耕作？

TEACHER AND PARENT GUIDE

学科知识
Academic Knowledge

生物学	切叶蚁有210种，其中有47种切割叶子、种植真菌；对热带雨林中切叶蚁采食的100多种叶子、果实和花朵的植物基质进行营养分析；牧蚁喜欢蚜虫；切叶蚁后会在离开它的家园时带走一小块菌种，藏在口中，建立新的家园；蚂蚁-真菌的共生关系；是蚂蚁种植真菌，还是真菌驯养蚂蚁；由蚂蚁种植的环蚁木属植物。
化 学	由蚂蚁采集的植物基质含有蛋白质、碳水化合物、钠、锌和镁；只有养殖真菌才产生氨基酸精氨酸；蚂蚁用有毒叶子来种植食用菌。
物 理	蚁巢控制湿度和温度，以确保持续耕种。
工程学	蚂蚁是令人瞩目的建筑师，能挖掘出复杂、气候可控的地下洞穴。
经济学	蚂蚁利用循环经济，效率极高。
伦理学	认为人类是最早的农民是很自大的，因为蚂蚁种植真菌和植物并驯养动物比人类早了几百万年。
历 史	蚂蚁的自给农业始于6600万年前恐龙灭绝后不久；牧蚁早在5000万年前就掌握了畜牧技术。
地 理	从草原到热带雨林，切叶蚁在不同的栖息地种植单一品种。
数 学	通过数学抽样法得出概率分布，其灵感来自蚂蚁，它们总是避开同伴走过的路径。
生活方式	从一个将食物运往世界各地的全球化经济，转变为一个能够利用社区可用资源满足所有基本需求的经济。
社会学	在另一个成功社区的启发下创建一个社区。
心理学	互利共生是指大家互相依赖，精诚合作，而不是像寄生虫那样，这样才能更好地工作，取得更多成就。
系统论	真菌蚂蚁喜欢真菌作物，为它们抵御害虫和病原体，并在地下巢穴中提供稳定的生长条件和有营养的"肥料"；人类和蚂蚁都经历了由狩猎采集到发现耕种好处的过程；蚂蚁取食真菌结节，让菌丝长成更多小球——类似于人们挖出一些土豆，让剩下的长出更多土豆。

教师与家长指南

情感智慧
Emotional Intelligence

种蘑菇的蚂蚁

种蘑菇的蚂蚁对他表亲种植植物的能力感到好奇。他对自己所拥有的无尽供应的食物感到满足。当被问及为什么他的饮食习惯几百万年来都是一样的,他肯定地说,只要食物供应源源不断,就没有必要改变。最重要的是,味道很好。他大方地分享防御外来入侵者的技巧,没有任何保留。他问他的表亲,她们家族是如何成功成为植物种植者的。一个又一个的问题揭示了一种独特的耕作方式。蚂蚁猜想,这种耕作方式人类只能梦想拥有了。

种植物的蚂蚁

种植物的蚂蚁对她表亲的毅力和工作风格表示钦佩。她知道她的客人想更多地了解她的耕作方式,便毫不保留地解释了她是如何从森林的地面上捡拾种子,并把它们带到树上安全的地方。尽管这极不寻常,特别是她们还能在植物提供的房间里养育幼蚁,储存垃圾,她却倾囊相授。这并不是吹嘘或夸大其词,只是对植物和蚂蚁如何从经济和社区中获益做出一个中肯的解释。

艺术
The Arts

让我们用美术用具来画一个蚂蚁农场:蚂蚁在地下照料蘑菇,在树上畜养蚜虫、种植开花植物。这是一个全新的视角,来展示蚂蚁如何将植物、蘑菇和动物作为其食物来源。你的绘画可能看起来有点夸张,但会让其他人知晓自然界循环食物生产的最佳形式,这已经存在了几千万年。现在做个变化,展示植物、蘑菇和蚜虫是如何反过来驯养蚂蚁,为蚂蚁提供所需的食物和住所。通过二者的可视化呈现,人们能以新颖有趣的视角来看待如何创造一个令人满意且有韧性的生活方式。

TEACHER AND PARENT GUIDE

思维拓展
Systems: Making the Connections

在我们之前，蚂蚁早已利用真菌、植物和动物进行耕种。这与我们用动植物耕种惊人的相似。有的蚂蚁扩张和管理真菌园，或者把开花植物的种子种到树上获取花蜜。还有像农场主养牛一样牧养蚜虫的牧蚁。作为回报，牧蚁为蚜虫遮风挡雨，有时还把它们从一棵植物带到另一棵植物。蚂蚁照料并保护蚜虫的卵，如同对待它们自己的卵那样，并在冬天将它们安置在蚁穴中。年轻的蚁后离开蚁穴时会将一只蚜虫放到口中，带到它的新家。就像人们以剥夺动物的"自由"为交换来为它们提供照料和保护一样，蚂蚁也是如此。蚂蚁不需要建造复杂的农业机械来减轻体力劳动，因为它们可以搬运5 000倍于自身重量的东西。人类要学习的重要一课是，蚂蚁可以养活几百万家族成员，它们在地球上的数量与人类相当，然而它们只从事有机耕作，只利用本地资源，并与自然王国的植物、真菌、细菌和动物（昆虫）协同合作。这种获取食物的机制往往令人吃惊。人们从未意识到这些昆虫早在几百万年前就拥有维持生命的智慧，这足以启示所有人。

动手能力
Capacity to Implement

让我们将现代农业中使用的耕作技术和蚂蚁使用的三种耕作技术作个比较。令人吃惊的是，蚂蚁和白蚁的总量与人类相当，然而蚁类却不像人类那样为了矿物质和营养去开采，而是给土壤施肥。用图表展示人们是怎么做的，然后再用另一张图表展示蚂蚁是怎么做的。你可以借用本册"艺术"部分的图画。接下来是最重要的部分，做一些关于数量级的计算。这将为我们如何采取有机和可持续的方式利用既有资源，如何运用更好的耕作手段提供创造性的建议。你还需要对不同的饮食做个调查，因为蚂蚁的食物种类并不多。在蚂蚁、植物、蘑菇和蚜虫的世界里几乎没有饥饿，请就此开展讨论。

教师与家长指南

故事灵感来自
This Fable Is Inspired by

尼可·吉拉尔多
Nicole Gerardo

尼可·吉拉尔多于1997年毕业于美国得克萨斯州休斯顿的莱斯大学，获得生态学和进化生物学学士学位。2004年，她在得克萨斯大学奥斯汀分校的帕特森实验室获得综合生物学博士学位。2008年，她加入了位于佐治亚州亚特兰大市的埃默里大学，担任生物系韦恩·罗林斯研究中心教授。尼可正在研究蚜虫、有益细菌和一系列蚜虫病原体（包括细菌和真菌）之间的相互作用。她还研究了几种昆虫系统，它们与有益的共生体有不同的联系，以及真菌蚂蚁及其相关的病原真菌。

图书在版编目（CIP）数据

冈特生态童书.第八辑：全36册：汉英对照 /
（比）冈特·鲍利著；（哥伦）凯瑟琳娜·巴赫绘；
何家振等译. —上海：上海远东出版社，2021
ISBN 978-7-5476-1773-1

Ⅰ.①冈… Ⅱ.①冈… ②凯… ③何… Ⅲ.①生态环
境 – 环境保护 – 儿童读物 – 汉、英 Ⅳ.①X171.1-49

中国版本图书馆CIP数据核字（2021）第249940号

策　　划	张　蓉
责任编辑	程云琦
封面设计	魏　来　李　廉

冈特生态童书

六足农夫

［比］冈特·鲍利　著
［哥伦］凯瑟琳娜·巴赫　绘
颜莹莹　译

记得要和身边的小朋友分享环保知识哦！
八喜冰淇淋祝你成为环保小使者！

Food 261

享用彩虹

Eat Your Rainbow

Gunter Pauli

[比] 冈特·鲍利 著
[哥伦] 凯瑟琳娜·巴赫 绘
郑旻 译

上海远东出版社

丛书编委会

主　任：贾　峰

副主任：何家振　闫世东　林　玉

委　员：李原原　祝真旭　牛玲娟　梁雅丽　任泽林
　　　　王　岢　陈　卫　郑循如　吴建民　彭　勇
　　　　王梦雨　戴　虹　翟致信　靳增江　孟　蝶

特别感谢以下热心人士对童书工作的支持：

匡志强　宋小华　解　东　厉　云　李　婧　陈　果
刘　丹　熊彩虹　罗淑怡　旷　婉　杨　荣　刘学振
何圣霖　廖清州　谭燕宁　韦小宏　李　杰　欧　亮
陈强林　王　征　张林霞　寿颖慧　罗　佳　傅　俊
胡海朋　白永喆　冯家宝

目录

享用彩虹	4
你知道吗？	22
想一想	26
自己动手！	27
学科知识	28
情感智慧	29
艺术	29
思维拓展	30
动手能力	30
故事灵感来自	31

Contents

Eat Your Rainbow	4
Did You Know?	22
Think about It	26
Do It Yourself!	27
Academic Knowledge	28
Emotional Intelligence	29
The Arts	29
Systems: Making the Connections	30
Capacity to Implement	30
This Fable Is Inspired by	31

一只熊和一只狐狸正在享受美餐。这儿有许多美味的食物。

　　"从漫长的冬眠醒来,就发现有这么多好吃的,它们能让我快快长胖,真是太快乐了。"熊说道。

A bear and a fox are enjoying a meal. There is a lot to eat.
"What a pleasure to wake up from a long winter sleep and find myself with lots of food. There is so much I can get fat fast," the bear comments.

……一只熊和一只狐狸正在享受美餐……

… a bear and a fox are enjoying a meal …

……可以拼凑成彩虹的颜色……

... in all the colours of the rainbow ...

"这可不是普通的食物,它们可以拼凑成彩虹的颜色哦,"狐狸补充道,"为了快乐和健康,我们每天都得吃下各种颜色的食物才行。"

"你说的对,"熊说道,"盘子里有一道彩虹,可真是漂亮啊。但说实话,我更喜欢红色的东西。"

"There is not just food, there is food in all the colours of the rainbow," the fox adds. "To be happy and healthy we should eat something of every colour every day."

"You are right. How beautiful to have a rainbow on your plate. But to be honest, I prefer whatever is red."

"你是说肉？"

"哦，不，肉不是唯一的红色食物，虽然我必须承认一块鲜嫩多汁的肉能让我心情愉悦。但我确实知道，红色和紫色的浆果也一样会让我保持快乐和健康。"

"啊，你让我想象到各种各样、颜色各异的食物。你知道的，我平时会吃很多棕色的东西。"

"Meat?"

"Oh no, meat is not the only red food. Although I must admit a juicy piece of meat makes me happy. But I do know that red berries, and purple ones too, will help keep me very happy and fit."

"Hm, you make me dream of rich food, of all possible colours. You know, I eat a lot of brown stuff."

你是说肉?

Meat?

棕色的食物富含纤维……

Brown food is rich in fibre ...

"棕色的食物富含纤维,有利于消化。但我相信你心里想的棕色食物并不是全谷物和巧克力。"

"你知道的,我平时会吃很多昆虫。昆虫大多数是棕色的。其实我有时候也会吃身边新鲜的绿色蔬菜,因为我觉得我还是需要它们的。"

"我觉得蔬菜好乏味啊,喜欢它们真是一点品味都没有。"熊说道。

"Brown food is rich in fibre. Great for digestion. But I am sure you are thinking of something else than whole grains and chocolate."

"You know, I eat a lot of insects and most are brown. I will also have greens when around and fresh. Then I feel I need them."

"I find greens boring. Lacks taste," Bear says.

"你最好忘了你刚刚说过的话。绿色蔬菜可以清洁你的身体,排出毒素。"

"那黄色呢?"

"哈哈哈!我总是偷别人的蛋,"狐狸咯咯地笑着说道,"那可是你能得到的最棒的黄色食物了。"

"那你吃香蕉或者玉米吗?它们也是很棒的黄色食物。"

"You better get over this. Greens clean the body, getting rid of toxins."

"What about yellow?"

"I steal eggs all the time," Fox giggles. "That's the best yellow you can get."

"And do you eat bananas or corn, another great yellow?"

那黄色呢?

What about yellow?

我会找一些橙色的食物……

I search for some orange food ...

"我干吗要去吃猴子喜欢吃的东西。香蕉和我一点都不般配。"

"好吧,其实我偶尔也会找一些橙色的食物来吃。在夏天,我的菜单上会多出杏子和芒果。它们实在是太可爱了!"

"橙色食物对心脏有好处,所以如果有可能,就去吃些当季的。那你平时吃白色的东西吗?"

"I must be desperate to eat what the monkeys do. Bananas are not for me."
"So, once in a while, I search for some orange food. In summer, apricots and mangos are on my menu. They are so sweet!"
"Orange is good for the heart, so enjoy it, in season, when you can. Do you eat anything that is white?"

"比如小绵羊?那是当然了!但我每次这么做,农场主就会气得发疯,恨不得宰了我。"

"不,不,不。我说的是洋葱和大蒜,或者豆腐,还有白萝卜啦。"

"唉!你干吗提这些,坏我胃口!我才不吃那种白色的东西。"

"You mean like sheep? Oh yes! But every time I do, the farmer gets mad at me and wants to kill me."

"No, no, no. I mean onions and garlic, or tofu and daikon."

"You are about to spoil my appetite. I don't want anything with that kind of white."

……农场主就会气得发疯……

... farmer gets mad at me ...

不是你喜欢吃什么……

It is not about what you like ...

"啊,真是对不起,我的朋友。但我还是不得不说一下,你认为白色的东西实际上是红色的!"

"什么?你可别糊弄我!我不想失去我的好胃口……"

"其实吃东西的时候,最重要的不是你喜欢吃什么,而是你的身体和大脑需要什么。如果你平时吃的食物里有彩虹的所有颜色,那你就更容易获得健康和快乐!"

"Sorry, my friend, but what you think is white – is actually red!"
"Now don't you confuse me, or make me loose my appetite…"
"It is not about what you like, it is about what your body and your brain need. If you get all the colours of the rainbow in your diet, then it will be easier to be happy and healthy."

"好吧,所有颜色的食物我都会吃的,但能不能红色比其他颜色更多一些?哈哈!当然,那得周围没有猎人才行。"

……这仅仅是开始!……

"Well, I will go for all colours on the condition that there is a little bit more red than any other colour. And of course, provided that there are no hunters around."

... AND IT HAS ONLY JUST BEGUN!...

……这仅仅是开始！……

... AND IT HAS ONLY JUST BEGUN! ...

Did You Know?

你知道吗？

Japanese cuisine, inspired by Buddhist culture, prescribes that food should have five colours: red, green, yellow, white and black. The Lebanese and Peruvian kitchen drops black but adds orange and purple.

受佛教文化的启发，日本料理规定食物应该有五种颜色：红、绿、黄、白、黑。黎巴嫩和秘鲁的菜肴中则去掉了黑色，取而代之的是橙色和紫色。

The blend of colours in food every day ensures that the different vitamins and minerals present in the food and needed by the body are part of your daily diet.

每日食用各种颜色的食物，能确保食物中存在各种不同的维生素和矿物质，满足每日人体所需。

Red colours are rich in lycopene, which is a powerful antioxidant. This includes tomatoes, guava, papaya, red grapefruit, watermelon, and strawberries.

红色食物中富含番茄红素，这是一种强抗氧化剂。这类食物有：西红柿、番石榴、木瓜、红葡萄柚、西瓜和草莓。

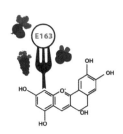

Purple (and blue) includes grapes, beetroot, and blueberries that are rich in anthocyanin, a strong antioxidant. Highly processed white foods form crystals that are hard to digest and should therefore be avoided.

像葡萄、甜菜根和蓝莓这样紫色（或蓝色）的食物中富含强抗氧化物花青苷。另外，精加工的白色食物会形成难以消化的结晶体，应避免食用。

Natural white food like leek, shallots, garlic, onions, daikon and tofu are best eaten raw to lower blood pressure and control cholesterol levels.

天然的白色食物，比如韭葱、青葱、大蒜、洋葱、白萝卜和豆腐，生吃可以降血压、控制胆固醇。

Greens, rich in folic acid and Vitamin B, help build blood and tissue, as well as purge toxins. Yellow is the colour of sunshine, and that gives us energy. Orange food is rich in carotenoids, which are key for the heart.

绿色蔬菜中富含叶酸和维生素B，有助于增强血液流动和组织机能，还可以帮助清除体内毒素。而黄色则是阳光的颜色，能给予我们能量。另外，橙色食物中富含对心脏有益的类胡萝卜素。

In addition to the colour, food requires a mix of preparation techniques: simmer, steam, grill, fry and eaten raw. Cooking releases nutrients. Cooking carrots makes 97% of the nutrients available.

除了颜色，摄取食物还需要各种烹调手段，比如炖、蒸、烤、炸或生吃。烹饪可以帮助食物释放营养物质，例如烹饪胡萝卜可释放其中97%的营养。

The colours are complemented with taste: salt, sour, sweet, bitter (which discourages appetite) and spicy (which entices appetite). There is the need to tickle all the senses: smell, taste, touch, sight and even sound!

颜色与味道是相辅相成的，这些味道有咸、酸、甜、苦（可抑制食欲）和辣（可提升食欲）。另外，在吃东西时，还有必要调动所有感官：嗅觉、味觉、触觉、视觉，甚至听觉！

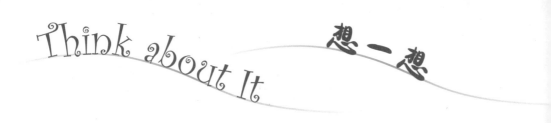

How many colours do you have on your plate every day?

每天你的餐盘里有多少种颜色?

Can you tell the content of the food from the colour?

你能通过颜色分辨出食物的成分吗?

How good is fat for you?

脂肪有什么好处?

Do you eat raw food?

你会吃生的食物吗?

Do not tell your mom what you are doing, but at every meal, take note of the colours on your plate. Do this for a week, at all three mealtimes. When you combine all the food, do you have a rainbow on your plate? Now see which colours are associated with the nutrition you need. You will find that you are now eating consciously, so this may a good time to point out the missing colours to your mom. Tell your family members about this, so they too can enjoy the rainbow on their plates, ensuring they are also getting all the nutrition they need.

嘘，别告诉妈妈。从现在起一周内，每日三餐你都要悄悄留意盘子里的颜色，看看每顿饭菜是否能组合成彩虹。哪些颜色和你需要的营养有关？你会发现你现在是在有意识地用餐，借机向妈妈指出饭菜中缺失了哪些颜色的食物。和家人分享，让他们也可以享受餐盘里的彩虹颜色，确保获得所需要的全部营养。

TEACHER AND PARENT GUIDE

学科知识
Academic Knowledge

生物学	蔬菜的不同营养价值：块根蔬菜、茎类蔬菜、叶类蔬菜、花类蔬菜、果类蔬菜。
化学	水果中的叶酸；维生素A、C、E；微量元素，钾，镁；氨基酸；抗氧化物；β-胡萝卜素和番茄红素；果胶；果实成熟时释放的挥发油所产生的香气。
物理	颜色是成熟度的一个标志，从绿色到黄色、橙红色和鲜红色；绿色未熟到过熟时出现黑点；梨的金黄色；色泽鲜艳，外观光滑；由接近球形的水滴造成的阳光色散和反射所形成的彩虹；人类的眼睛可以看见1 000万种不同的颜色。
工程学	果实中营养物质的保存，收获后营养物质迅速降解；可通过避免接触氧气、冷冻、加盐和糖或真空包装的方式来保存，延长其保质期。
经济学	温室的设计是为了延长当地新鲜蔬菜的种植季节，既可以防止高温，也可以防止霜冻。
伦理学	水果和蔬菜密植，很容易耗尽土壤中的营养；含有季节性营养物质的季节性食物已经常年供应，名不副实，世界各地都如此。
历史	埃及人在公元1500年就开始使用食用色素；公元前300年，葡萄酒曾被人为染红；在德国，伪造藏红花的人被判刑并活活烧死。
地理	红、黄、绿在中国是幸运色；非洲有缺乏微量营养素和缺碘造成的隐性饥饿现象，东南亚有缺铁性贫血，世界大部分地区都有人患维生素A缺乏症。
数学	食物和营养的计量不应以食物千克数和蛋白质、糖类和碳水化合物的含量为标准，而应以能量含量（每100克卡路里数）为标准，并辅以食物中微量营养素的密度。
生活方式	最好是吃当季的水果和蔬菜；食物的颜色给食物带来了情感，因此在生日聚会中，桌上所有的食物都显得格外夺目。
社会学	在不同文化中，颜色联想存在着根本性的区别：在西方国家，红色唤起兴奋、危险、紧急和爱；在中东，红色象征着危险；在中国，红色象征着幸运和幸福；对于日本人来说，橙色象征着爱、勇气和幸福。
心理学	"我们用眼睛吃饭"；看到彩虹的那一刻，你的心情就像天气一样由阴转晴，幸运和幸福也会与你永远相伴；蓝色代表平静，橙色代表冒险，紫色代表高品质，黑色代表优雅，白色会给人纯净和干净的感觉（即使它不是），红色则会带来刺激的感觉，让人感到充满活力，绿色代表平衡，会让人联想到大自然，黄色则代表有趣和有创意。
系统论	选择当季食物和营养成分正确搭配会增强人体免疫系统；将食物营养与农业质量联系起来。

教师与家长指南

情感智慧
Emotional Intelligence

熊

熊有着积极的心态,虽然冬眠瘦了不少,但相信补充食物马上就能胖回来。熊还乐于交流,非常了解狐狸心里的想法和取向,毫无保留地分享自己的喜好,甚至告诉对方自己不喜欢的东西。他性格十分开朗,像朋友和家人一样询问狐狸的个人喜好,有一颗好奇心。当狐狸承认自己常常偷蛋吃时,熊没有妄自评论,而是把注意力从蛋转移到了香蕉和玉米。另外,熊承认了他喜欢甜食,并且讨厌白色的食物。

狐狸

狐狸听懂了熊话语中的热情,并在情感交流中添加了一些智慧。因为周围有非常丰富的食物,熊不会感觉自己要与狐狸争夺食物,从而获得了熊的信任。狐狸毫不避讳表达自己的喜好,也会给熊一些更有智慧的引导,尤其是当熊说出自己不喜欢的东西时,狐狸能够快速建立起简单而清晰的论点。狐狸承认偷吃过蛋,同时明确表示自己不想像猴子一样吃香蕉而讽刺猴子。狐狸会直接问出心中所想,并寻找一些补充内容,这样熊就可以通过彼此的对话得知什么是健康的生活方式,从而增进彼此的友谊。

艺术
The Arts

烹饪是一门艺术。你会做饭吗?让我们从简单的烹饪开始,比如做一道有鸡蛋的菜肴。为了创造出彩虹色的餐盘,我们要往鸡蛋里加点什么?这样除了白色和黄色,餐盘还会有别的颜色。煮熟的鸡蛋已经有白色和黄色了,你可以在盘子里添加什么红色的食物?绿色的呢?还可以加些什么,让这道彩虹变得更完美、更美味呢?

TEACHER AND PARENT GUIDE

思维拓展
Systems: Making the Connections

食物也可以是药物。红色的水果和蔬菜含有番茄红素,这是一种抗氧化物,可能有助于预防前列腺癌、胃癌和肺癌。橙色食物含有隐黄质、α-胡萝卜素和β-胡萝卜素,这些营养素可转化为维生素A。一些橙色水果和蔬菜中的β-胡萝卜素可能有助于降低患肺癌、食道癌和胃癌的风险。褐色的扁豆、黑豆和鹰嘴豆在预防心脏病方面有重要作用。它们都富含叶酸,这是一种可以中和同型半胱氨酸的B族维生素,同型半胱氨酸是一种可以帮助血栓形成的氨基酸。另外,绿色蔬菜如花椰菜和卷心菜,富含异硫氰酸酯和吲哚。吃绿叶蔬菜也可降低患心脏病的风险,是营养的主要来源,同时还富含叶酸、维生素K、类胡萝卜素和ω-3脂肪酸。紫色和蓝色的水果蔬菜其颜色来自花青素,这是一种强抗氧化物。花青素的抗炎特性可能有助于降低患癌症和心脏病的风险,同时也能减轻关节炎带来的痛苦。这些知识都是十分明了的,当我们可以选择食用当地、当季的健康食品时,就更没有理由继续食用不健康的加工食品。营养价值非常依赖于种植食物的土地,如果土壤中的营养物质因过度耕作而耗尽,并且没有机会自我补充,那么耕作获得的食物将没有足够的营养可提供。这一切都将是我们需要面对和解决的巨大挑战。

动手能力
Capacity to Implement

如何提高食物中的营养含量?想要纯天然且营养丰富的食物,我们就必须关注土壤。土壤的健康,即土壤中含有足够的可以被水果和蔬菜吸收的矿物质和微生物,是一个先决条件。土壤中营养的补充依赖于微量元素如锌和碘的循环,它们丰富我们的日常饮食,帮助我们建立一个强大的免疫系统。我们要找出提高生产水果、蔬菜和谷物的土壤肥力的方法。我们的食物需要营养,同样,土壤也需要一定的营养。

教师与家长指南

故事灵感来自
This Fable Is Inspired by

德冈邦夫
Tokuoka Kunio

德冈邦夫生于1960年，是日本吉兆株式会社创始人汤木贞一的外孙。大学生时期，他曾是一个颇有抱负的乐队鼓手，但20岁时，他在接受禅宗训练后，决定成为一名厨师。1980年，他开始接受厨师培训，学习烹饪的基本原理。1995年，他顺利成为主厨，2009年，开始担任京都吉兆株式会社常务董事。从2004年开始，他开始参加各种海外烹饪活动。2008年，他为在北海道洞爷湖举行的八国集团峰会社交晚宴提供餐饮服务。在日本国内，他积极参与农业、林业和渔业等第一产业的区域振兴研究，并成为东京农业大学的客座教授和工业、文化和科学学会的名誉理事。在尊重传统的同时，德冈邦夫不断设计全新、多元的饮食文化方法，并致力于通过创新和提出建议，来传播日本美食的相关知识以及他对日本美食的热爱。

图书在版编目（CIP）数据

冈特生态童书.第八辑：全36册：汉英对照 /
（比）冈特·鲍利著；（哥伦）凯瑟琳娜·巴赫绘；
何家振等译. —上海：上海远东出版社，2021
ISBN 978-7-5476-1773-1

Ⅰ.①冈… Ⅱ.①冈…②凯…③何… Ⅲ.①生态环
境－环境保护－儿童读物—汉、英 Ⅳ.①X171.1-49

中国版本图书馆CIP数据核字（2021）第249940号

策　　划　　张　蓉
责任编辑　　程云琦
封面设计　　魏　来　李　廉

冈特生态童书
享用彩虹
[比]冈特·鲍利　著
[哥伦]凯瑟琳娜·巴赫　绘
郑　旻　译

记得要和身边的小朋友分享环保知识哦！
八喜冰淇淋祝你成为环保小使者！

Food
262

菜单上的金属

Metals on the Menu

Gunter Pauli

［比］冈特·鲍利 著
［哥伦］凯瑟琳娜·巴赫 绘
解天宇 译

丛书编委会

主　任：贾　峰

副主任：何家振　闫世东　林　玉

委　员：李原原　祝真旭　牛玲娟　梁雅丽　任泽林
　　　　王　岢　陈　卫　郑循如　吴建民　彭　勇
　　　　王梦雨　戴　虹　翟致信　靳增江　孟　蝶

特别感谢以下热心人士对童书工作的支持：

匡志强　宋小华　解　东　厉　云　李　婧　陈　果
刘　丹　熊彩虹　罗淑怡　旷　婉　杨　荣　刘学振
何圣霖　廖清州　谭燕宁　韦小宏　李　杰　欧　亮
陈强林　王　征　张林霞　寿颖慧　罗　佳　傅　俊
胡海朋　白永喆　冯家宝

目录

菜单上的金属	4
你知道吗？	22
想一想	26
自己动手！	27
学科知识	28
情感智慧	29
艺术	29
思维拓展	30
动手能力	30
故事灵感来自	31

Contents

Metals on the Menu	4
Did You Know?	22
Think about It	26
Do It Yourself!	27
Academic Knowledge	28
Emotional Intelligence	29
The Arts	29
Systems: Making the Connections	30
Capacity to Implement	30
This Fable Is Inspired by	31

一枚坚果挂在一棵腰果树上，享受着安静的下午。这时，几个来访的扁豆抬起头来，欣赏这棵树的景象。一个扁豆问这枚坚果：

"你倒挂在你的果实上，看起来真有趣。请告诉我，为什么人们不吃你的果实？"

"哦，他们会吃的。我们的果实非常多汁，是真正的美味。你有没有仔细观察过我们的坚果？"

A nut is hanging from a cashew tree, enjoying a quiet afternoon, when some visiting lentils look up and enjoy the sight of this tree. One lentil asks the nut,

"You look so funny hanging upside down from your fruit. But tell me, why don't people eat your fruit?"

"Oh they do. Our fruit is very juicy, and offers a real delicacy. Have you ever taken a closer look at our nuts?"

一枚坚果挂在一棵腰果树上……

A nut is hanging from a cashew tree ...

……世界各地的人都喜欢吃……

... people from around the world love to eat ...

"哦,是的,那些美味的小半月,世界各地的人都喜欢吃。"

"很高兴知道人们喜欢吃我们,谢谢。但很少有人知道,我们身上满是铜。"

"满是铜?你们离矿场或冶炼厂近吗?"

"Oh yes, those tasty little half-moons, that people from around the world love to eat."
"Good to know people love eating us, thanks. But few know that we are full of copper."
"Full of copper? Are you close to a mine or a smelter?"

"不,一点也不!我们没有使用高热和强大的力量,只是每次从土壤中取出一个铜原子,并将其放入我们的坚果中。"

"但是吃你的人知道你身上满是铜吗?他们肯定不想要身体里有任何铜线。"

"但铜对他们来说有益健康。"

"No, not at all! Instead of using high heat and a lot of force, we simply take one copper atom out of the soil at a time, and put it in our nuts."

"But do people who eat you know that you are full of copper? They surely don't need any copper wires in their body."

"But copper is so healthy for them."

……从土壤中取出一个铜原子……

... take one copper atom out of the soil ...

人们就不能制造红血球……

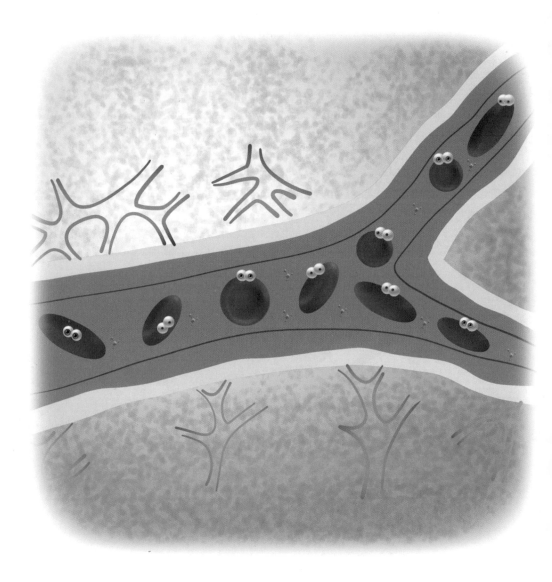

People cannot make red blood cells without …

"铜是有益健康的?"

"哦,是的!如果没有微量的铜,人们就不能制造红血球。"

"人们需要铁来制造血液。这就是红色的来源!"

"Copper is healthy?"
"Oh yes! People cannot make red blood cells without a tiny daily dose of copper."
"People need iron to make blood. That is where the red colour comes from!"

"你说的对,但仅靠铁是不行的,铜和铁一样需要。人体甚至需要微量的金。"

"人们的身体里有黄金?"惊讶的扁豆问道。

"不是开玩笑。每个成年人身上都有0.2克金,用来向身体周围输送电,这大约是一个矿场能从1吨岩石中提取出来的数量!"

"You are right, but iron alone won't do the job, copper is needed as much as iron. The body even needs a minute amount of gold."

"People have gold in their bodies?" the surprised lentil asks.

"No joke. Every adult has 0.2 grams of gold to send electricity around the body, and that is about how much a mine can extract from a ton of rocks!"

……人体甚至需要微量的金……

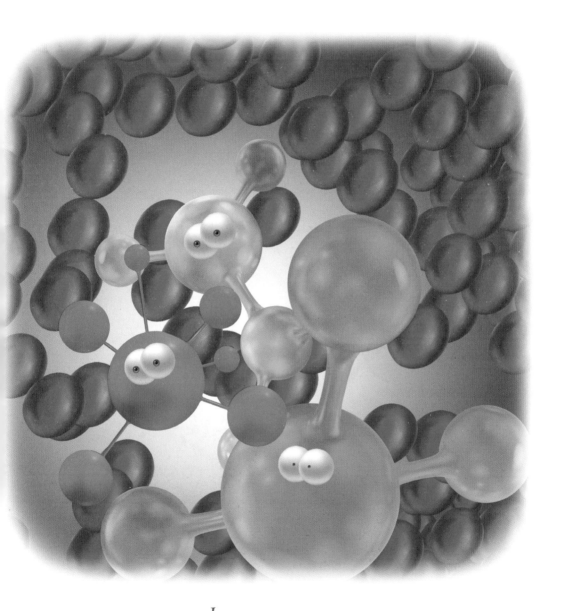

... body even needs a minute amount of gold ...

……扁豆也是伟大的矿工……

... lentils are great miners too ...

"我们扁豆也是伟大的矿工,但我们专注于锌。"

"扁豆汤中的锌?人身上到处都是生物学和化学,但你说得好像人身上都是金属和矿物质。"

"你的铜不是他们唯一需要的金属。如果他们的日常饮食不包括铬、铜、铁、锰、锌和钼,他们根本就活不长。"

"We lentils are great miners too, but we focus on zinc."
"Zinc in a lentil soup? The human body is all biology and chemistry, but you make it sound as if people are full of metals and minerals."
"Your copper is not the only metal they need. They wouldn't live long if their daily diet does not include chromium, copper, iron, manganese, zinc and molybdenum."

"目……什么？"

"钼。你知道花椰菜需要它来生长吗？你知道它还能使钢铁更坚固吗？"

"植物需要它来生长，钢铁也变得更坚固。我不知道金属有这么多好处。但告诉我，你如此重视的这种锌有什么用？"

"Molyb ... what?"

"Mo-lyb-de-num. Did you know that cauliflower needs it to grow? And that it makes steel stronger?"

"Plants grow with this, and steel turns stronger. I had no clue metals were so good. But tell me, what is the use of this zinc that you value so much?"

……花椰菜需要它来生长吗?

... cauliflower needs it to grow?

……它对人们的健康有奇效。

... it does wonders for people's health.

"锌能构建蛋白质。它可以治愈伤口。如果没有锌,免疫系统就不能保持活力,难以应对每天空气中的污染或寻找宿主的病毒所带来的压力。"

"但锌是一种金属,对吗?" 腰果问道。

"它是一种脆性金属,既不结实,也不算珍贵,但它对人们的健康有奇效。"

"Zinc builds protein. It heals wounds. And without zinc, the immune system would not maintain its vigour against daily stresses from pollution in the air, or viruses looking for a host."

"But zinc is a metal, right?" Cashew asks.

"It is a brittle metal, nothing strong or precious – but it does wonders for people's health."

"不管是脆弱的还是结实的,贵重的还是不贵重的,人们确实需要所有的东西来获得快乐和健康。"

"是的,除了蔬菜、蘑菇、海藻和肉类之外,还需要把金属列入菜单。"

……这仅仅是开始!……

"Whether brittle or strong, precious or not, people do need all of it to be happy and healthy."
"Yes, apart from vegetables, mushrooms, algae and meat, there is also the need to put metals on the menu."
... AND IT HAS ONLY JUST BEGUN!...

……这仅仅是开始！……

...AND IT HAS ONLY JUST BEGUN! ..

Did You Know?
你知道吗?

The number of people aged +90 is positively correlated with trace elements in soil and water. Deficiencies in these accelerate aging and cause retardation, abnormal pregnancies, and immunological abnormalities.

90岁以上的人口数与土壤和水中的微量元素呈正相关。缺乏这些元素会加速衰老,并导致发育迟缓、异常妊娠和免疫学异常。

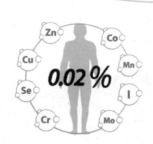

Essential metals, including zinc (Zn), copper (Cu), selenium (Se), chromium (Cr), cobalt (Co), manganese (Mn), and molybdenum (Mo), account for only 0.02% of the total body weight, but play key roles.

必需金属包括锌(Zn)、铜(Cu)、硒(Se)、铬(Cr)、钴(Co)、锰(Mn)和钼(Mo),只占人体总重量的0.02%,但发挥着关键作用。

Unbalanced food intake, too dense farming, excessive processing of crops, dieting practiced widely to reduce body weight, deficiency of zinc normally abundant in animal foods and crops turns into a health epidemic.

膳食摄入不平衡、耕作过于密集、农作物被过度加工、广泛控制饮食以减轻体重，使得缺锌成了一种流行病，而动物性食物和农作物中原本都富含锌。

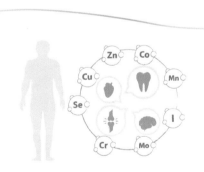

Trace minerals help in the formation of bones and teeth; they are essential for body fluids and tissues; they are part of the enzyme systems and are involved in the functioning of nerves.

微量矿物质有助于骨骼和牙齿的形成；它们是体液和组织所必需的成分；它们是酶系统的一部分并参与神经的运作。

Calcium is more than bone and teeth. It facilitates the transmission of information via the nervous system, the control of muscle contraction, especially the heart, and the avoidance of blood clotting.

钙不仅有益于骨骼和牙齿，它还能促进神经系统传递信息，控制肌肉收缩，特别是心脏，并避免血液凝固。

Zinc is part of all metabolic pathways for protein, lipids, carbohydrate and energy metabolism. It is essential for cell division and for the growth and repair of tissue, functioning of the immune system, and the skin.

锌是蛋白质、脂类、碳水化合物和能量代谢等所有代谢途径的一部分。它对细胞分裂、组织的生长和修复、免疫系统的运作和皮肤的生长是必不可少的。

Chromium promotes insulin, the hormone that controls glucose levels in the blood. The body needs copper to use iron efficiently, and is key for infant growth, brain development, the immune system and for strong bones.

铬能促进胰岛素分泌，这是一种控制血糖水平的激素。身体需要铜来有效利用铁，是婴儿成长、大脑发育、免疫系统和强壮骨骼的关键。

Molybdenum is found in the liver, kidneys, glands and bones. Since molybdenum is part of the chemical reactions linked to carbon, nitrogen and sulphur, molybdenum is important for human health and the ecosystem.

钼存在于肝脏、肾脏、腺体和骨骼中。钼与碳、氮和硫的化学反应有关，对人类健康和生态系统非常重要。

Think about It 想一想

Would you have a meal with metals on the menu?

你的食谱上的食物都含有金属吗？

Does it make sense that cashew nuts are full of copper?

腰果富含铜，这一点有意义吗？

Can you pronounce molybdenum? What is it good for?

你能读出钼的发音吗？钼有什么好处？

Do you have gold in your blood?

你的血液里有黄金吗？

Find out more about the metals you need for good health. Even though your body needs only a tiny amount, these metals are indispensable. Draw up a list of the ten most important metals and correlate these with your daily diet. Are you getting the trace elements that you need, or do you need more? And if you need more, then plan a menu consisting of dishes that include ingredients that contain all the metals and trace elements you need.

了解更多有益健康的金属。即使你的身体只需要很少的量,这些金属也是不可缺少的。列一份清单,将最重要的10种金属写下来,并说明它们与日常饮食的关联。你是获得了所需要的微量元素,还是摄入不足?如果你需要更多,那就拟定一份食谱,列出含有你所需所有金属和微量元素的菜肴。

TEACHER AND PARENT GUIDE

学科知识
Academic Knowledge

生物学	腰果含有非常丰富的铜和其他80多种营养物质；小扁豆是一种豆类；在所有豆类中，小扁豆的蛋白质与食物能量的比例最高；腰果树生长迅速，是一种常绿热带树种；腰果苹果的维生素C含量是橙子的5倍，钙、铁和维生素B_1的含量比其他水果如柑橘类、鳄梨和香蕉都要高。
化学	钙用于骨骼和牙齿；钙和磷合成羟基磷灰石；激活酶需要镁；氨基酸的代谢；腰果是多种营养物质的良好来源，包括硫胺、镁、铜、铁、锰、磷和锌；腰果壳油含有腰果酸和腰果间二酚。
物理	体内的金有助于电信号在整个身体传输。
工程学	钼是分解药物并将其转化为活性化学物质所必需的；腰果包括一种工业上用于制造清漆和杀虫剂的腐蚀性树脂；缺陷测试是通过等离子体-质谱法（ICP-MS）进行的，锌原卟啉的检测是通过定量的血液荧光法进行的。
经济学	腰果是一种顽强的植物，以生长在一般不适合其他果树生长的沙质土壤中而闻名，因此给人以经济发展的机会；腰果壳油占无壳坚果的四分之一，大约与内核的油相当；腰果树皮富含单宁酸，可用于皮革鞣制；人体内的金浓度等于甚至大于金矿中矿石的黄金浓度（每吨）。
伦理学	即使是微小的东西，它也能提供生计，即使它的作用不被认可，仍会提供独特的服务。
历史	在希腊发现的现存最古老的熟扁豆，可以追溯到公元前11000年；腰果树原产于巴西沿海地区，在17世纪被引入世界其他地区；扁豆起源于中东，并被首次人工种植。
地理	扁豆的一个分支（通常去壳）被称为木豆（印度的扁豆），在印度被煮成浓稠的咖喱饭菜，作为主食；加拿大的扁豆产量占全世界的33%；巴西是最大的腰果生产国；扁豆在沙土、黏土和肥土中都能生长；泥炭土和酸性土壤最容易缺铜。
数学	多变量系统建模，同时观察和分析一个以上统计变量，如微量矿物质对若干健康参数的影响；计量土壤学分析土壤的定量特征。
生活方式	扁豆可以浸泡、发芽、油炸、烘烤或煮着吃；易消化的淀粉含量低（5%），不易消化的淀粉含量高，使扁豆对糖尿病患者具有潜在价值。
社会学	扁豆只有1%的脂肪，因此可以减少患心血管疾病的风险；缺铁导致疲劳；扁豆中的高纤维会导致胃肠气胀。
心理学	来自扁豆等的膳食纤维，可以增加饱腹感，减少食欲和总热量的摄入。
系统论	野生扁豆具有抗病性和非生物胁迫耐性。

教师与家长指南

情感智慧
Emotional Intelligence

扁豆

扁豆喜欢看到倒挂的腰果。扁豆赞美腰果的味道，惊讶于腰果中充满了铜，而周围却没有任何工业或矿业。扁豆想知道为什么人类要吃铜。腰果回答说是为了健康，扁豆想知道这意味着什么，并希望腰果能更准确地回答。扁豆从一个又一个令人惊讶的事实中走出来，随后发现人体甚至需要金来运转。扁豆承认自己也在关注微量金属，并警告说每个人都需要它们。扁豆的主要贡献是锌，他高兴地解释了这一事实，自信而坦然。

腰果

腰果很乐意解释有关他自己的情况，并把重点引向了坚果。在指出果实的价值后，腰果透露了他所含铜的重要性，并解释了如何从土壤中收集铜。腰果非常浅显地描述了铜有益健康，进一步解释说铜对红血球的合成至关重要。腰果解释说，还需要铁和微量的金。腰果惊讶地发现，扁豆中含有大量的锌。腰果很奇怪为什么会有这么多关于金属的讨论。当腰果不知道"钼"这个词时，它有信心去问扁豆。腰果现在知道了金属对健康和幸福的重要性。

艺术
The Arts

岩石有各种有趣的颜色，红色、橙色、黄色、绿色、蓝色，看到这些，我们就会明白为什么艺术家最早的时候就用岩石，或者更准确地说，用其中所含的微量元素来制作颜料。准备好你的相机，无论是手机上的相机还是数码相机，拍一些你所在地区的岩石照片。拍下你看到的不同颜色，然后将它们与你在这个故事中了解到的微量矿物质相匹配。随着岩石被侵蚀，矿物质最终会进入土壤，被植物和动物吸收，并最终呈现在你的食物中。

TEACHER AND PARENT GUIDE

思维拓展
Systems: Making the Connections

　　物质不会凭空消失，也不会凭空产生，只是从一种形式转换成另一种形式。土壤中含有可供植物利用的微量矿物质。这种对植物的可用性是由土壤中的微量元素与周围的一切联系起来的方式来控制的，而这种联系又受到土壤酸度、排水和有机物含量的影响。人类的行为对土壤的肥力和食物中的微量元素的可得性产生了最普遍的影响。传统上，来自植物和动物的废物丰富了土壤。这就是为什么最肥沃的土壤通常位于城市周围的原因。对苏格兰土壤的详细分析表明，城市周围铜在消耗而铅在富集。现在，在土壤的构成和维护中还有一个主要因素，即与细菌、真菌、微藻和地衣的关联。地衣用酸来分解岩石矿物，即使在碱性很强、富含微量元素的岩石上也能生存。真菌在黏土、淤泥和沙子上释放的酸产生了可溶性硅、铁和锰。然后，根部和周围的微生物活动通过螯合和溶解作用提供营养物质，如磷酸盐和大多数的微量元素。如果我们想了解微量元素如何最终进入我们的食物，并决定我们的健康，就需要生物学家、化学家、物理学家和统计学家的跨学科研究。从地球的形成开始对整个系统进行了解，是了解我们健康的关键。

动手能力
Capacity to Implement

　　到一块耕地去看看。仔细观察土壤。你能分辨这是哪种土壤吗？是沙土、淤泥、白垩土、泥炭土还是黏土？先不要下定论。再看一下土壤的颜色。从颜色可以看出含有哪些矿物质？再看一下土壤的结构：它是干燥的，很容易地从你的指间滑过，还是粘在一起有黏性？这些基本观察有助于你决定在这里种什么最好。最后，看看这片土地上产生了什么废弃物，这些废弃物能否被土地循环利用。然后你就会知道这块土地产出的食物会不会富含营养，为你提供所需的微量矿物质。把你的结论和建议写下来，并且与朋友和家人分享。

教师与家长指南

故事灵感来自
This Fable Is Inspired by

玛丽·弗朗西斯·皮恰诺
Mary Frances Picciano

　　玛丽·弗朗西斯·皮恰诺在美国宾夕法尼亚州立大学获得科学硕士学位，然后在1973年在同一所大学获得博士学位。1974年，她担任伊利诺伊大学农学院助理教授，1984年成为正教授。1989年，她加入宾夕法尼亚州立大学，担任营养学教授。她对母婴和幼儿营养进行研究。她在母乳的组成和生理学以及微量元素的代谢方面做了大量的工作。2002年她加入美国国家卫生研究院，担任高级营养研究科学家。她撰写了100多篇关于饮食和营养需求的文章以及创新的研究论文。

图书在版编目(CIP)数据

冈特生态童书.第八辑:全36册:汉英对照/
(比)冈特·鲍利著;(哥伦)凯瑟琳娜·巴赫绘;
何家振等译.—上海:上海远东出版社,2021
ISBN 978-7-5476-1773-1

Ⅰ.①冈… Ⅱ.①冈…②凯…③何… Ⅲ.①生态环境-环境保护-儿童读物—汉、英 Ⅳ.①X171.1-49

中国版本图书馆CIP数据核字(2021)第249940号

策　　划	张　蓉
责任编辑	程云琦
封面设计	魏　来　李　廉

冈特生态童书

菜单上的金属

[比]冈特·鲍利　著
[哥伦]凯瑟琳娜·巴赫　绘
解天宇　译

记得要和身边的小朋友分享环保知识哦!
八喜冰淇淋祝你成为环保小使者!

Food 263

假的食物？

Fake Food?

Gunter Pauli

[比] 冈特·鲍利 著

[哥伦] 凯瑟琳娜·巴赫 绘

闫宇昕 译

上海远东出版社

丛书编委会

主　任：贾　峰

副主任：何家振　闫世东　林　玉

委　员：李原原　祝真旭　牛玲娟　梁雅丽　任泽林
　　　　王　岢　陈　卫　郑循如　吴建民　彭　勇
　　　　王梦雨　戴　虹　翟致信　靳增江　孟　蝶

特别感谢以下热心人士对童书工作的支持：

匡志强　宋小华　解　东　厉　云　李　婧　陈　果
刘　丹　熊彩虹　罗淑怡　旷　婉　杨　荣　刘学振
何圣霖　廖清州　谭燕宁　韦小宏　李　杰　欧　亮
陈强林　王　征　张林霞　寿颖慧　罗　佳　傅　俊
胡海朋　白永喆　冯家宝

目录

假的食物?	4
你知道吗?	22
想一想	26
自己动手!	27
学科知识	28
情感智慧	29
艺术	29
思维拓展	30
动手能力	30
故事灵感来自	31

Contents

Fake Food?	4
Did You Know?	22
Think about It	26
Do It Yourself!	27
Academic Knowledge	28
Emotional Intelligence	29
The Arts	29
Systems: Making the Connections	30
Capacity to Implement	30
This Fable Is Inspired by	31

一只鳄梨和一个甜菜根正在一家供应美味水果和蔬菜的三星餐厅里浏览菜单。在那里用餐的客人并不知道他们盘子里的食物是用什么做成的。

鳄梨说道："看看这鞑靼牛排！我都馋得流口水了。"

An avocado and a beetroot are looking at the menu of a three-star restaurant serving tasty fruit and vegetables dishes. The guests dining there have no idea what the food on their plates is made of.

"Look at the steak tartare! It makes my mouth water," the avocado says.

……浏览菜单……

... are looking at the menu ...

……一块肉也没有……

... not a single piece of meat in ...

甜菜根指出:"这肉看起来像生牛肉,但并不是!整盘菜里甚至连一块肉也没有。"

"你在开玩笑吗?肉是如此红嫩多汁。这是地道的法国菜,味道非常棒。"

"你觉得什么东西能呈现出这种大红色?"

"That may look like raw beef – but it is not! There is not a single piece of meat in that whole dish," the beetroot points out.

"You've got to be kidding? There is nothing that looks as red and juicy as meat. This dish is a French masterpiece. It smells very well seasoned."

"Who do you think can make this grand red colour?"

"鲜红色肯定来自最新鲜的肉啦。"

"我告诉过你,这些菜里丝毫没有动物性成分。为什么你不相信我?"

"你看一看,再闻一闻!"

"That bright redness can only come from the freshest meat."
"I told you that there are no ingredients from any animal involved in preparing any of these dishes. Why don't you believe me?"
"Just look at it! And smell it!"

再闻一闻!

And smell it!

……能仔细看看我吗？

... take the time to look at me?

甜菜根问道："你能仔细看看我吗？我是什么颜色？"

"哦，你是深红色，接近紫色。人们只要吃一小块甜菜根，他们的小便就会从黄色变成粉红色，这让他们惊慌。"

"其实，那块鞑靼牛排是用甜菜根做成的。肉之所以有这种质感，是因为人们精心烹饪和切块。我的颜色和肉的颜色非常匹配。"

"Would you take the time to look at me? What colour am I?" Beetroot asks.

"Well, you are dark red, nearly purple. And when people eat just one piece of you, their urine turns from yellow to pink, making them panic."

"Well, that steak tartare is made out of me. Thanks to smart cooking and cutting, it has that texture. My colour is a perfect match to imitate meat."

"你为什么要这么做？"鳄梨问道。

"有些人不喜欢杀死动物作为食物，但仍然喜欢肉类美妙的滋味。"

"但这不是真正的肉啊！"

"Why would you ever want to do that?" Avocado asks.
"Some people don't like to kill animals for food, but still like to enjoy that great taste."
"But that is fake meat!"

但这不是真正的肉啊!

But that is fake meat!

你看到巧克力甜点吗……

Did you see that chocolate dessert ...

"没错,那是超级健康的食物。如果你不愿意,你可以选择不杀动物去吃鲜肉。大自然的美丽之处在于它为每个人都提供了让他们满意的选择。"

鳄梨问:"你看到刚送来的巧克力甜点吗?"

"No, that is super healthy food. If you don't want to, you don't need to have an animal killed to eat its raw flesh. The beauty of Nature is that it offers options to please everyone's choices in life."

"Did you see that chocolate dessert that just came by?" Avocado asks.

"噢，怎么会没有注意到泡沫丰富的慕斯呢？一定很好吃！"

"那份慕斯不含任何巧克力和牛奶！"

"你是在说真正的假货？没有巧克力怎么能做出巧克力慕斯呢？没有牛奶怎么会有奶油泡沫呢？"

"Well, it is impossible not to notice that foamy mousse. It must be so tasty!"

"That mousse has no chocolate or milk!"

"Now you are talking about real fake stuff. How can you ever have a chocolate mousse without chocolate? How can you ever have a creamy foam without milk?"

那份慕斯不含任何巧克力和牛奶!

That mousse has no chocolate or milk!

我很自豪能用来制作……

I am so proud to be the one they use for ...

"我很自豪能用来制作慕斯。没有牛奶,没有鸡蛋,也没有巧克力——但是味道依然很好。吃过一次的人还想再吃一次!"

"但你是绿色的,而且是固体!我知道有些人不想吃任何动物制品。但是为什么不吃巧克力呢?那是一种植物啊!"

"成熟的鳄梨与椰奶和角豆混合,赋予甜点的颜色和质地可以骗过任何人。"

"I am so proud to be the one they use for making that mousse. It has no milk, no eggs and no chocolate – and yet it tastes great. People who have it once, want it again!"

"But you are green and solid! I understand that some people don't want to eat anything of animal origin. But why eliminate chocolate? That is a plant!"

"Ripe avocado, mixed with coconut milk and carob, gives a dessert the colour and texture that would fool anyone."

"角豆？那是一种豌豆啊！谁会笨到把豌豆错当成巧克力？"

"有时候不是被愚弄的问题。通常它是指享受美味的同时不会摄入脂肪！"

……这仅仅是开始！……

"Carob? But that is a pea! Who will be fooled into mistaking peas for chocolate?"

"Sometimes it is not about being fooled. Often it's about enjoying a great tasting dish – without the fat!"

... AND IT HAS ONLY JUST BEGUN!...

······这仅仅是开始！······

... AND IT HAS ONLY JUST BEGUN! ...

Did You Know?
你知道吗?

Beetroot originated in North Africa in prehistoric times, and grew wild along Asian and European seashores. In these earlier times, people exclusively ate the beet greens and not the roots.

甜菜根起源于史前时期的北非,在亚洲和欧洲的海岸生长。在早期,人们只吃甜菜叶,而不吃甜菜根。

The Romans were the first to cultivate beets to eat their roots. The tribes that invaded Rome spread farming and eating of beets throughout northern Europe, both as animal fodder and for human consumption.

罗马人是第一批种植甜菜并且食用其根部的人。除了人食用之外,甜菜还作为动物饲料,入侵罗马的部落将甜菜种植和食用方式传播到整个北欧。

In the 19th century, it was discovered that beets were a concentrated form of sugar. The first sugar factory was built in Poland. Napoleon decreed that beets be used as the key source of sugar, catalysing its popularity.

19世纪，人们发现甜菜中糖分的浓度很高。波兰建成了第一家制糖厂。拿破仑下令将甜菜作为糖的主要来源，从而促进了糖的普及。

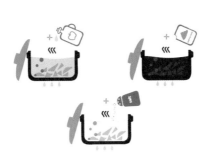

Beets' colour modifies during cooking. Adding lemon juice or vinegar will brighten the colour, while an alkaline substance, such as baking soda, will often cause them to turn a deeper purple. Salt bleaches the beets.

甜菜的颜色在烹饪过程中会改变。加入柠檬汁或醋会使其颜色变亮，而加入碱性物质如小苏打，会使其变成深紫色。盐会使甜菜变白。

Avocados were the staple food in Mexico, with the oldest traces dating back to 10,000 years ago. Its cultivation dates back 5,000 years ago. It was first called an alligator pear.

鳄梨在墨西哥是主食，最早可以追溯到1万年前。鳄梨的种植可以追溯到5 000年前，最初被称为鳄鱼梨。

Mexico remains the largest producer of avocados, harvesting more than 2 million tons per year. Mexico started processing the seeds as a highly nutritional additive to bread.

墨西哥现在仍然是最大的鳄梨生产国，每年收获超过200万吨鳄梨。墨西哥很早就开始将鳄梨种子加工成面包的营养添加剂。

Avocado seeds were dispersed by large mammals, like the ground sloth, which are now extinct. No animal today has a large enough digestive system to disperse the seeds. Domestication saved the avocado from extinction.

鳄梨种子是通过大型哺乳动物传播的，比如现在已经灭绝的地懒。现在没有任何动物有足够大的消化系统来传播鳄梨的种子。人工种植使得鳄梨免遭灭绝。

The carob is member of the pea family. It is rich in calcium, but has only one-third of the calories of cacao and 17 times less fat. The plant is very resistant to drought and grows around the Mediterranean.

角豆是豌豆科的一种，富含钙，热量仅是可可的三分之一，脂肪相较可可减少17倍。角豆十分耐旱，生长在地中海地区。

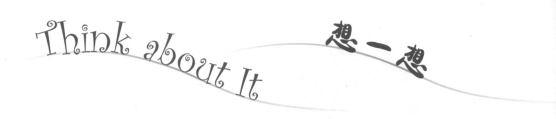

Would you stop eating meat to save animals from pain and slaughter?

为了使动物免受痛苦和屠杀，你会不再食用肉类吗？

Would you mistake a pea for a chocolate?

你会把豌豆错当成巧克力吗？

Can a vegetarian restaurant ever be awarded three Michelin Stars?

素食餐厅能被授予米其林三星吗？

When someone proposes something that seems impossible, are you ready to learn more, or do you dismiss it to avoid wasting time?

当有人提出似乎不可能的事情时，你是准备去深入了解，还是为了避免浪费时间而打消这个念头？

Do It Yourself! 自己动手！

Find some ripe avocados, ones with firm, green flesh. Scoop the flesh out and place in a blender, and blend it with some agar-agar, a seaweed extract that is used instead of gelatine. Let it settle for a few minutes, then place it a cup or bowl in the refrigerator, to cool down and set. In about two hours you will have made an avocado mousse. If you want to add carob powder, you can do so after blending. Add the concentrated powder and you will be surprised with the result. Does this sound too easy? Using this basic recipe, you can now start to get creative, adding some mint, a sweetener or even a hot red chilli pepper. Enjoy creating a dessert worthy of three stars!

找一些有着结实的绿色果肉的成熟鳄梨。把果肉挖出来放入搅拌机，和一些琼脂混合。琼脂是一种用来代替明胶的海藻提取物。静置几分钟后倒入杯子或碗里，再放入冰箱冷却定型。大约2小时后，鳄梨慕斯就做好了。如果你想加入角豆粉，可以在搅拌后加入。加入浓缩粉末的效果一定会让你感到惊讶。这听起来是不是太简单了？基于这个基本配方，你可以发挥创意，加入一些薄荷、甜味剂，甚至是红辣椒。享受创造米其林三星甜点的过程吧！

TEACHER AND PARENT GUIDE

学科知识
Academic Knowledge

生物学	鳄梨既指树也指水果；鳄梨是一种浆果，只有一粒种子；角豆树是豆科常绿乔木或灌木。
化学	鳄梨富含单不饱和脂肪；鳄梨叶是一种香料；鳄梨富含维生素B和维生素K；鳄梨对鸟类和马有毒；甜菜苷是从甜菜根中提取的一种着色剂，用于改善番茄酱、果酱、冰淇淋和糖果的颜色和味道。
物理	鳄梨成熟时果皮呈绿色、棕色、紫色或黑色，外观可能是梨形、蛋形或球形；鳄梨油的冒烟点较高；长时间烹饪使得鳄梨无法食用。
工程学	慕斯的制作过程包括起泡，这个过程后来因使用二氧化碳烹饪食物而变得复杂，位于加泰罗尼亚的罗萨斯的巴塞罗那El Bulli餐厅因这种工艺而闻名。
经济学	甜菜根是一种食物、着色剂和药物；全球鳄梨市场规模达750万吨；葡萄牙出产全世界30%的角豆。
伦理学	有些人希望改善动物福利，因此不吃任何动物制品，甚至不吃鸡蛋或奶酪。
历史	鳄梨与现已灭绝的更新世巨型动物共同进化，后者的消化系统可以让大种子穿过其肠道；甜菜在古代中东变成人工种植；墨西哥人和秘鲁人分别在1万年前和4 000年前开始食用鳄梨；人们在1750年将鳄梨引入印度尼西亚；在19世纪人们用甜菜根为葡萄酒着色；鞑靼人会将肉放在马鞍下使其软化。
地理	鳄梨生长在热带和地中海地区；墨西哥的鳄梨产量占世界总量的32%；鳄梨原产于特瓦坎山谷，并向南传播到秘鲁；在东欧，人们会制作甜菜汤来改善颜色和味道；角豆原产于地中海和中东，十分抗旱且耐盐性强。
数学	吃一年素食可以节约1 500吨水，少砍伐1 000平方米森林，减少3.3吨二氧化碳排放。
生活方式	吃肉、吃素或素食饮食；年轻一代很关心动物权益。
社会学	英语中"鳄梨"一词来自西班牙语aguacate；在秘鲁的盖察，它被称为帕尔塔；甜菜一词起源于凯尔特语。
心理学	"寻找新鲜事物的刺激感"和"成为小众群体的一部分"驱使人们购买这些"假"的食物，包括看起来像鞑靼牛排或巧克力慕斯的食物，而实际上它们不是。
系统论	只有减少肉类消费，我们才能拥有一个可持续发展的社会。

教师与家长指南

情感智慧
Emotional Intelligence

鳄梨

鳄梨一看到食物就兴奋，根本不相信这道红色多汁的菜里没有肉。尽管对这道菜"负责"的甜菜已经确认这道牛排是素的，鳄梨还是不想听到这一事实。鳄梨固执己见的依据是食物的外观和味道。鳄梨想要知道为什么有人想用甜菜代替肉。即使逻辑上说得通，鳄梨坚持认为这种食物是"假的"。鳄梨意识到这件事已经没有讨论的空间，就把注意力转移到巧克力甜点上。鳄梨表示这份甜点不含巧克力和牛奶，并且为能够用来制作这道甜点感到自豪，还花时间来解释制作过程。鳄梨谈论的并非愚弄谁，而是有关为了减少脂肪摄入而特意做出的选择。

甜菜根

甜菜根一开始就表明这道菜不是肉。因为鳄梨坚持己见，所以甜菜根提出问题，引导鳄梨尝试去理解。当甜菜根的观点不被接受时，甜菜根自信地直接问为什么鳄梨不相信。甜菜根指出替代红肉的正是甜菜根。甜菜根并不想让鳄梨相信蔬菜替代肉类是最好的解决方案，而是提供一个可行的选择。甜菜根注意到甜点。当鳄梨说这份慕斯不含巧克力或牛奶时，这款甜点就被认为是假的。甜菜根最初试图说服鳄梨相信牛肉是替代的，现在又质疑坚硬的绿色鳄梨如何变成慕斯。甜菜根坚持认为巧克力不会被淘汰，豌豆也无法取代巧克力。

艺术
The Arts

甜菜根能提供很多种颜色，不仅仅是一种果汁。学习如何用不同的方法处理甜菜根以得到不同深浅的红色和紫色。关键是要认识到植物不仅是肉类的替代品，而且有着自己的独特作用。

TEACHER AND PARENT GUIDE

思维拓展
Systems: Making the Connections

　　吃肉会产生一系列问题，不仅对我们的健康不利，还会破坏地球环境。畜牧业是造成环境问题的元凶之一，排放的温室气体占世界排放总量的18%。这相当于地球上所有汽车、火车、飞机和轮船的排放量。在美国，家畜养殖造成的森林砍伐和温室气体排放加剧了气候变化。联合国政府间气候变化专门委员会（IPCC）指出了这些问题，并证实，除非改变我们的食物系统，包括动物农业实践和改变饮食，全球气温还会继续升高，这将带来更多的干旱、野火、昆虫泛滥、传染病和极端天气。几乎每个人都可以做到停止或减少肉类消费。如此小的转变却能产生很大的影响。Drawdown教育项目的研究结果显示，人类转变到以素食为主在100个最优解决方案中排在第四位。没有人必须在一夜之间变成完全不吃肉的状态。我们知道，想要成为素食主义者，首先要成为弹性素食主义者。"周一无肉"运动鼓励人们在饮食中加入蔬菜，这是一个循序渐进的过程，也是一个减少个人碳排放的好方法。改变我们的饮食习惯能产生深刻影响。我们不仅要顾及动物和环境，还要考虑到我们的同胞和后代的利益。当替代品的味道与在巴黎米其林三星餐厅能享用的晚餐一致时，吃素便成为既让人接受又令人愉悦的做法。

动手能力
Capacity to Implement

　　我们不想让社会割裂。我们并非争论究竟是赞成还是反对，我们也不是让所有人承认他们的所作所为对环境产生了什么不良影响，但现在确实有必要改变我们食用的种类。我们唯一能做的就是尽可能准备最好的食物，并且相信这将激励人们吃更多不用肉就能烹制的美味佳肴。这能够说服更多人。所以，请罗列一份菜单，涵盖所有最好吃的素食。不要纠结于这些菜里有没有肉，而是要夸赞其美妙的味道。

教师与家长指南

故事灵感来自
This Fable Is Inspired by

阿兰·帕萨德
Alain Passard

　　阿兰·帕萨德从他的祖母那里得知，烹饪可以是一种庆祝活动。帕萨德还没读完高中就进入烹饪行业。他最初是在法国布里多尼之星米歇尔·凯弗的厨房里当学徒。帕萨德在4年的学习中积累了许多经验，然后到兰斯的米其林三星餐厅 La Chaumiere 工作。后来，他换到了巴黎的阿克拉斯特拉特。3年后，帕萨德接手巴黎昂吉恩公爵厨房。26岁时，他成为法国最年轻的米其林二星级厨师。接着他又去了布鲁塞尔的卡尔顿餐厅，先后给餐厅带来一星和两星。1986年10月，他在巴黎开设了琶音餐厅。在6个月内，餐厅获得了一星，一年后获得了两星。1990年1月，《戈米氏餐饮指南》给琶音餐厅打了19分（满分20分）。1996年，在这家餐厅十周年纪念日，阿兰·帕萨德和他的餐厅获得了米其林三星评级。阿兰·帕萨德用他的创新烹饪方法及简单而又完美的菜肴改变了烹饪世界。他是素食餐厅界第一个获得米其林三星的人。

图书在版编目（CIP）数据

冈特生态童书.第八辑：全36册：汉英对照 /
(比)冈特·鲍利著；(哥伦)凯瑟琳娜·巴赫绘；
何家振等译. —上海：上海远东出版社，2021
ISBN 978-7-5476-1773-1

Ⅰ.①冈… Ⅱ.①冈… ②凯… ③何… Ⅲ.①生态环
境-环境保护-儿童读物—汉、英 Ⅳ.①X171.1-49

中国版本图书馆CIP数据核字(2021)第249940号

策　　划	张　蓉
责任编辑	程云琦
封面设计	魏　来　李　廉

冈特生态童书
假的食物？
[比]冈特·鲍利　著
[哥伦]凯瑟琳娜·巴赫　绘
闫宇昕　译

记得要和身边的小朋友分享环保知识哦！
八喜冰淇淋祝你成为环保小使者！

Health 278

发烧树
The Fever Tree

Gunter Pauli
［比］冈特·鲍利 著
［哥伦］凯瑟琳娜·巴赫 绘
章里西 译

上海远东出版社

丛书编委会

主　任：贾　峰

副主任：何家振　闫世东　林　玉

委　员：李原原　祝真旭　牛玲娟　梁雅丽　任泽林
　　　　王　岢　陈　卫　郑循如　吴建民　彭　勇
　　　　王梦雨　戴　虹　翟致信　靳增江　孟　蝶

特别感谢以下热心人士对童书工作的支持：

匡志强　宋小华　解　东　厉　云　李　婧　陈　果
刘　丹　熊彩虹　罗淑怡　旷　婉　杨　荣　刘学振
何圣霖　廖清州　谭燕宁　韦小宏　李　杰　欧　亮
陈强林　王　征　张林霞　寿颖慧　罗　佳　傅　俊
胡海朋　白永喆　冯家宝

目录

发烧树	4
你知道吗?	22
想一想	26
自己动手!	27
学科知识	28
情感智慧	29
艺术	29
思维拓展	30
动手能力	30
故事灵感来自	31

Contents

The Fever Tree	4
Did You Know?	22
Think about It	26
Do It Yourself!	27
Academic Knowledge	28
Emotional Intelligence	29
The Arts	29
Systems: Making the Connections	30
Capacity to Implement	30
This Fable Is Inspired by	31

一棵柳树正在周游世界，寻找志同道合者来协助他使人们获得健康，减轻痛苦。抵达秘鲁后，柳树受到了金鸡纳树的欢迎。金鸡纳树说道：

"感谢你长途跋涉来到我们位于高地的寒舍。你是止痛大师，我们应该去拜访你，而不是你来看我们！"

A willow tree is travelling around the world, looking for friends to help him make people healthy and relieve their pain. Arriving in Peru, the willow is welcomed by a cinchona tree, saying,

"Thank you for travelling all the way to our modest region, here in the highlands. You are the master of pain relief, and we should be visiting you, instead of you visiting us!"

一棵柳树正在周游世界。

A willow tree is travelling around the world.

……很荣幸你把我也当作大师。

... generous of you to consider me a master too.

"我这位止痛大师很荣幸能踏上这场朝圣之旅,最终得以找到你这样的宗师。"柳树回答道。

"很荣幸你把我也当作大师。我很钦佩你这位宗师,不过我们似乎正在失去我们在社会中的作用。"

"This master of pain relief is honoured to undertake this pilgrimage, and find grandmasters, like you," the willow tree replies.

"It is generous of you to consider me a master too. I do admire you as a grandmaster, as we seem to be losing our role in society."

"我们永远不会失去我们的作用。大师之所以是大师，最终变成宗师，是因为我们始终为人类的福祉而工作，不断分享我们的智慧，还有优秀的学生与世人分享他们的发现。"

"只要我们还能继续活下去，你说的就能实现。现在我们只剩下1000多棵了。几百年前，不少国家为了金鸡纳树的控制权还发动了战争。"金鸡纳树说。

"We will never lose our role. Masters are masters, and turn into grandmasters because we work for the common good and share our wisdom and have great students who share with us their discoveries."

"As long as we have a chance to live on. There are perhaps only a thousand of us left. A couple of hundred years ago, countries waged war over our family of trees to control us," Cinchona Tree says.

大师之所以是大师……

Masters are masters ...

你的树皮用来缓解高烧……

Your bark is used to reduce a high fever ...

"你是'发烧树'。当有人患上疟疾,一种由蚊子传播的可怕疾病时,你的树皮一直被人们用来缓解疟疾引起的高烧,人们当然需要你。"

"你知道携带疟疾的蚊子其实就是被白人殖民者带到美洲来的吗?这种疾病以前在这里从未出现过,但我们早就备好了天然疗法。"

"You are 'The Fever Tree'. Your bark is used to reduce a high fever when a person suffers from malaria, the dreaded disease carried by mosquitoes. Of course people need and want you."

"Did you know that the mosquitoes that carry malaria were brought here, to the Americas, by the white colonizers? This illness never existed here before, and yet we already had a natural cure waiting."

"这算是深谋远虑还是先见之明呢?你们在控制疟疾发烧方面确实很出色。"

"不过请注意,我无法真正治愈疟疾。寄生虫一旦进入人体,我就无法杀死它,但我可以缓解疾病带来的身体不适,"金鸡纳树说。

"当人们制造出一种化学物质作为你的替代品,并开始研究疫苗时,他们就对你失去了兴趣。"

"Foresight or farsighted? You are outstanding in controlling malarial fever."

"Please note that I cannot cure malaria. I cannot kill the parasite once it is in a person's body, but I can ease the discomfort," Cinchona says.

"People lost interest in you when they made a chemical to replace you, and started working on their vaccines."

……开始研究疫苗。

...started working on their vaccines.

血液里的这些寄生虫变化非常快……

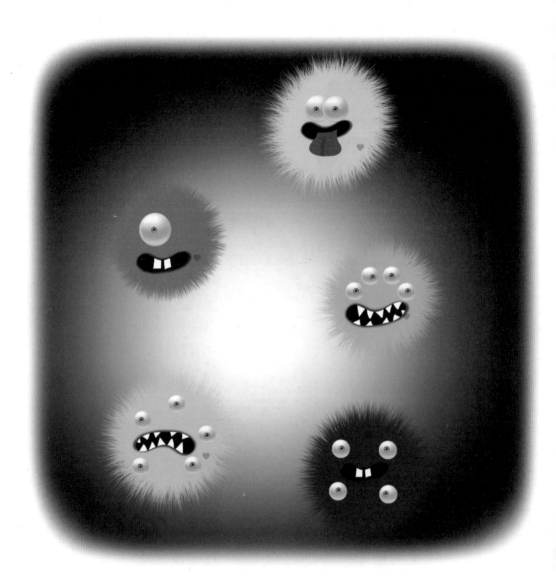

The bugs in their blood change so quickly ...

"我知道。知道血液里的这些寄生虫有多聪明吗?他们变化非常快,瞬间就对人工药物产生了抗药性,而疟疾疫苗也因此效果欠佳。"

"是的,这些宛如小僵尸般的寄生虫,数以万亿地生活在我们体内和周围,无时无刻都在变化。无论我们尝试什么手段,他们都会迅速变异,适应力也会越来越强。人类必须奋起直追,不过他们从来没有追上过。"

"I know. How smart was that? The bugs in their blood change so quickly they become resistant to artificial medicine in no time. Vaccines against malaria don't work well."

"Yes, all these tiny zombies, living by their trillions in and around us, are always changing. Whatever we try, they quickly mutate and become more and more resilient. People have to play catch-up without ever catching up."

"那讲讲你们的故事吧？你们可是著名的'止痛树'。"

"世界上有几百种不同的柳树，几乎所有种类的柳树都可以用来做止痛药。"

"你们也被人造化学品取代了吗？"

"哦，是的。我们甚至被橙子取代了！"

"And what about you? You are known as 'The Painkiller Tree'."

"There are hundreds of different willows around the world, and nearly all our family members can be used to make painkillers."

"And did you get replaced with artificial chemicals?"

"Oh yes. We were even replaced by oranges!"

我们甚至被橙子取代了!

We were even replaced by oranges!

维生素C非常重要……

Vitamin C is very important ...

"橙子？橙子能止痛吗？"

"不，不，不。但橙子含有丰富的维生素C。"

"确实。维生素C对冬季预防流感非常重要。"

"是的，不过我叶子里含有的维生素C比橙子多10倍。我能在许多橙子无法生长的地方茁壮成长。那人类为什么要抛弃我呢？"

"Oranges? Do oranges kill pain?"
"No, no, no. But oranges are sold as a rich source of Vitamin C."
"And they are. Vitamin C is very important to keep the flu away in the winter."
"Of course, but my leaves have ten times more Vitamin C than oranges. I thrive in many places where oranges cannot grow. So why discard me?"

"是啊,他们总是被那些新奇的玩意儿吸引。不过我有信心人类很快就会注意到在自己眼皮底下早就有了他们一直想要的东西。剪树枝和种树会让他们很开心。"

"是啊!你真正需要的东西其实就在你生活的地方。希望人类能重新发现自己生活中已经拥有的奇迹。"

……这仅仅是开始!……

"Well, people are always fascinated by anything exotic. Yet I am confident that they will soon want what they already have, in their area. Making cuttings and growing a tree will make them happy."

"Yes! All one needs is right where one lives. Let's hope people rediscover the marvels of what they already have at home."

... AND IT HAS ONLY JUST BEGUN!...

……这仅仅是开始!……

... AND IT HAS ONLY JUST BEGUN! ...

Did You Know?
你知道吗?

世界上有29种金鸡纳树，秘鲁有其中的20种。不幸的是，由于森林砍伐、土壤退化和使用有害化学物质的单一栽培农业的发展，许多金鸡纳树正处于灭绝的边缘。

Peru has 20 of the world's 29 cinchona species. Unfortunately, many are on the brink of extinction due to deforestation, degradation of the soil and the growth of monoculture agriculture with harmful chemicals.

金鸡纳树生长在海拔1300—2900米的潮湿森林中，高度可达15米，大部分在秘鲁西北部，但也有在中部的。

Cinchona trees grow up to 15 metres in height, in humid forests between 1,300-2,900 metres above sea level, mostly in the northwest but also in the centre of Peru.

Pre-Columbian people of Peru, Colombia, Ecuador, and Venezuela used cinchona to treat fever and pain. Now cinchona is used to make tonic water and angostura bitters, used in Peru's national cocktail, the pisco sour.

哥伦布发现美洲前，秘鲁、哥伦比亚、厄瓜多尔和委内瑞拉人用金鸡纳树来治疗发烧和疼痛。金鸡纳树现在被用来制作奎宁水和安古斯图拉树皮制剂，还用于调制秘鲁的国民鸡尾酒皮斯科酸酒。

Simon Bolivar, liberator of South America, and the Peruvian Congress decided, in 1825, soon after independence from Spain, to include the cinchona in the national coat of arms, in recognition of its medicinal benefits.

1825年秘鲁脱离西班牙独立后不久，南美解放者西蒙·玻利瓦尔和秘鲁国会决定，将金鸡纳树纳入国徽，以认可其药用价值。

Almost all willows readily take root from cuttings or broken branches on the ground. Willows are planted on overflowing meadows and stream borders. They grow from the Arctic Circle to the tropics.

几乎所有的柳树插枝都很容易生根。柳树种植在河漫滩草甸和溪边，从北极圈到热带地区都可生长。

All 400 species of willows have medicinal properties, recognised by the Egyptians, Greek, Native American and Chinese cultures. Hippocrates described the effectiveness of the willow in the 5th century BC.

400多种柳树都有药用价值，在埃及、希腊、美洲原住民和中国文化中都认可这一点。希波克拉底在公元前5世纪记录了柳树皮的药效。

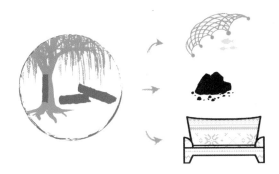

Early fishing nets, ropes and strings were made from willow bark. Its wood is a source of charcoal. Having twigs that bend easily makes it a renewable resource for the traditional woven furniture of Scandinavia.

早期的渔网、绳子和线都是用柳树皮做的。柳树可制成木炭。柳树枝容易弯曲，这使它成为斯堪的纳维亚传统编织家具的可再生资源。

The Inuits eat young willow leaves raw, with seal oil, for its Vitamin C content. The inner bark is scraped out and cooked in strips, like spaghetti. It is used for wound healing, and mixed with cranberries, as a cramp remedy.

因纽特人吃生嫩柳叶和海豹油，以获取其中的维生素C。刮下柳树的内层皮切成条煮，就像意大利面一样。柳树用于伤口愈合，并与蔓越莓混合，用于治疗抽筋。

Would you like to have your medicine grown at home?

你想在家里种自己需要的药吗?

Since bacteria and viruses mutate, plants change as well, but chemicals do not.

由于细菌和病毒会变异,植物也会变异,但化学物质不会。

What makes a master, and a grandmaster?

是什么造就了大师和超级大师?

Would you rather drink orange juice or eat willow spaghetti?

你是喜欢喝橙汁还是吃柳树意大利面?

When you have headache, what do you do? Do you first drink a glass of water to see if perhaps it is the result of dehydration? A few glasses of water usually reduce pain. If the pain does not subside, do you smell some camphor? Or do you take a painkiller? The question is what works best. Or should the question be what is best for you? Ask others what they do for a headache, and talk about the solutions that people have been using for centuries with great success.

头痛时,你会怎么做?你是否会先喝一杯水,看看是不是因为脱水?喝几杯水通常可以减轻疼痛。如果疼痛没有消退,你是闻樟脑,还是吃止痛药?问题是哪种效果最好,或者哪种方法最适合你。问问别人他们是如何治疗头痛的,然后聊一聊几个世纪以来人们一直在使用并取得巨大成功的治疗方案。

TEACHER AND PARENT GUIDE

学科知识
Academic Knowledge

生物学	柳树也叫燕柳树，有400多种；柳树是落叶树；在北极和阿尔卑斯地区发现的矮柳只有6厘米高；柳树花称作柳絮；柳树上有100多种蚜虫；柳树是通过插入土里的插条或断枝长出来的；29种金鸡纳树都来自安第斯山脉；抗寄生虫免疫和抗毒性免疫。
化 学	柳树汁富含由阿司匹林的前体水杨苷代谢而成的水杨酸；非甾体类抗炎药；金鸡纳树皮含有奎宁和生物碱。
物 理	对蚊子进行足够的照射以减少疟原虫。
工程学	在中国北方，人们用柳树来造头木林，用头木法获得薪柴或织造材料，不消耗土壤或森林；玻利维亚物种树皮的奎宁产量为8%—13%；疟疾疫苗使用来自疟疾寄生虫、乙型肝炎病毒的一部分和化学辅助剂来增强免疫反应，疫苗有效率仍然很低。
经济学	柳树作为生物能的来源；单宁、纤维、纸张、绳索和线可以用柳树木生产；奎宁的人工合成根除了种植栽培柳树的大型产业；为了维持金鸡纳树的垄断地位，秘鲁及其周边国家从19世纪开始就禁止出口金鸡纳种子和树苗；本土生长，适应性强。
伦理学	疟疾每年造成超过3亿人罹患急性疾病，导致至少100万人死亡。疟疾造成的死亡有90%发生在撒哈拉以南的非洲；除了需要有效的药物，还需要通过良好的食物和丰富的微量矿物质来增强免疫系统。
历 史	在亚述、苏美尔、埃及、希腊和美洲土著的古代文献中都提到药用柳树；用柳树制成的渔网可以追溯到公元前8300年；1853年，安徒生写了一篇名为《在柳树下》的故事；英国每年因疟疾直接损失6 000万英镑。
地 理	世界范围内柳树有520种，主要分布在北半球温带，在寒带、热带和南半球极少，且大洋洲无野生种；金鸡纳树是秘鲁的国树。
数 学	从基本的罗斯模型开始，通过包括宿主、携带者、寄生虫相互作用的关键特征来描述疟疾发病率的建模策略。
生活方式	出于美观和减碳，公园里经常种植柳树；基于金鸡纳树皮测试的顺势疗法；分享智慧；人们总是被异国情调所吸引。
社会学	中国古代有折柳告别的习俗；早在殖民之前安第斯的盖丘亚文化中就有使用奎宁的记载；大师和宗师；为共同利益而工作。
心理学	深谋远虑是预见未来或明智地为未来做准备的能力；先见之明是指有向前看的能力；任何来自远方的都被认为是更好的。
系统论	蝴蝶效应是指在一个动态系统中，初始条件的微小变化会导致整个系统大的长期连锁反应；在居民区，柳树根部到处伸展并侵略性地寻找水分，这成为一个问题；随着对金鸡纳树皮需求的增加，森林遭到了破坏；寄生虫变异成多种耐药寄生虫，使疫苗失效。

教师与家长指南

情感智慧
Emotional Intelligence

金鸡纳树

金鸡纳树谦逊，生活简朴，而实际上他的同类只剩下几千棵了。他对柳树的来访表示感谢，并满怀信心地赞美柳树为宗师。在金鸡纳树所剩无几的时代，他也在寻找自己的亲缘关系。金鸡纳树让柳树想起了历史事实。他强调自己角色的局限性：无法治愈疟疾。金鸡纳树对合成药物的发展和寄生虫的变异以及变得具有抗药性感到好奇。他向同样被化学物质取代的柳树解释说，人类更喜欢新奇的产品。金鸡纳树仍然相信，人类会重新发现他们已经拥有自己想要的东西，包括身边那些大师级的树。

柳树

柳树自尊心很强，自己是大师，也大方地称金鸡纳树为宗师。柳树相信自己永远不会失去作用，因为她为解除人类的痛苦而工作。她指出了一个痛苦的现实：化学物质取代了天然药物。然而，柳树又指出，寄生虫和病毒的变异速度快于人类寻找治疗方法的速度。柳树承认她也被化学物质所取代，失去了作为维生素C供应者的角色。柳树想知道人类为什么要丢弃她，并且希望人们能够重新发现他们周围生长的奇迹。

艺术
The Arts

柳树是世界各地艺术作品中常见的著名形象。你会怎样描绘柳树？是在晨雾中，还是在夕阳下？用柔和的颜色画出你对柳树的印象，只需表现柳树的形状，不要有太多的细节。柳树的轮廓即使模糊也很容易认出，看起来有点神奇。再试着用其他的方法，艺术地描绘柳树吧。

TEACHER AND PARENT GUIDE

思维拓展
Systems: Making the Connections

顺势疗法是从金鸡纳树研究中演化而来的。金鸡纳工业发展较快，但缺少对采集和出口的限制，导致金鸡纳物种濒临灭绝。直到1971年，科学家才开始真正理解阿司匹林作为消炎药的作用机制，然而几千年来，人类一直在用它治病。从某种角度来说，这是自然医学的胜利。我们的许多药物最初来自植物，我们还在继续寻找有潜力的新药。许多治疗方法都是在对相关植物有效成分进行临床试验或调查之前就有的。真正能治疗疾病的植物进入药物开发之后，就以药丸、片剂和胶囊的形式出现。但要强调的另一方面是免疫系统。我们有必要加强自身的免疫系统。与其把重点放在治疗疾病上，倒不如尽我们所能把预防作为优先事项。这意味着加强和抵御细菌与病毒攻击的能力，这是通过选择生活方式来实现的，包括锻炼和吃健康的饮食，而不是服药。第一种选择是将预防医学作为优先事项，重点关注每个人的健康，包括地球的健康，这些都至关重要。这意味着要选择像药物一样的食物，以及一种提高身体机能的生活方式，同时确保孩子们能够在有清洁空气和优质饮用水的环境中成长。增强免疫系统的天然补充品需要森林和其他生态系统欣欣向荣，需要生物多样性再次走上进化的道路。

动手能力
Capacity to Implement

你家乡有多少种自然疗法？去药房看一看，做个表格，列出哪些是天然的，哪些是化学合成的。看看你和朋友、家人服用的化学药物中有哪些能找到天然替代品。当你找到天然替代品时，看一下哪些是本地就有的，哪些是工厂制造的。最后，问问那些生产药品的人，他们是本地公司还是跨国公司。现在开始注重预防医学，了解你可以在家里种植的天然药物。和你的朋友、家人讨论一下，这有助于引导你们将来使用更多的当地疗法。

教师与家长指南

故事灵感来自
This Fable Is Inspired by

安妮·麦金太尔
Anne McIntyre

安妮·麦金太尔曾在英国坦布里奇韦尔斯的草药学院和英国阿育吠陀学院学习。多年来，她学习顺势疗法和芳香疗法，并成为一名专攻草药的专家。安妮有将近 40 年的临床实践，使用的是东西方医学的结合。她是一位多产的作家，并通过她的出版物分享她的经验，出版了《女性草药大全》《治疗性饮料》《花疗》《儿童草药：西医和阿育吠陀观点》《草药指南》《阿育吠陀圣经和传统配药：使用印度和西方草药的阿育吠陀疗法指南》。她在英国格洛斯特郡的科茨沃尔德有一个美丽的药用植物园。

图书在版编目（CIP）数据

冈特生态童书.第八辑：全36册：汉英对照 /
（比）冈特·鲍利著；（哥伦）凯瑟琳娜·巴赫绘；
何家振等译. —上海：上海远东出版社，2021
ISBN 978-7-5476-1773-1

Ⅰ.①冈… Ⅱ.①冈…②凯…③何… Ⅲ.①生态环境－环境保护－儿童读物—汉、英 Ⅳ.①X171.1-49

中国版本图书馆CIP数据核字（2021）第249940号

策　　划　张　蓉
责任编辑　程云琦
封面设计　魏　来　李　廉

冈特生态童书
发烧树
［比］冈特·鲍利　著
［哥伦］凯瑟琳娜·巴赫　绘
章里西　译

记得要和身边的小朋友分享环保知识哦！
八喜冰淇淋祝你成为环保小使者！

Health 277

臭臭的？健康的！

Smelly but Healthy

Gunter Pauli

[比]冈特·鲍利 著
[哥伦]凯瑟琳娜·巴赫 绘
章里西 译

上海远东出版社

丛书编委会

主　任：贾　峰
副主任：何家振　闫世东　林　玉
委　员：李原原　祝真旭　牛玲娟　梁雅丽　任泽林
　　　　王　岢　陈　卫　郑循如　吴建民　彭　勇
　　　　王梦雨　戴　虹　翟致信　靳增江　孟　蝶

特别感谢以下热心人士对童书工作的支持：

匡志强　宋小华　解　东　厉　云　李　婧　陈　果
刘　丹　熊彩虹　罗淑怡　旷　婉　杨　荣　刘学振
何圣霖　廖清州　谭燕宁　韦小宏　李　杰　欧　亮
陈强林　王　征　张林霞　寿颖慧　罗　佳　傅　俊
胡海朋　白永喆　冯家宝

目录

臭臭的？健康的！	4
你知道吗？	22
想一想	26
自己动手！	27
学科知识	28
情感智慧	29
艺术	29
思维拓展	30
动手能力	30
故事灵感来自	31

Contents

Smelly but Healthy	4
Did You Know?	22
Think about It	26
Do It Yourself!	27
Academic Knowledge	28
Emotional Intelligence	29
The Arts	29
Systems: Making the Connections	30
Capacity to Implement	30
This Fable Is Inspired by	31

生姜和小豆蔻两种香料正在厨房里讨论健康食品。他们对能够帮助人们保持健康而感到非常自豪。生姜说道：

"我很敬佩你跟我一样能帮助人们保持身材。但你告诉我，你吃了多少大蒜？你的口气……唉，实在让人难以忍受。"

"我？臭吗？怎么可能！人们明明是用我来掩盖他们吃大蒜的味道。"

Two spices, ginger and cardamom, are in the kitchen discussing healthy food and how proud they are of helping people stay fit, when the ginger says,

"I am impressed that you too help people stay in shape. But tell me, how much garlic have you eaten? Because your breath … well, it is quite unbearable."

"Me? Smelly? Impossible! People use me to cover up the smell of the garlic they eat."

……生姜和小豆蔻正在厨房里……

... ginger and cardamom, are in the kitchen ...

我就是喜欢大蒜!

I just love garlic!

"对对对,你和我都有消除口臭的功效。但我必须告诉你,今天你嘴里的蒜味太难闻了。"

"对不起,但我就是喜欢大蒜!而且我真的很需要它。三天不吃蒜,走路打蹿蹿!"小豆蔻说。

"那为什么我们都得忍受这种气味呢?因为你,整个房间都散发着大蒜味。我们能不能打开窗户透透气?"

"Yes, you and I are able to control bad smells. But I must tell you, your garlic breath is rather overpowering today."

"My apologies, but I just love garlic! And I really need it. Can't start my day without it," Cardamom says.

"But why should we all suffer the smell? The whole room is reeking of garlic because of you. Could we please open a window and let in some fresh air?"

"哦，得了吧！我肯定不是唯一喜欢大蒜的人。大蒜很容易种植和收割，所以到处都在种，很容易高产。"

"不过，你还是需要多体谅你周围的人。有些人真的受不了这种气味。请问，你吃完后会用冷水漱口吗？"

"Oh, come on! I am certainly not the only one who loves garlic. It is grown everywhere, as it is easy to plant and harvest, and it grows so abundantly."

"Still, you need to be more considerate of those around you. Some of us really cannot stand the smell. And may I ask, do you rinse your mouth with cool water after eating it?"

……很容易种植……

... it is easy to plant ...

我还会刮舌头……

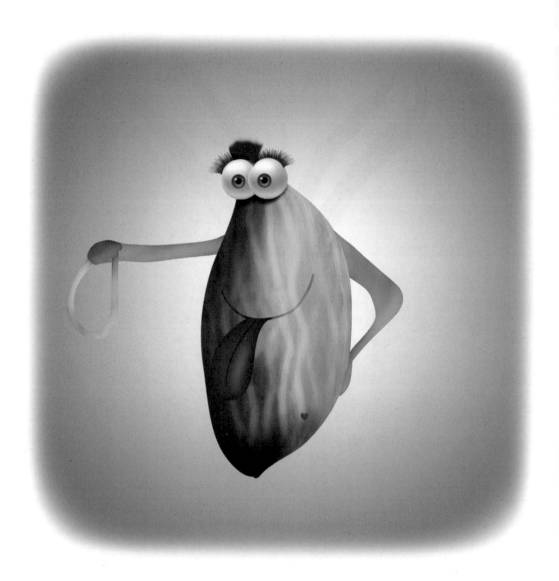

I scrape me tongue ...

小豆蔻回答说:"当然,而且我还会刮舌头。"

"口臭和大蒜味还是有区别的。哦,不应该叫蒜味,应该叫那种从你嘴里冒出来的'毒气'。"

"我可没口臭!我只是很喜欢吃蒜。对我来说,它闻起来就像久旱之后的雨水一样让人舒服。"

"Of course I do, and I scrape me tongue," Cardamom replies.

"There is a difference between bad breath and the smell of garlic. Or, should I say, those awful fumes coming out of your mouth."

"I don't have bad breath! I just enjoy plenty of garlic. To me, it smells as good as the rain does after a long drought."

"要真是这么好闻就好了……你想让朋友们留在你身边的话,至少嚼一些你自己的种子,或者柠檬皮。以你现在散发的味道,没人愿意靠近你。"

"谢谢你给我建议,帮我控制不好的气味。你真好,大多数人都会直接强迫我不再吃大蒜。"

"If only it really did… Chew some of your own seeds, or some lemon rind, at least, if you want your friends to stick around. The way you smell now, no one will want to come near you."

"Thank you for making some suggestions to help me control the odour. That is very kind of you. Most people would try and force me to stop eating garlic."

嚼一些你自己的种子……

Chew some of your own seeds ...

……对抗流感？

... fights the flu?

"我绝不会那么做,因为我知道这对你很有好处。但是,我建议你不要再文绉绉地叫它'不好的气味'了。这明明就是恶臭,我可受不了!"

"但大蒜营养丰富,几乎没有热卡。你知道它还能降低血压和对抗流感吗?"

"I would never do that as I know it's very good for you. But, I suggest you stop calling it an odour. It's a really bad smell, and one I can't stand!"

"But garlic has so much nutrition, and hardly any calories. And did you know that it decreases blood pressure and fights the flu?"

"嗯，如果我是病毒或细菌的话，绝不会冒险接近你这么臭的人。难怪你的肺会这么健康。"

"更重要的是，这种气味可以防止昆虫落在我的皮肤上，吸我的血，往我体内灌满致病的寄生虫，"小豆蔻夸耀道。

"Well, if I were a virus or bacteria, I would never venture anywhere near someone as smelly as you. No wonder your lungs are so healthy."

"And what is more, it keeps insects from landing on my skin, sucking my blood and filling me with parasites that make me sick," Cardamom boasts.

……可以防止昆虫落在我的皮肤上。

... it keeps insects from landing on my skin,

……把我当茶喝。

... serve me as a tea.

"你需要更多科学依据来证明这些说法。我们生姜也能够预防许多疾病。这就是为什么人们把我加到他们的食物里，或者把我当茶喝。但我不会到处宣称我可以治愈所有的病。"

"你知道在古希腊奥运会上，大蒜被推荐给运动员以提高他们的成绩吗？而且事实证明，如果你的血液中含有重金属，大蒜可以保护你的器官免受其害，而且……"

"You'll need more science to prove such claims. We ginger plants also help prevent many illnesses. That is why people add me to their food or serve me as a tea. But I do not go around claiming I can cure anyone."

"Did you know that garlic was recommended to athletes to improve their performance at the Olympic Games in Greece? And it has been proven that if you have heavy metals in your blood, garlic protects your organs, and …"

"我明白了!很显然你是大蒜的忠实粉丝,也是一名超棒的推销员。我决定给鼻子塞上塞子,然后每天吃几瓣大蒜。"

"太好了!我们俩一起臭臭的不是更好吗?"

……这仅仅是开始!……

"I get it! You clearly are garlic's greatest fan, and a great salesperson too. I will simply put a peg on my nose and start eating a few cloves every day myself."

"Good! Isn't it better if we smell bad together?"

... AND IT HAS ONLY JUST BEGUN!...

……这仅仅是开始！……

...AND IT HAS ONLY JUST BEGUN!...

你知道吗?

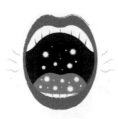

大蒜素

Bad breath is caused by bacteria proliferating in cavities in the mouth. Garlic's antimicrobial activity has been recognised for many years. The active component is allicin, which creates a unique odour.

口腔异味是由口腔内细菌增殖引起的。大蒜的抗菌活性已被认可多年。其活性成分是大蒜素,能产生一种独特的气味。

Egyptians, Babylonians, Greeks, Romans and Chinese cultures all included garlic into their cuisine for its health and medicinal properties.

在埃及、巴比伦、希腊、罗马和中国文化中都将大蒜纳入饮食,因为大蒜具有健康和药用的特性。

Chopped garlic has antimicrobial effects in ground beef and raw meatballs, reducing the colonies of bacteria, yeast, and moulds during storage. Garlic protects organs against damage from heavy metals.

碎牛肉和生肉丸中的碎大蒜有抗菌作用，可以在储存过程中减少细菌、酵母和霉菌的菌落。大蒜可以保护器官免受重金属的损害。

Ginger is used as an ingredient in food, pharmaceutical products, and cosmetics. Garlic and ginger extracts in hot water have the best anti-fungal activities.

生姜被用作食物、药品和化妆品的原料。大蒜和生姜热水提取物的抗真菌活性最好。

Several spices, such as clove, oregano, thyme, cinnamon, and cumin exhibit antimicrobial activities, reducing food spoilage bacteria. Some are even effective at controlling antibiotic resistant microorganisms (MRSA).

一些香料如丁香、牛至、百里香、肉桂和孜然，具有抗菌活性，可以减少食物中的腐败细菌。有些甚至能有效控制耐抗生素微生物（MRSA）。

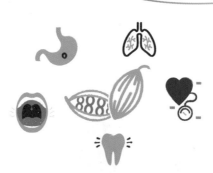

Cardamom reduces blood pressure, helps with ulcers, controls bad breath, and even prevents cavities, while improving breathing. It originally comes from India but Guatemala is now the world's leading exporter.

小豆蔻可以降低血压，辅助治疗溃疡，控制口臭，甚至防止蛀牙，同时改善呼吸。小豆蔻最初来自印度，但危地马拉现在是世界上最大的出口国。

Garlic was used to provide strength and increase the work capacity of labourers building the pyramids in Egypt. Garlic was used in ancient Greece, as one of the earliest "performance enhancing" agents.

埃及建造金字塔时，大蒜被用来增强劳动者的体力，提升其劳动能力。在古希腊，大蒜被用作最早的"性能增强剂"。

The use of garlic as a food and as a medicinal agent in Asia dates back to 2000 BC. Garlic was in wide use in China, particularly when consumed together with raw meat.

在亚洲，大蒜作为食物和药品使用可追溯到公元前2000年。大蒜在中国被广泛使用，特别是与生肉一起食用。

Are you ready to smell bad to be healthy?

你准备好为了保持健康而去忍受难闻的气味吗?

Do you like the smell of garlic?

你喜欢大蒜的味道吗?

Would you continue doing something you like, even when your friends do not like it?

即使朋友不喜欢,你也会坚持做自己喜欢的事情吗?

Is there a difference between bad breath and a garlic smell?

口臭和大蒜味有区别吗?

Ask around to see who likes the taste of garlic in their food, and who likes the smell that lingers on one's breath. Also speak to those who do not like the taste or smell. Listen to their arguments, and list the positive as well as negative comments, before explaining the health benefits to them. Next, find out if anyone is flexible and accommodating, and willing to change their opinion or follow a new line of thought, once they are made aware of the health benefits of consuming garlic. Summarise your findings and share the information with friends and family members.

问问周围的人，谁喜欢食物中的大蒜味，谁喜欢呼吸中那股挥之不去的大蒜味。也可以和那些不喜欢大蒜味的人交流。听听他们的观点，列出积极和消极的评论，然后解释大蒜对他们健康的好处。接下来，看看有没有人在知道吃大蒜对健康的好处后，能够灵活适应，愿意改变自己的观点或尝试新的思路。总结你的发现，并与朋友和家人分享。

TEACHER AND PARENT GUIDE

学科知识
Academic Knowledge

生物学	大蒜是一种世界各地都种植的草本植物；大蒜与洋葱、大葱和韭菜有关；小豆蔻有一种强烈独特的味道，具有浓郁的树脂芳香；生姜是一种多年生草本植物，和姜黄、小豆蔻一样都属于姜科；小豆蔻味道辛辣，有一种甜美带有刺激性的香气，还带有一点柠檬和薄荷的味道。
化 学	生姜中具有抗菌活性的挥发性化合物包括蒎烯、冰片、莰烯和芳樟醇等；大蒜含有大蒜素，即二烯丙基硫代亚磺酸酯；大蒜中强烈的化合物已经演化成一种防御机制，阻止鸟类、昆虫和蠕虫啃食；在汗液和呼吸中发现大蒜味；小豆蔻籽油能杀死细菌和真菌。
物 理	当大蒜的细胞被剥碎、咀嚼或捣碎时，细胞中储存的酶会引起细胞液中几种含硫化合物的分解。
工程学	气味工程包括气味排放采样、气味样本分析、环境气味监测；嗅觉技术是处理嗅觉表征的工程学科；环境流体力学。
经济学	中国的大蒜产量占全球80%；由于古希腊和古罗马的需求，小豆蔻贸易发展成为一种奢侈品行业；小豆蔻的种植是高度劳动密集型的，每公顷土地都需要高度的维护，因此是一个创造高就业机会的产业；小豆蔻是目前世界第三昂贵的香料，仅次于藏红花和香草。
伦理学	每个人对不同气味都有不同的记忆，因此我们需要尊重喜欢或不喜欢某些气味的人；气味基因工程可以人工改变人的情绪。
历 史	5 000多年前大蒜就在世界各地广泛种植；人们在图坦卡蒙（公元前1325年）的坟墓中发现了保存完好的大蒜；小豆蔻出现在苏美尔和阿育吠陀文献中；公元前3000至公元前1500年，生姜跨太平洋运输；琳达·巴克和理查德·阿克塞尔因对嗅觉系统的研究而获得2004年诺贝尔生理学或医学奖。
地 理	大蒜起源于西伯利亚；如果种植适当，可以把大蒜种到北至阿拉斯加的边远地区；酸奶黄瓜是一种混合了大蒜和盐的酸奶，是东地中海菜肴中常见的酱汁；沙特阿拉伯和科威特是世界上最大的豆蔻进口国。
数 学	有一种算法可以对分子的气味进行还原，根据分子自身的结构推算出它的气味。
生活方式	在橄榄油中乳化大蒜可以制作蒜泥蛋黄酱；嗅觉缺失症或嗅觉丧失；从1980年开始，嗅觉艺术被认定为一种艺术。
社会学	嚼小豆蔻可以清新口气，还可以帮助消化；小豆蔻作为药用已有几千年历史。它最常用来治疗消化不良、哮喘和口臭；在中东烹饪中，小豆蔻常被混入咖喱粉中，也是典型香料混合物中使用的香料之一；在亚洲，小豆蔻经常被用来制作一种传统饮料——小豆蔻茶。
心理学	嗅球负责处理气味，直接连接大脑中负责情感反应和记忆的杏仁核和海马体；事实上，人们更喜欢能引起强烈记忆的气味，因为嗅觉与记忆紧密相连，每个人的偏好都不一样。
系统论	嗅觉系统不仅仅是指嗅球，还包括鼻孔、骨头、鼻腔以及覆盖在黏膜、薄膜、腺体、神经元和纤维上的组织层。

教师与家长指南

情感智慧
Emotional Intelligence

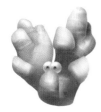

生姜

生姜对大蒜的气味有强烈的反应,她知道小豆蔻控制气味的能力,但希望小豆蔻更体贴,对他人的感受更在意些。生姜想为小豆蔻提供减少他人不适的方法。她认为,更好地控制大蒜的气味将有利于小豆蔻,否则人们不愿接近他,也不愿与他交往。生姜的方法不是规定性的,也没有多大说服力。她只是提出了一些建议。当讨论到健康时,生姜说小豆蔻令人讨厌的大蒜味给她留下深刻的印象。生姜提醒小豆蔻不要做出没有科学依据的结论。最后生姜承认,证明大蒜诸多好处的论据很充分。

小豆蔻

小豆蔻表示歉意,承认吃大蒜是他日常生活的一部分,但他不准备改变这一点。小豆蔻赞美大蒜,对这种容易生长且高产的植物有一种特殊情感。小豆蔻解释了他为减少大蒜味而采用的步骤。小豆蔻很清楚,需要把大蒜的气味与它的健康特性和驱蚊能力区分开。当生姜认可小豆蔻是最棒的推销员时,他的自尊心得到提升。小豆蔻最后说最好两个人都有大蒜味,这样他们之间就不会相互影响。

艺术
The Arts

让我们来看看嗅觉艺术。气味有不同的种类,对它的鉴赏非常个人化。你能否为棋盘上的每个棋子赋予不同的气味,一边享受这些气味一边玩游戏?或者可不可以制作一个盒子,装入12种不同气味的东西?玩一个识别气味的游戏,让被试者告诉你喜欢或讨厌哪种气味。科学家已经确定了10种不同的气味类别。去了解一下,并列出你认为可以创造愉快气氛的气味。这是一门新兴的气味艺术。

TEACHER AND PARENT GUIDE

思维拓展
Systems: Making the Connections

　　气味和嗅觉科学与光和声学区别非常大。世界上有3种原色和5种基本味道，研究人员提出，气味可分为10种基本类型，但对它们的定义还远不够精确。我们很难知道一种分子的哪些部分造成了它的气味。松树或新割的草的木质气味属于一类。甜味如焦糖、巧克力、香草、花香和香水是另一类。更复杂的香气，比如烤面包或刚煮好的咖啡，最好是被描述为两种或两种以上香气的组合。我们知道光的波长决定了它的颜色，但是一种分子的气味可能取决于它所含的碳原子数量，它的稳定程度，以及它伸出的支链。气味以化学感知的形式对大脑进化产生了巨大的影响。嗅觉研究是理解大脑进化的重要途径，将其局限于食物和香水，就好比是要理解颜色视觉而只研究红色，对蓝色和绿色均不闻不问一样。所需的研究显然是多学科的，从心理学到化学、农业和人工智能。有关植物气味的生物化学仍然是一个相对较新的研究领域。气味和嗅觉是在嗅球中处理的。这与杏仁核和海马体有直接联系，是大脑中提供情感反应和记忆的部分。人们更喜欢能唤起强烈回忆的气味。因此，气味的"基因工程"有一个主要的道德缺陷：它可以通过人工手段控制人们的记忆和情绪。这是一个需要仔细研究的研究领域。

动手能力
Capacity to Implement

　　雨的味道！下次久旱之后下雨时，出去在雨中跑跑步。先仔细闻一闻房子里所有东西的味道，然后穿着雨衣在外面淋雨时，再特别注意一下雨的味道。是不是让人愉快？是什么引起了雨的气味？观察一下尘土、岩石和小草，辨别出三种不同的气味，以及让你最愉悦的气味，庆祝期盼已久的雨的到来。写下你的观察并与家人和朋友分享。

教师与家长指南

故事灵感来自
This Fable Is Inspired by

伊莎贝尔·乔伊·贝尔
Isabel Joy Bear

伊莎贝尔·乔伊·贝尔17岁时开始在科学与工业研究组织（CSIRO）的化学实验室担任实验室助理。她参加了墨尔本技术学院（现在的皇家墨尔本理工大学）夜校的学习，在那里获得应用化学和应用科学的文凭。由于没有达到加入CSIRO的标准，她被迫换了工作。后来，伊莎贝尔搬到英国，1978年她获得维多利亚学院固态化学博士学位。

1953年，她重新加入了CSIRO，并在接下来的40年里一直致力于矿物化学研究。1964年，伊莎贝尔·乔伊·贝尔和她的同事迪克·托马斯成为第一个用科学方法描述雨水气味的人。她注意到雨水的气味与一种淡黄色的油有关，这种油被称为"潮土油"。科学家们随后将这一情况可视化，展示了降雨期间释放的气味。

图书在版编目（CIP）数据

冈特生态童书.第八辑：全36册：汉英对照 /
(比)冈特·鲍利著；(哥伦)凯瑟琳娜·巴赫绘；
何家振等译.—上海：上海远东出版社，2021
ISBN 978-7-5476-1773-1

Ⅰ.①冈… Ⅱ.①冈… ②凯… ③何… Ⅲ.①生态环境－环境保护－儿童读物—汉、英 Ⅳ.①X171.1-49

中国版本图书馆CIP数据核字(2021)第249940号

策　　划　张　蓉
责任编辑　程云琦
封面设计　魏　来　李　廉

冈特生态童书

臭臭的？健康的！

[比]冈特·鲍利　著
[哥伦]凯瑟琳娜·巴赫　绘
章里西　译

记得要和身边的小朋友分享环保知识哦！
八喜冰淇淋祝你成为环保小使者！

Health 276

大地之肠

The Earth's Intestines

Gunter Pauli

[比]冈特·鲍利 著
[哥伦]凯瑟琳娜·巴赫 绘
章里西 译

上海远东出版社

丛书编委会

主 任：贾 峰

副主任：何家振 闫世东 林 玉

委 员：李原原 祝真旭 牛玲娟 梁雅丽 任泽林
　　　　王 岢 陈 卫 郑循如 吴建民 彭 勇
　　　　王梦雨 戴 虹 翟致信 靳增江 孟 蝶

特别感谢以下热心人士对童书工作的支持：

匡志强 宋小华 解 东 厉 云 李 婧 陈 果
刘 丹 熊彩虹 罗淑怡 旷 婉 杨 荣 刘学振
何圣霖 廖清州 谭燕宁 韦小宏 李 杰 欧 亮
陈强林 王 征 张林霞 寿颖慧 罗 佳 傅 俊
胡海朋 白永喆 冯家宝

目录

大地之肠	4
你知道吗?	22
想一想	26
自己动手!	27
学科知识	28
情感智慧	29
艺术	29
思维拓展	30
动手能力	30
故事灵感来自	31

Contents

The Earth's Intestines	4
Did You Know?	22
Think about It	26
Do It Yourself!	27
Academic Knowledge	28
Emotional Intelligence	29
The Arts	29
Systems: Making the Connections	30
Capacity to Implement	30
This Fable Is Inspired by	31

一条蚯蚓和一个海参正在争论谁是真正的"大地之肠"。

"好吧，今天就让我们一了百了吧！伟大的古希腊智者亚里士多德谈到'大地之肠'时，他说的是我。他甚至不知道你的存在，"蚯蚓说。

An earthworm and a sea cucumber are debating the issue of what the real intestines of the Earth are.
"Well, let's settle this once and for all! When the great Greek sage Aristotle talked about 'The Earth's Intestines', he talked about me. He didn't even know you existed," Earthworm says.

……古希腊智者亚里士多德谈到……

... Greek sage Aristotle talked about ...

瞧，你生活在陆地，我生活在海洋……

Look, you live on land and I live in the sea ...

"你们的'智者'显然也不是无所不知。瞧,你生活在陆地,我生活在海洋。我们做着同样的工作,由于海洋覆盖地球的面积比陆地大得多,'大地之肠'这个名字当然就是我的了!"

"Your wise man clearly did not know everything. Look, you live on land and I live in the sea. We do the same job, and as the sea covers much more of the globe than land does, surely the name 'The Earth's Intestines' is mine!"

"或许你是对的。但是几千年前，中国人在他们的医学名著中就把我描述为味咸性寒。那时他们就知道蚯蚓对心脏和血液循环有好处。"

"有意思……"海参回答道。"中国人在同一本书里也写到了我。我被形容为味咸性温，是治疗肾和关节疾病的良药。"

"You may be right. But thousands of years ago, the Chinese described me as being salty and cold, in their famous book on medicine. They knew that I am good for the heart and for blood circulation then already."

"Interesting…" Sea Cucumber replies. "The Chinese wrote about me too, in the same book. I'm described as salty and warm, and therefore good medicine for the kidneys and the joints."

中国人把我描述为……

Chinese described me as ...

……具有独特的断后重生能力……

... unique ability to regrow body parts ...

"你不要以为我的好处就只是肥沃土地和提供药物。我还是为数不多的具有独特的断后重生能力的生物之一，"蚯蚓自豪地说。

"哦，了不起！但是你听听这个：当捕食者出现的时候，我可以把我的一些器官从肛门排出，很快这些器官又会重新长出来。"

"And you mustn't think that all I am good for is enriching the soil and providing medicine. I am one of the few creatures with the unique ability to regrow body parts," Earthworms boasts.

"Oh, that's great! But listen to this: When a predator appears, I can push some of my organs out through my guts, and then regrow them, in no time."

"这太令人震惊了！你知道吗？我可以轻易改变我的饮食习惯，甚至吃不好的食物，但仍然可以正常消化任何食物。"

"什么都吃吗？人类扔掉的石化产品或有毒化学品也吃吗？"

"That is astounding! And did you know that I can easily change my diet, and even eat bad food, and still get the right mix to digest anything?"

"Eat anything? Even the petroleum or toxic chemicals people throw away?"

......甚至吃不好的食物......

... even eat bad food ...

10代人?

Ten generations?

"嗯，这么做也许要花些时间，但并非不可能。这可能需要我们10代人的努力，但最终我们将会把污染最严重的地方也清理干净，帮助环境恢复正常。"

"10代人？这意味着，在你的有生之年，可能看不到自己努力工作的成果……"

"Well, it may take us a while to so, but it is not impossible. It could take ten generations of us, but we will clean up even the most polluted places, and help the environment get back to normal."

"Ten generations? That means you won't see the results of your hard work in your own lifetime…"

"但我的后代可以看到!我们一年到头不停地在自己的茧里产卵。我只能活5年左右,但这足以让我为更美好的世界添砖加瓦了。"

"这么说,你现在不仅想要'大地之肠'的称号,还想被称为'土壤废物管理大师'?"

"But my offspring will! We lay lots of eggs all year around, inside of their own cocoon. I only live for about five years, but that is long enough for me to help make the world a better place for everyone."

"So, now you not only want the title of 'The Earth's Intestines', but also want to be known as 'Master Waste Manager of the Soil'?"

"土壤废物管理大师"……

"Master Waste Manager of the Soil"...

我们不管理废物！

We don't manage waste!

"我们不管理废物！我们是改进者，我们是创造者，我们提供空气，我们生产食物。我们要确保在这个人类污染得如此厉害的世界上，所有生命形式都有更好的生存机会。"

"这真的很棒。但我们也扮演着重要的角色。我可以提议把我们并称为'大地之肠'吗？海洋里的海参，陆地上的蚯蚓！"

"We don't manage waste! We improve, we create, we provide air, and our castings are food. We ensure that all life forms have a better chance of survival in this world where people pollute so much."

"That is very impressive. But we play an important role too. May I suggest that we declare both of us as 'The Earth's Intestines'? The sea cucumber for the ocean, and the earthworm on land!"

"给不给头衔无所谓了,由于我们过去的贡献而得到认可并不重要。我们需要清楚,为了能给所有物种创造一个更美好的世界,我们现在还能做些什么。"

……这仅仅是开始!……

"It is time to be generous with titles, as in the end the recognition we receive for what we did in the past does not matter – we need to be clear about what we can do now to secure a better world for all."

... AND IT HAS ONLY JUST BEGUN!...

……这仅仅是开始!……

... AND IT HAS ONLY JUST BEGUN! ...

Did You Know?
你知道吗?

Sea cucumbers extract oxygen from water in a pair of "respiratory trees" that branch in the cloaca just inside the anus. Sea cucumbers "breathe" by drawing water in through the anus and then expelling it.

海参通过一对"呼吸树"从水中提取氧气,这对"呼吸树"位于肛门内的泄殖腔内。海参"呼吸"的方式是先从肛门吸入水,然后再将水排出体外。

Sea cucumbers have a water vascular system that provides hydraulic pressure to its tentacles, allowing them to move. This is connected to over 100 small miniature hearts to pump blood.

海参有一个水血管系统,可以为其触手提供水压,使其能够移动。它连接着100多个小型心脏来输送血液。

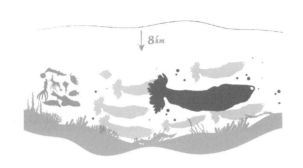

Sea cucumbers are found in great numbers on the deep seafloor, where they make up the majority of the animal biomass. At depths of +8 km, sea cucumbers comprise 90% of the total mass of the macro-fauna.

海参在深海中大量存在，在那里它们构成了动物生物量的大部分。在海洋 8 000 多米深处，海参占大型动物种群的 90%。

In New Zealand, the strawberry sea cucumber lives on rocky walls on the southern coast of the South Island, where populations reach densities of 1,000 animals per square meter, earning the name "strawberry fields".

在新西兰，草莓海参生活在南岛南海岸的岩壁上，这里的种群密度达到每平方米 1000 只，因此得名"草莓田"。

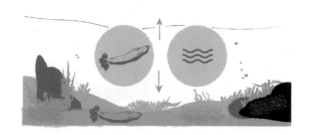

Some sea cucumbers have nearly the same density as seawater, so they make long "jumps" of up to 1,000 m high, before slowly falling back to the ocean floor.

有些海参的密度几乎和海水一样，所以它们可以"跳"1000米高，然后慢慢地落回海底。

Sea cucumbers scavenge on debris and can process 19 kilograms of sediment per square meter per year, ensuring that dissolved oxygen remains widely available.

海参以垃圾为食，每年每平方米可以处理19千克沉积物，确保溶解氧仍然广泛可用。

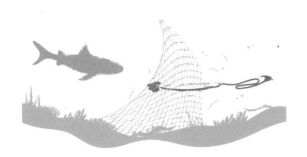

Sea cucumbers can discharge sticky threads to entangle predators. Others mutilate their bodies by violently contracting muscles and expel internal organs out of their anus. These parts are quickly regenerated.

海参可以释放出黏性的丝线来缠绕捕食者。另一些海参则通过剧烈收缩肌肉将内脏通过肛门排出的方式肢解自己的身体。被排出的部分很快就再生恢复了。

Sea cucumbers are bottom feeders eating plants, algae, and animals such as shellfish, crabs, crayfish, sea anemones, starfish, snails, and bristle worms. All feed comes from the sea floor.

海参是底栖动物，以植物、藻类和动物为食，如贝类、螃蟹、小龙虾、海葵、海星、蜗牛和毛足虫。所有的食物都来自海底。

Would you like to eat all the waste from the floor?

你会把地上的垃圾都吃光吗?

How important is it to keep your room clean?

保持房间整洁有多重要?

Shall we manage waste or improve the quality of leftovers?

我们是仅仅管理垃圾,还是应该将它们变废为宝呢?

Is it important to recognise the best?

识别出最好的是否重要?

Let's make a calculation that shows the real importance of the sea cucumber. The Earth consists of 30% land and 70% sea. Earthworms are found only up to 30 cm deep in the ground. Sea cucumbers form 90% of all living biomass 8 kilometres below the surface of the sea. So the question is: how many more biomass weight of sea cucumbers than earthworms are there around the globe? The conclusion you draw from this calculation may astound you!

让我们做一个计算来显示海参真正的重要性。地球表面由30%的陆地和70%的海洋组成。蚯蚓生活在地下30厘米深的地方。海参在海平面以下8千米处占生物总量的90%。所以问题是：在全球范围内，海参的生物量比蚯蚓多多少？从这一计算中得出的结论会让你大吃一惊！

TEACHER AND PARENT GUIDE

学科知识
Academic Knowledge

生物学	海参的身体可以围绕着一个中轴被分成五个几乎相同的部分。
化 学	海参蛋白质含量高（高达40%），脂肪含量极低（1%）；甘氨酸是海参的主要氨基酸；海参增加海水pH值；海参的碳酸钙有助于珊瑚的形成，而它的氨则促进珊瑚的生长，就像肥料一样。
物 理	海参有一个液压系统，用于移动、饮食和废物运输以及呼吸；海参通过调节足部的水压来移动。
工程学	液压吸盘，灵感来自海参；以海参为核心的综合多营养水产养殖（IMTA），整合不同营养水平的生物混合培养，让各类水产养殖废弃物创造价值。
经济学	对新药和农药的研究正在从陆地转向海洋，由于海洋生物种类繁多，生活在极其恶劣的环境中，产生了多种多样的生物活性化合物；海参是一种具有生物活性的天然产物，如具有抗真菌和抗病毒特性的三萜苷。
伦理学	我们怎么能过度捕捞到令生态系统崩溃的地步？协调好海参保护和海参社会经济重要性之间的关系。
历 史	古代海参可以追溯到5亿年前。
地 理	印度政府在2020年建立了世界上第一个海参保护区；大堡礁海洋公园和加拉帕戈斯群岛国家公园（都是联合国教科文组织世界遗产）见证了海参数量的锐减。
数 学	海参呈放射状对称；海参一次释放多达13万枚卵；利用基于个体的生物能量模型估计海参废弃物生物修复的潜力。
生活方式	如何理解海参的名称：法语的bêche-de-mer，葡萄牙语的bicho do mar（字面意思是"海洋动物"）。
社会学	海参在东亚和东南亚被认为是一道美味。
心理学	不断增长的人口和清洁饮用水供应不足是影响消费者行为的资源困境或公共困境的一部分；人们对海鲜消费的态度是通过对味道、营养价值、易于制作、熟悉和新鲜度的信念而形成的；大多数只注重信息的方法通常只会提高知识水平，并不能带来所期望的行为改变；自我意识需要掌握和认识自我，理解自我是大脑的一个功能部分；炫耀是大脑试图改善自身形象的举动，只有当你觉得自己的形象有问题时，你才会努力去改善它。
系统论	海参在海洋生态系统中起着关键作用，它们分解碎屑，循环营养物质；过度开采导致海参养殖业的发展，尤其是中国和日本市场；全球海参存量调查表明，在世界许多地区，海参种群正面临着巨大的捕捞压力，需要采取有效的养护措施；从养殖场重新放养海参是对过度捕捞的昂贵补救措施。

教师与家长指南

情感智慧
Emotional Intelligence

蚯蚓自信满满，希望被认为是最好的。她引用古希腊哲学家的观点来证明自己的立场。她还引用中国古代医书作为依据，展示自己擅长的领域。她坚持说，自己做的不仅仅是清洁土壤，还可以做很多。她承认海参的贡献，又举例证明自己做得比海参更多更好。蚯蚓对自己为公共利益和改善所有物种生活所做出的贡献感到自豪。在对话结尾，蚯蚓通过总结所有自己贡献的方式获得认可，这是她自尊心的基础。最后，蚯蚓赞同海参提出的建议，并呼吁海参要专注于行动，而不仅仅是为了获得认可。

海参寻找证据，论证古希腊哲学家可能并不知道海参的存在。当蚯蚓引用中国古代医书肯定自己的地位时，海参提醒她，海参也因其药用价值而被中国古代医学所认可。当蚯蚓继续显示自己的能力时，他也不示弱，补充介绍了自己独特的防御方式。当蚯蚓意识到海参的独特特征时，她又告诉海参她有吃生活垃圾的能力。海参试着将讨论引向更深的程度。他意识到蚯蚓拥有令人印象深刻的能力，遵循"知之为知之，不知为不知，是知也"的逻辑，克制自己不多提及自己非凡的能力。他在结束辩论时表示，蚯蚓和海参都应该得到高度认可。

艺术
The Arts

海参有着如此奇怪和有趣的形状，很容易被认为是外星物种。让我们通过拍摄来制作一张海报，展示至少24种不同的海参。你可以从最受欢迎的10种用于食材和药物的海参开始。接下来，用令人印象深刻的颜色和照片填充海报的其余部分。向人们展示你的海报，以证明海参对人类健康和环境的宝贵贡献。

TEACHER AND PARENT GUIDE

思维拓展
Systems: Making the Connections

海参是海洋生态系统的重要组成部分。它们清洁海底，控制海洋的酸度，并与许多其他海洋生物共享共生关系。大量被开发的海参种类是沉积物食用者，它们从海底收集有机碎屑和沉积物。和蚯蚓一样，海参吞食沉淀物，并在沉淀物穿过肠道时提取有机物。这个"清洗"沉积物中的各种碎片和微生物的过程改善了海洋环境。如果这种活动停止，海底将加速呈现缺乏维持生命的氧气的局面，环境污染就会加重，将不适合生物生存。海参也被称为重要的"生物扰动器"。生物扰动是指生物对沉积层进行颠覆和改造的过程。它混合了表层和次表层沉积物层，增加了渗透性，从而使得沉积物变得更加多孔和松散，进而加强了水—沉积物界面的交换，促成较深沉积层的氧气供应增加。在局部范围内，海参可以通过它们的粪便增加海水的碱性，这意味着海水不会变酸。海参还是许多小物种的宿主，这些小物种要么生活在海参体壁上，要么生活在海参体内。如果没有海参作为宿主，许多这些小生物将无法生存。反过来，这些与海参的共生关系增加了生态系统的生物多样性。海参提供了许多生态系统服务：改善沉积物质量，改变水的化学性质，增加生物多样性，并在海洋食物网向更高营养层次的能量转移中提供链接。过度捕捞导致的海参大规模耗竭影响了海洋的生产力和多样性。保护海参种群所获得的经济收益将大大超过不可持续的开发所获得的利润。

动手能力
Capacity to Implement

设计一个利用蚯蚓清理污染土壤的系统。关键是要相信蚯蚓可以提供这种有价值的服务，而且是石化工业解决方案所不能做到的。详细计算净化土壤的最佳方法，并估计需要多少条蚯蚓才能完成这项工作。

教师与家长指南

故事灵感来自
This Fable Is Inspired by

斯维特拉娜·鲍里索夫纳·恰钦
Svetlana Borisovna Chachina

斯维特拉娜·鲍里索夫纳·恰钦是俄罗斯鄂木斯克国立技术大学生物学博士。她一直在领导一项有关蚯蚓在清理有毒废物中作用的研究，特别是针对来自石油的有毒废物。她的研究证明，被石油污染的土壤中碳氢化合物的含量在蚯蚓和细菌存在的22周后下降了95%。微生物制剂提高了蚯蚓在被石油污染的基质中的存活率。在被石油污染的土壤中，5个月后碳氢化合物含量下降了97%。

图书在版编目（CIP）数据

冈特生态童书.第八辑：全36册：汉英对照 /
（比）冈特·鲍利著；（哥伦）凯瑟琳娜·巴赫绘；
何家振等译.—上海：上海远东出版社，2021
ISBN 978-7-5476-1773-1

Ⅰ.①冈… Ⅱ.①冈…②凯…③何… Ⅲ.①生态环
境－环境保护－儿童读物—汉、英 Ⅳ.①X171.1-49

中国版本图书馆CIP数据核字（2021）第249940号

策　　划	张　蓉
责任编辑	程云琦
封面设计	魏　来　李　廉

冈特生态童书
大地之肠
[比]冈特·鲍利　著
[哥伦]凯瑟琳娜·巴赫　绘
章里西　译

记得要和身边的小朋友分享环保知识哦！
八喜冰淇淋祝你成为环保小使者！

Health
275

无眼 ≠ 无助
Eyeless, not Helpless

Gunter Pauli

[比] 冈特·鲍利 著
[哥伦] 凯瑟琳娜·巴赫 绘
章里西 译

上海远东出版社

丛书编委会

主　任：贾　峰
副主任：何家振　闫世东　林　玉
委　员：李原原　祝真旭　牛玲娟　梁雅丽　任泽林
　　　　王　岢　陈　卫　郑循如　吴建民　彭　勇
　　　　王梦雨　戴　虹　翟致信　靳增江　孟　蝶

特别感谢以下热心人士对童书工作的支持：

匡志强　宋小华　解　东　厉　云　李　婧　陈　果
刘　丹　熊彩虹　罗淑怡　旷　婉　杨　荣　刘学振
何圣霖　廖清州　谭燕宁　韦小宏　李　杰　欧　亮
陈强林　王　征　张林霞　寿颖慧　罗　佳　傅　俊
胡海朋　白永喆　冯家宝

目录

无眼 ≠ 无助	4
你知道吗？	22
想一想	26
自己动手！	27
学科知识	28
情感智慧	29
艺术	29
思维拓展	30
动手能力	30
故事灵感来自	31

Contents

Eyeless, not Helpless	4
Did You Know?	22
Think about It	26
Do It Yourself!	27
Academic Knowledge	28
Emotional Intelligence	29
The Arts	29
Systems: Making the Connections	30
Capacity to Implement	30
This Fable Is Inspired by	31

一条蚯蚓钻出地面去拜访英国农场里的一只兔子。兔子的12只兔宝宝的眼睛仍然闭着,他们浑身无毛,看起来光溜溜的。

"恭喜恭喜!!你这一窝生的真不少啊。"

"谢谢!现在食物充足,是生育的好时机。"

An earthworm makes a trip to the surface to visit a rabbit on an English farm. Her twelve kittens' eyes are still closed, and they look naked without any hair.

"Congratulations! That is quite a brood you have there."

"Thank you. This is a good time to breed, as we have a lot of food around."

……一条蚯蚓钻出地面……

… earthworm makes a trip to the surface …

……我们的确是被罗马人带到这里的……

... we were brought here by the Romans ...

"我听说你有意大利血统。"

"哈哈,你真幽默!是的,我们的确是被罗马人带到这里的,他们饲养我们作为食物。你知道兔子一年可以生12窝吗?"

"这么说来,每年每对兔子可以生144只小兔子!难怪罗马人饲养你们。有你们在,就不会挨饿。"

"I am told you are of Italian descent."
"Oh, that's very funny! Yes, we were brought here by the Romans, who raised us for food. You know we can have twelve litters a year?"
"That means one hundred and forty-four kittens, for every pair, every year! No wonder the Romans bred you. No-one goes hungry with you around."

"不过,喂好管好这些小宝宝是一项相当繁重的工作。幸运的是,他们现在还什么也看不见,所以……"

"你的小宝宝生来就看不见吗?"

"哦,他们能看见,但他们的眼皮还张不开。他们刚出生的时候还没长毛,又冷又渴,就闭着眼睛四处摸索,通过感知我的体温来找到我。"

"It is quite a job to keep these kittens fed and under control, though. Fortunately, they cannot see a thing yet, so …"

"Your babies are born blind?"

"Oh, they can see, but their eyelids are still closed. And as they are still without any hair, and want to stay warm and drink, they feel their way around, finding me by sensing my body heat."

你的小宝宝生来就看不见吗?

your babies are born blind?

我捕捉到鼹鼠的振动……

I pick up a mole's vibrations ...

"对小宝宝们来说，各个感官的发育都至关重要。有些人只知道单纯仰仗他们的眼睛和耳朵生活，却忘记了我们还有其他的感官。"

"你一定也进化出了一些有趣的感官。听说你没有耳朵，却能听见，这是真的吗？"

"啊哈。是的，放心好了，当我捕捉到鼹鼠的振动时，我就会迅速挪到安全的地面上。"

"It is so important for little ones to develop all their senses. Some rely on their eyes and ears only, and forget we are blessed with other senses."

"But you must have developed some interesting senses yourself. Is it true that you have no ears, but you can hear?"

"Ah, rest assured, when I pick up a mole's vibrations, I quickly move to safety on the surface."

"但我曾见过海鸥拍打地面,于是你们蚯蚓被引诱出来然后被吃掉,是这样吗?"

"是的,因此我们是会失去一些伙伴。但如果我们能在一个月内生下几千个宝宝,就会有很多蚯蚓能存活下来尽其职责。"

"每个月生1000个小宝宝!你赢了!但是你没有腿怎么走路呢?你的表皮就像我那些没毛的宝宝一样滑。"

"But I have seen gulls tap on the ground, and then you earthworms come out and get eaten?"

"Yes, we lose some worms that way. But as we can have thousands of babies in just a month there will be many earthworms left, offering their services."

"One thousand little worms every month! You beat me! But how do you get around without any legs? You are as slippery as my little ones with no hair."

我曾见过海鸥拍打地面……

I have seen gulls tap on the ground ...

……你会发现我有很多毛发。

... you'll see I have a lot of hair.

"嗯，我有一个惊喜给你。如果你仔细看，你会发现我有很多毛发。"

"你是用你的毛发走路吗？从来没听过这么魔术的玩意……"

"这不是魔术，而是实实在在的能力。你猜怎么着？我甚至能通过皮肤尝到味道！"

"Well, I have a surprise for you. If you look very closely, you'll see I have a lot of hair."

"Are you walking on your hair? Never heard of such a magical trick…"

"It is not magic. It is very practical. And guess what? I can taste through my skin!"

"显然是的,因为你没有舌头。但你能尝出食物是好是坏吗?"

"当然!即便我的大脑尺寸有限,我也能知道什么是有毒的,什么是安全的。"

"那没有鼻子,你怎么呼吸?"兔子问。

"Obviously, as you have no tongue. But can you taste if food is good or bad?"
"Sure! Even with my tiny brain, I know what is toxic and what is safe."
"And not having a nose, how do you breathe?" Rabbit asks.

......你怎么呼吸?

... how do you breathe?

……我皮肤上的湿东西能吸收空气……

... wet stuff on my skin absorbs air ...

"我皮肤上黏糊糊的湿东西能吸收空气。"

"所以这就是你呆在地下的原因,这样你就不会干燥和窒息。你知道如何在各种条件下生存,我明白了。但是如果下雨怎么办?你不会被淹死吗?"

"我们已经准备好应对遇到的任何事情。我们甚至可以透过皮肤看东西。"

"The slimy wet stuff on my skin absorbs air."
"So that is why you stay underground, so you won't dry out and suffocate. You know how to survive under all conditions, I see. But what if it rains? Won't you drown?"
"We are ready to deal with whatever comes our way. We can even see through our skin."

"你在跟我开玩笑吧!透过皮肤看东西?"

"这是大自然的又一个奇迹!我们可能没有眼睛,但我们肯定不是无助的。"

……这仅仅是开始!……

"You've got to be kidding me! Seeing through your skin?"

"That is just another marvel of Nature! We may be eyeless, but we are certainly not helpless."

... AND IT HAS ONLY JUST BEGUN!...

……这仅仅是开始！……

... AND IT HAS ONLY JUST BEGUN! ...

Did You Know? 你知道吗？

Rabbits breed and grow so quickly that one pair can produce more than 300 kg of meat in a year. Compare that to the yield of 200 kg for an average year-old beef steer.

兔子繁殖和生长非常快，一对兔子一年可以产出300多千克的肉。相比之下，一头普通的一岁肉牛的产量是200千克。

A rabbit needs 4 kg of feed to make 1 kg of meat. In comparison, beef cattle need 7 kg of feed, or more, to create 1 kg of meat. Rabbits have less cholesterol and fat than chicken, beef, lamb or pork.

一只兔子吃4千克饲料产1千克肉。相比之下，肉牛需要7千克甚至更多的饲料才能产出1千克肉。兔肉的胆固醇和脂肪比鸡肉、牛肉、羊肉和猪肉少。

Rabbits have an ideal fatty acid ratio of 4∶1 (omega-6 to beneficial omega-3 fatty acids). Rabbit manure, high in nitrogen and phosphorus, builds soil. There are 50 rabbit species in the world.

兔子的脂肪酸比例是理想的4∶1（ω-6脂肪酸和有益的ω-3脂肪酸之比）。兔粪富含氮和磷，可以滋养土壤。世界上有50种兔子。

Rabbits easily convert protein from cellulose-rich plants, which are not economical to feed to chickens and turkey. Grains and soya fed to poultry put them in direct competition with man for food. Rabbits do not compete.

兔子很容易从富含纤维素的植物中转化蛋白质，而用这些植物来喂鸡和火鸡并不经济。用谷物和大豆喂养家禽意味着与人类竞争食物。而兔子不会竞争。

Rich fertile land has up to 700,000 earthworms per hectare, meaning that the weight of earthworms beneath a farmer's soil could be greater than that of livestock on it. There are 3,000 different species.

肥沃的土地上每公顷有多达70万条蚯蚓，这意味着农场土壤下蚯蚓的重量可能比地里牲畜的重量还大。有3 000种不同的蚯蚓。

Charles Darwin studied earthworms for over 30 years. Darwin published his best-selling book, *The Formation of Vegetable Mould – and the Role of Earthworms*, in 1881, and sold more copies than *The Origin of Species*.

查尔斯·达尔文研究蚯蚓超过30年。1881年达尔文出版了他的畅销书《腐殖土的形成和蚯蚓的作用》，销量超过《物种起源》。

Earthworm's bodies are covered in hair that grip the soil and, through muscle action, allow them to move through their burrows. They have light receptors detecting when it is light or dark to avoid drying out in the sun.

蚯蚓的身体上覆盖着毛发，这些毛发能抓住土壤。蚯蚓的身体通过肌肉运动，让它们在洞穴中移动。它们有光感受器，可以探测到何时亮或暗，以避免在阳光下遭到曝晒。

Earthworms breathe oxygen and release carbon dioxide, just like humans do, but they don't have lungs. As they cannot breathe through their mouth, and don't have a nose, they breathe through their skin.

蚯蚓吸入氧气，释放二氧化碳，就像人类一样，但它们没有肺。由于它们不能用嘴呼吸，也没有鼻子，它们只能通过皮肤呼吸。

Rabbits have 12 litters a year. Does this mean overpopulation or lots of food?

兔子一年生12窝。这意味着种群过密还是食物充足？

Seeing with your skin and breathing through your skin?

你可以用皮肤看和用皮肤呼吸吗？

Are you helpless when you are blind?

当你失明时你觉得无助吗？

Have you developed all your senses?

你所有的感官都发达吗？

Calculate how many people can you be fed with rabbits, and how much food you would need to produce to feed the rabbits. Be careful, as this is not a linear calculation. You will need to take into account the efficiency of the conversion of non-edible food that people will not eat, and the fact that other farmed animals, such as chickens, eat soy and corn that could also be eaten by people. So how much more efficient is the rabbit?

算算你能用兔子养活多少人,以及你需要生产多少饲料来喂兔子。注意,这不是线性计算。你需要考虑到人们不会去吃的非食用食物的转化效率,还要考虑到其他养殖动物(如鸡)吃的大豆和玉米是人们也吃的。那么兔子的效率高出多少?

TEACHER AND PARENT GUIDE

学科知识
Academic Knowledge

生物学	蚯蚓有两组环形和纵向肌肉，用于在土壤上下移动；蚯蚓的三种类型：深层土壤生活、表层取食、土壤取食；蚯蚓是雌雄同体；蚯蚓有能力再生失去的部分；兔子和野兔的区别；兔子会产生两种粪便。
化学	氮肥会产生对蠕虫致命的酸性土壤；蚯蚓会在肠道中积累镉、汞、锌、铜等重金属。
物理	安哥拉羊毛的直径为14—16微米，细度和柔软度与羊绒相似；空气溶解在蚯蚓皮肤上的黏液里，它们必须保持湿润才能呼吸，干燥时就会窒息；蚯蚓有表皮感受器、颊感受器和光感受器；蚯蚓有方向感；兔子的视野接近360度，只是在鼻梁处有一个小盲点；兔子能调节体温。
工程学	蚓粪的有效氮含量、有效磷含量、有效钾含量分别是人工合成肥料的5倍、7倍和11倍；蚯蚓在土壤中挖洞会形成许多通道，帮助维持土壤结构、通风和排水。
经济学	对所有有机"废物"进行蚯蚓堆肥，可产生健康的土壤并提高生产力；蚯蚓通过物理研磨和化学消化将植物碎屑破碎、混合，加速土壤–植物系统中的养分循环；蚯蚓为鱼、家禽和猪提供了极好的蛋白质来源。
伦理学	为什么农场要去饲养将植物蛋白转化为动物蛋白效率低得多的家畜，而且其饲料会与人类口粮相竞争；仅仅是为了口味而放弃高效的养兔去养牛？为什么不改进烹饪方式呢？
历史	大约在公元前1000年，腓尼基人驯养了兔子，并把这项技术带到西班牙；罗马人把兔子传播到整个罗马帝国，他们吃新生的兔子，称之为劳瑞斯。
地理	俄罗斯是世界上最大的兔子生产国，其次是法国；安哥拉兔起源于土耳其；蚯蚓被认为是新西兰毛利人首领的美味佳肴；欧洲兔子已被引入到除南极洲以外的每一个大陆；兔子是中国十二生肖之一；在日本的传统故事中，兔子住在月球上。
数学	异速生长数学：部分身体的大小和身体整体的大小之间的比例关系，因为两者都在发育过程中生长。
生活方式	安哥拉羊毛是从安哥拉兔身上收集的，它不会引起过敏。
社会学	19世纪初，养兔是贵族的特权，废除封建特权后，西欧乡村和城市郊区都出现了养兔现象；20世纪50年代，随着其他更美味的肉类出现，兔肉产量大幅下降；兔子通常被当作生育或重生的象征。
心理学	生活在黑暗中，用你所有的感官来感受，你会有什么感觉？当长时间陷于黑暗中，逃避和回避学习。
系统论	通过掺杂死亡植物，蚯蚓能提高土壤肥力及其碳储量；偏离了有机栽培技术，以及使用人工合成化肥和杀虫剂，蚯蚓受到了严重伤害，目前至少有三种蚯蚓已被列为灭绝物种，更多的蚯蚓处于濒危状态；蚯蚓是土壤健康的环境指标；蚯蚓在将大块有机质转化为腐殖质、提高土壤肥力方面发挥着重要作用。

教师与家长指南

情感智慧
Emotional Intelligence

蚯蚓

蚯蚓表达了对一窝新生小兔的祝贺,自信地问了一个关于兔子家族的问题。他还和兔子谈论敏感话题:兔宝宝是否天生就看不见。蚯蚓鼓励兔子提升感官。当被问到他自己的感官时,蚯蚓自信地说他有感觉的能力。蚯蚓喜欢展示令人惊讶的事实,比如他有很多毛发。他不断地分享蚯蚓一个又一个让人吃惊的能力,兔子最终被这些她不知道的蚯蚓独特的特征所困惑。蚯蚓坚韧而自信。他不想自己被认为是无助的生物,他表现出知道如何照顾自己,即使人们认为蚯蚓有很多局限性。

兔子

兔子想让蚯蚓觉得她有责任心,因为她只在食物充足的时候繁殖。兔子解释说,小兔子刚出生时看不见,虽然会被认为是负面的,但实际上是一种优势,这更易于她照顾新生儿。她庆幸兔子的其他感官可以正常发育,然后同情地质疑蚯蚓的感官,因为他没有耳朵。兔子意识到蚯蚓更多产,但仍然为蚯蚓没有耳朵、头发、舌头和鼻子而感到遗憾。不过她还是乐于了解蚯蚓分享的那些令人惊讶并且没有让蚯蚓感到无助的事实。蚯蚓积极的生存态度让兔子钦佩不已。

艺术
The Arts

找一幅丢勒画的《野兔》复制品,仔细看一看。再来看看版画《沃尔珀丁格》,沃尔珀丁格也叫鹿角翼兔,是德国巴伐利亚高山森林中的一种神话生物。尤其还要看看沃尔特·莫尔斯画的兔子,他模仿了丢勒的风格,但更加超现实主义。你觉得艺术家画笔下的沃尔珀丁格有意思吗?有创意吗?现在轮到你画兔子了,请采用不同的超现实主义印象派画风。你画的兔子可能看起来像蚯蚓。你自己都会大吃一惊!

TEACHER AND PARENT GUIDE

思维拓展
Systems: Making the Connections

我们很容易就认为蚯蚓是无助的，经常关注它们没有鼻子和肺的生活是多么困难，却没有花时间去发现蚯蚓是一种多产的生命形式。蚯蚓有能力断后重生，改善土壤，可持续地产生超过牧场奶牛体重的总生物量。蚯蚓提供了如此广泛的生态系统服务，这一事实启发我们思考，每个人都有能力用我们所拥有的有限资源为改善社会做出贡献。相比于蚯蚓，人们对兔子的了解更多，并且兔子作为多产的象征出现在著名的艺术作品中，但它们已不再是动物蛋白质的主要来源之一，尽管兔子在将植物转化为肉类方面比鸡或鱼更有效。养牛作为重要的生产方式之一，有着多种负面影响，如排放大量温室气体和破坏生态系统。有一种新的用养兔子和蚯蚓替代养牛业的商业模式。想象一下这样的食物链：一对兔子每年生产300千克的肉，它们的粪便喂养蚯蚓，用蚯蚓喂养鱼，然后用含有丰富天然肥料的鱼塘水灌溉农田。这是一个模仿大自然的最佳模式，为我们提供了一个消除饥饿的机会，同时创造了就业机会，并确保所有这一切都能小规模实施，即使是那些仅靠几立方米空间维持生计的家庭。看起来，现代的牛肉养殖方式使我们无助和依赖，而"无助"的蚯蚓和高效的兔子却确保了我们的生存。

动手能力
Capacity to Implement

你知道在哪儿能找到蚯蚓吗？把新鲜的厨余垃圾装满一个盒子，然后放入几条蚯蚓。不知不觉间，你会发现盒子里的蚯蚓远比你想象的多。遵循蚯蚓养殖的规则，继续你的研究，观察那些被认为是废物的东西是如何通过没有眼睛、鼻子或肺的小生物的辛勤工作变成肥土的！

教师与家长指南

故事灵感来自
This Fable Is Inspired by

珍·斯威尼
Jean Sweeney

  珍·斯威尼担任3M公司的首席可持续发展官直到2017年。从产品开发到制造，她曾在多个领域工作。她曾任多项领导职务，如3M公司的澳大利亚和新西兰制造总监，以及环境、健康、安全和可持续发展副总裁。作为首席可持续发展官，珍负责3M的全球环境、健康、安全和可持续发展项目。珍致力于回馈社会，她把时间集中在一些支持环境、教育和动物福利的非营利性组织工作上。她热衷于鼓励年轻女性从事科学、技术和工程方面的职业。珍强烈倡导目的驱动创新和商业利益的可持续发展。

图书在版编目（CIP）数据

冈特生态童书.第八辑：全36册：汉英对照 /
（比）冈特·鲍利著；（哥伦）凯瑟琳娜·巴赫绘；
何家振等译. —上海：上海远东出版社，2021
ISBN 978-7-5476-1773-1

Ⅰ.①冈… Ⅱ.①冈…②凯…③何… Ⅲ.①生态环
境–环境保护–儿童读物—汉、英 Ⅳ.①X171.1-49

中国版本图书馆CIP数据核字（2021）第249940号

策　　划　　张　蓉
责任编辑　　程云琦
封面设计　　魏　来　李　廉

冈特生态童书
无眼≠无助
[比]冈特·鲍利　著
[哥伦]凯瑟琳娜·巴赫　绘
章里西　译

记得要和身边的小朋友分享环保知识哦！
八喜冰淇淋祝你成为环保小使者！

Health
274

跳 蛛
Jumping Spiders

Gunter Pauli

[比]冈特·鲍利 著
[哥伦]凯瑟琳娜·巴赫 绘
章里西 译

上海远东出版社

丛书编委会

主　任：贾　峰
副主任：何家振　闫世东　林　玉
委　员：李原原　祝真旭　牛玲娟　梁雅丽　任泽林
　　　　王　岢　陈　卫　郑循如　吴建民　彭　勇
　　　　王梦雨　戴　虹　翟致信　靳增江　孟　蝶

特别感谢以下热心人士对童书工作的支持：

匡志强　宋小华　解　东　厉　云　李　婧　陈　果
刘　丹　熊彩虹　罗淑怡　旷　婉　杨　荣　刘学振
何圣霖　廖清州　谭燕宁　韦小宏　李　杰　欧　亮
陈强林　王　征　张林霞　寿颖慧　罗　佳　傅　俊
胡海朋　白永喆　冯家宝

目录

跳蛛	4
你知道吗？	22
想一想	26
自己动手！	27
学科知识	28
情感智慧	29
艺术	29
思维拓展	30
动手能力	30
故事灵感来自	31

Contents

Jumping Spiders	4
Did You Know?	22
Think about It	26
Do It Yourself!	27
Academic Knowledge	28
Emotional Intelligence	29
The Arts	29
Systems: Making the Connections	30
Capacity to Implement	30
This Fable Is Inspired by	31

一只跳蛛看着远处有一大群蝗虫正向自己迫近。成千上万只甚至是上百万只蝗虫正蜂拥而至。他抓住其中一只问道：

　　"你们是要来这里大快朵颐，吃个片甲不留吗？"

　　"对不起！难道你是唯一有权利在这里吃东西的人吗？"

A jumping spider is watching a swarm of grasshoppers approaching. Thousands, perhaps even millions, of them are arriving, so he asks one,
"Are you coming to eat everything and leave nothing?"
"Excuse me! Are you the only one around here with the right to eat?"

一只跳蛛看着远处……

A jumping spider is watching ...

食物不代表一切……

Food isn't everything...

"当然不是,但至少我们吃完不会让土地荒芜,让居民绝望。"

"听着,我们和你们一样都想要快乐和健康,对我们来说,这意味着得填饱肚子。"

"你知道,食物不代表一切。如果你整天只会吃吃吃,而不做足够的运动,你就会变得又胖又懒。"

"Of course not, but at least we do not leave the land bare, and its inhabitants in despair."

"Look, you and I both want to be happy and healthy, and for us that means having a full tummy."

"Food isn't everything, you know. If all you ever do is eat, and you don't move around getting enough exercise, you will get fat and lazy."

"你和我都是动物世界里最好的跳跃运动员。这不仅使我们保持健康,而且对我们的胃、心脏和眼睛也有好处。"

"在所有的蜘蛛里,我的视力是最好的,甚至还具备运动探测器。没有人能悄悄接近我,即使是从后面也不行。"

"是的,你的四对眼睛给你带来了极大的优势。"

"You and I are some of the best jumpers in the animal world. It not only keeps us fit, it is also good for our stomach, heart and eyes."

"I already have the best eyesight of all spiders, and am even equipped with motion detectors. No one can come close, not even from behind."

"Yes, your four pairs of eyes must give you a tremendous advantage."

……所有蜘蛛里视力最好的……

... the best eyesight of all spiders ...

我的腿和膝盖的工作原理跟弹弓很像……

My legs and knees work like a catapult ...

"我们都有一些独特的技能,让我们有机会在某些方面做到最好。你跳得比大多数昆虫都高。你是怎么做到的?"

"我的腿和膝盖的工作原理跟弹弓很像,"蝗虫说。

"所以你肯定有肌腱和肌肉。我可没有这些,但是我也能跳得很高。"

"We all have something unique that gives us the chance to be the best at something. You jump higher than most insects. How do you do that?"

"My legs and knees work like a catapult," Grasshopper says.

"So, you must have tendons and muscles to use. I don't have any, but I can still jump very high too."

"要跳20倍于你身长的距离,你的腿必须有弹性。"

"不,我只需要泵出体液、改变压力就行了,"跳蛛吹嘘道。

"我不太懂这是怎么回事。你能更详细地解释一下吗?"

"To leap twenty times your body length, you must have elastic in your legs."

"No, all I have to do is pump liquid and change pressure," Spider boasts.

"I have no idea how that works. Could you please explain in a bit more detail?"

……泵出体液、改变压力……

... pump liquid and change pressure ...

……给气球充气时……

... when you fill a balloon with air ...

"好的。你先说说给气球充气时，会发生什么？"

"气球鼓起来了。"

"如果你把气球吹到快要爆，然后忽然放开呢？"

"Okay. So, when you fill a balloon with air, what happens?"

"It expands."

"And if you fill it until it is about to burst, and let it go?"

"它会到处乱飞。"

"我的腿就是这样。我在腿中加满'油',然后找准时机放开起跳!"

"难以置信!那这个过程你能重复多少次?"

"It will fly around like crazy."
"That is what I do with my legs. I pump them full of oil, and then I let go to blast off!"
"Incredible! And how many times can you repeat that?"

……它会到处乱飞……

It will fly around like crazy …

……加"油"多快跳得就多快……

... jump as fast as I can pump ...

"哦,我加'油'能有多快我跳得就能有多快,"蜘蛛回答说。"但我并不会像气球一样失去准星,我总是清楚地知道自己想在哪儿落地。"

"好吧,不过有些事我能用我的腿做,而你不能。你猜得到是什么吗?"

"Oh, I can jump as fast as I can pump," Spider replies. "But I don't go crazy, I always know exactly where I want to land."

"Well, there is something I can do with my legs, that you can't. Can you guess what?"

"哈哈，这次你想炫耀什么？"

"事实上，我可以用双腿摩擦翅膀来创作音乐。"

"用腿在翅膀上拉小提琴？然后用肚子上的耳朵来听？你真是一个十足的怪人！"

……这仅仅是开始！……

"What are you showing off now?"

"The fact that I can make music, by rubbing my legs against my wings."

"Using your legs to play violin on your wings? And using the ears on your belly to listen to it? You are a strange creature, indeed!"

... AND IT HAS ONLY JUST BEGUN!...

……这仅仅是开始！……

... AND IT HAS ONLY JUST BEGUN! ...

Did You Know? 你知道吗?

Of the over 11,000 spider species, over 10% are jumping spiders. These spiders have four pairs of eyes and the best vision amongst arthropods.

在 11 000 多种蜘蛛中,超过 10% 是跳蛛。这些蜘蛛有 4 对眼睛,是节肢动物中视力最好的。

The jumping spider has full 3-D vision, allowing it to very precisely estimate range and direction, and to identify potential prey. It uses a silk thread as a safety line when jumping.

跳蛛有完整的三维视觉,可以非常精确地估计距离和方向,并识别潜在的猎物。它在跳跃时用一根丝线作为安全绳。

Some jumping spiders have been found at high altitudes, such as on the slopes of the Mount Everest. They are found in deserts as well as tropical forests.

一些跳蛛在高海拔地区被发现，比如在珠穆朗玛峰的斜坡上。在沙漠和热带森林中也都能找到它们。

Jumping spiders have motion-detecting eyes, which pick up movement on the sides and behind them, giving them near 360-degree field of vision.

跳蛛的眼睛有运动探测功能，可以捕捉身体两侧和背后的动静，使它们拥有近360度的视野。

Jumping spiders are capable of learning, recognising, and remembering colours, and of adapting their hunting behaviour accordingly. Their mechanism for jumping is hydraulic, and independent of muscle use.

跳蛛能够学习、识别和记忆颜色，并相应地调整它们的狩猎行为。它们的跳跃机制是液压传动，与肌肉活动无关。

Grasshoppers and locusts are the same insects, with different names. Both are short-horned members of the order Orthoptera.

蝗虫和蚱蜢是同一种昆虫，只是名字不同。它们都是直翅目的短角成员。

Studies show that jumping is good for a child's heart, respiration, bones, connective tissue and muscles. As jumping requires the use of both brain hemispheres for sensory-motor integration, it improves coordination.

研究表明，跳跃对孩子的心脏、呼吸、骨骼、结缔组织和肌肉都有好处。由于跳跃需要使用两个大脑半球来进行感觉和运动的整合，它提高了协调性。

When jumping, trillions of body cells move up and down, energising the mitochondria in the cells. Jumping helps the digestive track. Eye muscles are strengthened through the continuous change of focus.

当跳跃时，数以万亿计的体细胞上下移动，激活了细胞内的线粒体。跳跃有助于消化。眼部肌肉通过焦点的持续变化得到加强。

Do you like jumping on a trampoline?

你喜欢在蹦床上跳吗?

Does anyone have the right to eat everything that is available?

每个人都有权利吃掉所有能吃到的东西吗?

What would you do with four pairs of eyes?

如果你有4双眼睛，你会做什么?

Which is stronger: hydraulic power or muscle power?

液压动力和肌肉的力量，哪个更强?

How do hydraulics work? Time to find out. Look around, at home and at school, to find out where hydraulics are used in daily life – car brakes, an elevator, the dishwasher, and even your hairdresser's chair. Study the different ways in which hydraulic systems are used, and how hydraulic operating systems are maintained. Share your newly acquired knowledge with your friends and family members.

液压系统是如何工作的？是时候揭晓答案了。看看周围，在家里和学校，了解液压在日常生活中的应用——汽车刹车，电梯，洗碗机，甚至理发店的椅子。研究使用液压系统的不同方式，以及如何维护液压操作系统。与你的朋友和家人分享你新获得的知识。

TEACHER AND PARENT GUIDE

学科知识
Academic Knowledge

生物学	跳蛛不用网来捕捉猎物；蝗虫生长分为卵、若虫和成虫三个阶段；蝗虫学研究蝗虫，蝗虫有11 000种，是现存最古老的咀嚼食草昆虫群；蚱蜢是夜行性的，可以游泳和在水上滑行；蚱蜢有三对腿、两对翅；蚱蜢体内的血淋巴，用于愈合伤口、热传递和提供静水压力；蚱蜢模仿叶子。
化学	蚱蜢的胸腹由几丁质组成；血清素是蚱蜢的兴奋剂，信息素则是吸引蚱蜢成群结队的物质。
物理	跳蛛对蓝光和紫外线很敏感；液压机对少量流体施压，以产生大量的动力；蚱蜢有全方位的视觉，3只单眼用来辨别明暗；蚱蜢用后腿上的一列乳头状突起摩擦前翅边缘，称为发音器；肌肉不能同时以高强度和高速度收缩。
工程学	采用液压代替楔子、斜面、滑轮并做手动升降；用液压压实垃圾；"半机械蝗虫"能够准确探测爆炸物。
经济学	成群的蝗虫会造成毁灭性的后果，并导致饥荒；一个大型蝗虫群可拥有多达3.5万亿只蝗虫。
伦理学	有什么吃什么，不给别人留下任何食物。
历史	蝗虫进化于大约2.5亿年前的三叠纪早期；最古老的跳蛛化石来自始新世（5000万年前），蝗虫来自石炭纪（3亿年前）；蚱蜢是雅典的象征；1663年，法国数学家布莱斯·帕斯卡发现了流体压力传递原理；拉封丹重述了伊索寓言《蚂蚁和蚱蜢》。
地理	在墨西哥、印度尼西亚以及乌干达、赞比亚、津巴布韦、尼日利亚和南非等非洲国家，蝗虫是一种食物来源。
数学	比例的数学：当肺扩张时，空气充满5亿个微小的肺泡，每个直径不到1毫米，呼气时，数百万个微小的呼吸通过越来越大的气道毫不费力地融合成一次最终的呼吸。这一概念也适用于蝗虫群。
生活方式	学习我们从未想象过的东西；日常锻炼的需要；花时间互相学习，倾听不同观点；孩子们天生渴望跳跃，享受跳跃，这有利于细胞、心脏和消化系统的健康。
社会学	会改变颜色和行为并成群结对的蝗虫种类；由于蝗虫能够跳和飞得很远，鸣声可以产生必要的群居凝聚力，这对限制分散和引导同伴到更有利的栖息地是必要的。
心理学	梦见蝗虫代表着自由、独立、精神觉悟，也代表着无法安定下来或做出决定。
系统论	大自然运用肌肉、弹射器和液压，为不同的物种提供了不同的解决方案，让一切都像一个系统一样运作。

教师与家长指南

情感智慧
Emotional Intelligence

跳蛛

跳蛛毫无顾忌向蝗虫提问。她自信地告诉蝗虫，生活不仅仅是为了食物，锻炼也是至关重要的。当蝗虫友善地提到他们都有非凡的跳跃能力时，她开始吹嘘自己拥有超强的视力。当蝗虫提到这给她带来了极大的优势时，她承认每个物种都有自己的独特之处，这让他们有机会在某方面成为最优秀的。跳蛛对蝗虫跳跃的原理表现出了兴趣，但只是把他的回答当作另一个炫耀自己的机会。当蝗虫询问细节时，她耐心地向他解释她特殊的跳跃机制，但当蝗虫提到他的腿还有其他用途时，她就生气了，并指责蝗虫是在炫耀。她对蝗虫发出声音的能力感到惊讶，认为他是一个奇怪的生物。

蝗虫

蝗虫被跳蛛直截了当的话吓了一跳，他反问跳蛛他是否也有吃东西和生存的权利。他试图安抚这只跳蛛，说他们都是最棒的跳跃运动员，而且运动对健康很有好处。当跳蛛夸耀自己的视力时，蝗虫承认她的四只眼睛带来的特殊优势。当被问到蝗虫为什么比一般的昆虫跳得高时，他解释说，蝗虫的关节被设计成像弹弓一样。他以提问题作为对她的回应，并坚持了解更多的细节。他对跳蛛的能力印象深刻，在谈话结束时告诉跳蛛，蝗虫还有另一种不寻常的能力，那就是发声。

艺术
The Arts

让我们用回收来的金属做蝗虫和蜘蛛。收集一些用过的钉子、螺丝、垫圈、滚珠轴承、电线、金属刨花和其他你可以在车库或者车间找到的金属碎片。找一种有趣的方法将这些材料组合起来，创造出两个金属雕塑，一个是蜘蛛，一个是蝗虫。从材料的颜色、形状和质地中获得灵感。

TEACHER AND PARENT GUIDE

思维拓展
Systems: Making the Connections

　　生活中，每个人、每个物种都会在有机会的时候表现其最好的能力。我们应该始终对新的科学发现保持开放的态度。跳蛛和蝗虫都是跳跃冠军，它们进化出了不同的技术。这不是一种比另一种更好的问题，而是每一种都利用其独特的能力来确保生存。孩子们喜欢跳跃只是为了享受跳跃的乐趣，已经证明跳跃对孩子们的健康有很多方面的有益影响，特别是对于骨骼、肌肉、消化系统和心脏健康。当我们看到现在的孩子们花了多少时间坐着或睡觉时，难道我们不应该鼓励他们多运动吗？教他们蜘蛛跳、蝗虫跳，激励他们走出家门，去参加更多的锻炼。孩子们喜欢模仿，像青蛙和兔子一样跳跃，大自然激发他们去运动和探索。在床或蹦床上蹦跳，能激活线粒体，训练眼睛在移动时集中注意力，有助于清除体内废物，增强关节（尤其是膝盖和臀部），增加骨密度，增强淋巴排毒，改善消化和平衡，增加肺活量，也能促进心理健康。我们对动物和周围自然世界的了解与观察激发了我们的想象力。孩子们喜欢运动，尤其是跳跃，这是很自然的事情。因为跳跃对他们的健康有很多好处，至少可以确保他们在一天结束时感到疲惫，晚上可以好好睡觉。

动手能力
Capacity to Implement

　　让我们做一个弹弓来演示蝗虫膝关节的机理。弹弓被用作武器，用它来练习射击是很有趣的，只不过不能对着人，或任何动物或鸟。首先，做一些研究，了解一下让物体飞起来的物理原理，只需用到一些小树杈、弹性材料和一点点力量。有不同种类的弹弓，所以不要只设计一种。当你设计和完成了三种不同的类型时，你足以证明已经了解它们是如何工作的，这时你也掌握了蝗虫运动的机制。

教师与家长指南

故事灵感来自
This Fable Is Inspired by

凯西·冈特
Kathy GUNTER

1992 年，凯西·冈特获得美国华盛顿州贝灵汉的西华盛顿大学文学学士学位。在那里她继续深造，并于 1996 年获得教育学硕士学位。2003 年，她以一篇关于成年人跌倒和骨折风险的论文获得了俄勒冈州立大学科瓦利斯分校的博士学位。目前她是俄勒冈州立大学生物与人口健康科学学院的教授。她也是研究儿童和家庭健康的哈里·福特中心的健康饮食和积极生活研究所主任。她研究蹦床跳跃对儿童骨骼健康的影响，并发表了大量相关文章。

图书在版编目（CIP）数据

冈特生态童书.第八辑：全36册：汉英对照 /
（比）冈特·鲍利著；（哥伦）凯瑟琳娜·巴赫绘；
何家振等译. —上海：上海远东出版社，2021
 ISBN 978-7-5476-1773-1

Ⅰ.①冈… Ⅱ.①冈…②凯…③何… Ⅲ.①生态环
境-环境保护-儿童读物—汉、英 Ⅳ.①X171.1-49

中国版本图书馆CIP数据核字（2021）第249940号

策　　划　　张　蓉
责任编辑　　程云琦
封面设计　　魏　来　李　廉

冈特生态童书
跳蛛
［比］冈特·鲍利　著
［哥伦］凯瑟琳娜·巴赫　绘
章里西　译

记得要和身边的小朋友分享环保知识哦！
八喜冰淇淋祝你成为环保小使者！

Energy 273

光之花蜜

Nectar of Light

Gunter Pauli

[比]冈特·鲍利 著
[哥伦]凯瑟琳娜·巴赫 绘
贾龙智子 译

上海远东出版社

丛书编委会

主　任：贾　峰

副主任：何家振　闫世东　林　玉

委　员：李原原　祝真旭　牛玲娟　梁雅丽　任泽林
　　　　王　岢　陈　卫　郑循如　吴建民　彭　勇
　　　　王梦雨　戴　虹　翟致信　靳增江　孟　蝶

特别感谢以下热心人士对童书工作的支持：

匡志强　宋小华　解　东　厉　云　李　婧　陈　果
刘　丹　熊彩虹　罗淑怡　旷　婉　杨　荣　刘学振
何圣霖　廖清州　谭燕宁　韦小宏　李　杰　欧　亮
陈强林　王　征　张林霞　寿颖慧　罗　佳　傅　俊
胡海朋　白永喆　冯家宝

目录

光之花蜜	4
你知道吗？	22
想一想	26
自己动手！	27
学科知识	28
情感智慧	29
艺术	29
思维拓展	30
动手能力	30
故事灵感来自	31

Contents

Nectar of Light	4
Did You Know?	22
Think about It	26
Do It Yourself!	27
Academic Knowledge	28
Emotional Intelligence	29
The Arts	29
Systems: Making the Connections	30
Capacity to Implement	30
This Fable Is Inspired by	31

一朵被森林里的所有人称为火狐的蘑菇看到了一片芥菜，向其中的一位提问道：

　　"真的有人打算把你们变成能在夜晚发光以照亮街道的植物吗？"

　　"你怎么知道的？这是个秘密！"惊讶的芥菜问道。

A mushroom, known by everyone in the forest as the firefox, sees a family of mustard plants and asks one,

"Is it true that someone is planning to turn you into plants that will light up at night to illuminate the streets?"

"How did you know? It is a secret!" The surprised mustard plant asks.

一朵蘑菇……

A mushroom ...

我是自然界最好的光源之一。

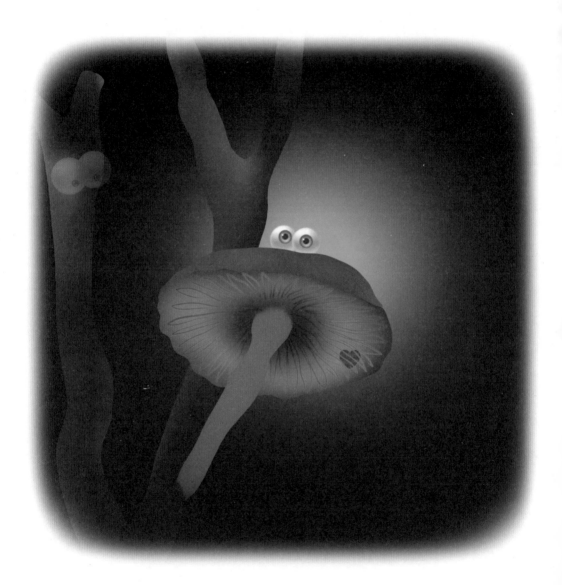

I'm one of the best lights in Nature.

"啊,公开的秘密就不再是秘密了……"

"你真是消息灵通。这是怎么办到的?"

"嗯,我是自然界最好的光源之一。所以我必须留意其他想和我做一样的事的家伙。我的建议是不要费力去模仿我们。"

"Ah, a secret out in the open is not a secret anymore…"

"You are very well informed. How come?"

"Well, I'm one of the best lights in Nature. So I must keep an eye on others wanting to do what I do. My advice is not to even bother imitating us."

"有些穿白大褂的聪明人把细菌放进了我体内来制造光。想象一下,小僵尸和大植物的组合可能并不容易。"

"嘿,当你想要新奇、大胆的东西时,就选最好、最有创意的。向那些把不可能变为可能的人寻求建议。"

"你在夜晚降临时开始发光,并且整夜闪耀——这是真的吗?"

"Some smart people in white coats put bacteria into my body that make light. Imagine, the mix of tiny zombies and a big plant may not be easy."

"Look, when you want something new and daring, go with the best and most creative. Ask advice from those who did the impossible before."

"Is it true that you start glow at the same at night, and shine all night?"

……穿白大褂的人……

... people in white coats ...

……昆虫会来寻找食物。

... insects come to look for food.

"这是肯定的,我当然能。"

"你为什么在晚上'开灯'呢?你是想给人指路,还是想引起很多人的注意,炫耀自己可以发光?"

"还有更好的理由。当我'开灯'的时候,昆虫会来寻找食物。然后我的小种子会粘在它们的腿上,这些苍蝇和蚊子把我的孢子带到枯树和树枝上——那些我靠风可够不着的地方。"

"You bet I can, and of course I do."

"Why do you turn on your lights at night? Do you want to show the way, or get a lot of attention and show off that you can make light?"

"There is a better reason. When I turn on my lights, insects come to look for food. Then my tiny seeds stick to their legs and these flies and mosquitos carry my spores to dead trees and branches I could never reach with wind."

"这么说，你的光就像花蜜，吸引蜜蜂授粉？"

"没错！但是，当有太多饥饿的苍蝇时，我会稍微改变光线，来吸引喜欢吃苍蝇的虫子。"

"什么？我真不敢相信。你根据自己的想法，利用你发出的光调节着你周围的生命？"芥菜问道。

"So, your light is like a flower's nectar, attracting bees to pollinate it?"

"Exactly! But, when there are too many hungry flies, I change my light slightly – and attract the bugs that like to eat those flies."

"What? I can not believe this. Your light regulates life around you, as you see fit?" Mustard Plant asks.

……当有太多饥饿的苍蝇时……

… when there are too many hungry flies …

……给松鼠吃的蘑菇？

... mushrooms that squirrels eat?

"你以为我只是那种给松鼠吃的蘑菇吗?我们没有大脑,但比你想象的要聪明。"

"我也没有大脑,但我想了解你发光的秘密。你知道我的秘密,所以请和我分享你的秘密吧。"

"哦,不用电进行照明是非常简单的。"

"Did you think I was just one of those mushrooms that squirrels eat? We do not have a brain, but are smarter than you think."

"I have no brain either, but I would like to understand your secret of making light. You know my secret, so please share yours?"

"Oh, making light without electricity or batteries is very simple."

"那你是怎么做的?"

"我需要氧气,以及长时间干旱后的一点雨。"

"就这些吗?"芥菜问道。

"如果旱季过后天气又暖和又潮湿,我会持续照耀好几天——多亏了水和氧气。"

"So how do you do it?"

"I need oxygen, and a bit of rain after a long period of drought."

"Is that all?" Mustard Plant asks.

"If the weather is warm and damp after a dry season, I keep on shining for days and days – thanks to water and oxygen."

我需要氧气,以及一点雨……

I need oxygen, and a bit of rain ...

……比萤火虫做得好。

...better than the fireflies.

"你比萤火虫做得好。干得不错!"

"连我的根都会发光,让它们看起来像光纤电缆……"

"别告诉我你是通过位于地下,把你和树木连接起来的奇妙网络线路发送信息的。"

"You do better than the fireflies. Well done!"
"Even my roots make light, making them look like fibre optic cables…"
"Don't tell me you are sending information through your wonderful network of tiny wires that connect you with the trees underground?"

"当然了！或者说，你认为只有人类知道如何通过线路传输数据吗？"

"所以你就是未来的互联网，可以食用，还可以生物降解。太棒了！"

……这仅仅是开始！……

"You bet we do! Or did you think that people are the only ones who know how to transmit data through wires?"

"So you are the internet of the future, edible and biodegradable. Amazing!"

… AND IT HAS ONLY JUST BEGUN!…

……这仅仅是开始！……

... AND IT HAS ONLY JUST BEGUN! ...

Did You Know?
你知道吗？

Aristotle documented the glowing of rotting tree bark, and described bioluminescence. Scientists now know that there are three genes necessary to generate luminescence.

亚里士多德记载了腐烂树皮的发光现象，并描述了这种生物发光现象。科学家现在知道了有三种基因是发光所必需的。

Bioluminescence is a natural phenomenon caused by a substance called luciferin, one that oxidises with the help of the enzyme luciferase, to emit a light that is cold, bright and long lasting.

生物发光是由一种叫作荧光素的物质引起的自然现象，这种物质在荧光素酶的帮助下氧化，发出明亮、持久的冷光。

Bioluminescence is found in many species, mainly in the sea, from glowing worms to deep-sea fish and bacteria. They use bioluminescence to attract prey.

在许多物种身上都发现了生物发光现象,主要是在海洋中,从发光的蠕虫到深海鱼类,再到细菌。它们通过生物发光吸引猎物。

Night-time luminescence attracts beetles, flies, wasps, ants and moths. These insects are key for distributing the mushroom's spores, so the mushroom can reproduce and colonise new food sources.

蘑菇在夜间发光以吸引甲虫、苍蝇、黄蜂、蚂蚁和飞蛾。这些昆虫是传播蘑菇孢子的关键,蘑菇以这样的方式繁殖和开拓新的食物来源。

There are no bioluminescent plants, except for the dinoflagellates, plant-like protists, living in the ocean. These single-celled, microscopic organisms float near the surface and can produce big light shows.

除了像植物一样的原生生物甲藻外，没有生活在海洋中的发光植物。这些单细胞的微生物漂浮在水面附近，可以产生强烈的光照。

The firefox mushroom is also commonly known as the bitter oyster mushroom. It grows on all continents, on deciduous trees like oak and birch. It has the unique ability to detoxify contaminated soil.

火狐蘑菇也被称为苦酸平菇（学名*Panellus stypticus*，鳞皮扇菇）。它分布于所有的大陆，生长在橡树和桦树这样的落叶树上。它具有为被污染土壤解毒的独特能力。

The bitter oyster mushroom has been used in Traditional Chinese Medicine (TCM) to stop bleeding, and also a purgative to improve bowel movement and treat constipation.

苦酸平菇在传统中医中被用来止血，也是一种改善肠道运动和治疗便秘的通便剂。

The mushrooms have an internal clock for turning on the light. Even when cultivated in laboratories, in continuous light or darkness, they will still display the strongest bioluminescence between 18:00 and 21:00.

这种发光蘑菇有一个内部的时钟，以此控制"开灯"时间。即使在实验室培养，处于持续的光照或黑暗环境，它们仍然会在晚上6点到9点之间发出最强的光。

Think about It

Can there be a biodegradable internet?　　能否有一个可生物降解的互联网？

Do you study your competition?　　你研究过你的竞争对手吗？

Can someone with no brain still be smart?　　没有大脑的生物可能聪明吗？

Why develop genes when Nature already has a solution that is proven to work?　　既然自然界已经有了一个行之有效的解决方案，为什么还要研发基因技术？

Time to learn more about bioluminescence. Make a list of at least ten plants and animals that can produce light. And notice how none of these have any batteries or lamps. Research this ability, as it could give you some great ideas for creating energy-efficient lights, even better than LED. Also look into the reasons for there to be light. Make a presentation of your findings on light from Nature, and offer some ideas on how this could be used as an energy-efficient source in your home, school or community.

是时候学习更多关于生物发光的知识了。列出至少10种能发光的动植物。注意这些生物体内都没有电池或灯。研究这种能力，因为它可以启发你创造节能灯具，甚至比LED更节能。同时也要探究光存在的原因。介绍一下你对自然界的光的发现，并谈谈如何让它们成为家庭、学校或社区的节能光源。

学科知识
Academic Knowledge

生物学	已知有80种蘑菇能够在黑暗中发光；生物发光只受基因控制；火狐蘑菇生长在硬木上；异宗配合蘑菇；蘑菇在夜间发光以吸引昆虫传播蘑菇孢子，否则在黑暗中昆虫可能无法找到蘑菇，就像花朵利用形状、颜色、花蜜和气味来吸引潜在的传粉者一样；发光蘑菇中的荧光素与萤火虫体内的相同；北方飞鼠食用真菌。
化学	咖啡酸是生物发光的代谢产物；火狐蘑菇含有止血成分，并有通便作用；木质素、纤维素和半纤维素被酶分解；生物发光需要pH值在3.0—3.5之间；超氧化物歧化酶。
物理	火狐蘑菇的孢子呈椭圆形或腊肠形；生物发光发出短波紫外线；通过昼夜节律调节生物发光。
工程学	火狐蘑菇用于生物修复；基因编译软件在计算机上组装DNA，用于创造新的生命形式，然后将技术参数发送给DNA组装公司来构建实际的DNA；打印DNA。
经济学	推广现成的解决方案，比如以用可生物降解的化学物质生产的冷光灯取代传统公共路灯，或者推广用可生物降解的电缆连接的互联网。
伦理学	合成生物学和转基因生物（GMO）对人类构成的风险从未被详细评估过，转基因生物可能成为没有天敌的入侵物种；事先不会知道一个新物种被释放后会作出什么行为；生物黑客。
历史	亚里士多德在公元前4世纪描述了蘑菇，将其中一种称为火狐，称它发出的光为冷光；在17世纪，斯堪的纳维亚人在黑暗的冬夜用发光真菌照明；"火狐"这个词是17世纪北美洲东部和南部的早期殖民者在看到森林里的亮光时创造出来的。
地理	火狐蘑菇生长于温带气候；一些生活在海岛上的民族将发光真菌融入仪式服装和面部彩绘中。
数学	照明问题是一类数学问题——研究装有镜面墙的房间的照明，允许光线在镜面墙间重复反射；第一个多边形不发光房间有4条边和2个固定边界点；固定在墙边的灯具，其光线会在墙上形成非常清晰、尖锐的曲线，明暗区域的边界呈双曲线。
生活方式	路灯作为一种让人感到舒适和安全的手段是很重要的。
社会学	城市照明有利于安全驾驶和预防犯罪；照明增加了市民对城市的信任感。
心理学	光可以引发诸如快乐、乐意、不确定、恐惧、喜爱、迷恋、愉悦、鼓舞、惊讶、蔑视、妄想、满足等情绪以及审美偏好。
系统论	将富含有毒多酚的工业废物转化为中性成分。

教师与家长指南

情感智慧
Emotional Intelligence

火狐蘑菇

火狐蘑菇想知道芥菜是否会成为夜间照明的竞争对手。在发光这件事上,火狐蘑菇担任了导师的角色,并以自信的态度向芥菜提供了一条建议:不要做蘑菇做的事,如果要做,和专家谈谈。火狐蘑菇确信自己的能力,并分享了自己发光的原因。他透露了自己能够控制周围昆虫的数量。虽然火狐蘑菇认识到了他的局限性,但他证明了自己对周围的生物有很强的影响力。火狐蘑菇使一切看起来简单易行,但他保留了技术细节。然而,火狐蘑菇给芥菜带来了一些额外的惊喜,这引起了芥菜的关注,再次体现了他的自信和强烈的自尊心。

芥菜

芥菜无法掩饰他的惊奇。火狐蘑菇竟然知道本应是秘密的事,这激起了他的兴趣。芥菜承认,在植物中植入细菌被视为一个重大突破,但最终成为了一项挑战(而且失败了)。芥菜想知道火狐蘑菇的光线有什么用,并问了一些涉及火狐蘑菇的自我的问题。芥菜很快就抓住了本质:光代替了花蜜。当火狐蘑菇承认自己没有大脑时,芥菜表示自己也没有。对于蘑菇发光的原理,他坚定地寻求更多的解释。芥菜被火狐蘑菇的解释深深打动,并表示了自己的感受。当他们谈到传输数据的线缆时,芥菜惊呆了。

艺术
The Arts

让我们用荧光笔画一片有树木和蘑菇的森林。请使用白色、黄色和绿色,如果方便的话,再用点蓝色。现在画两张图,一张用普通颜料,一张用荧光笔,然后给其他人看。哪幅画更能打动人?学会用少量的荧光展现大大的效果。

TEACHER AND PARENT GUIDE

思维拓展
Systems: Making the Connections

　　自然界中没有什么东西是仅仅因为一个理由而存在的。我们应当超越单个物种，去观察一个物种如何与其他物种相关联，并互相影响。每个物种的所有能力中最重要的是促进生命的延续。火狐蘑菇也不例外。它已经进化出在夜间发光的独特能力。将近有100种蘑菇具有这种特征，还有一些昆虫和植物也有类似的能力，这些都基于相同的化学反应。被认为"没有大脑"的物种利用生化物质在水里和水外、地面和空中产生光，这证明了它们拥有共同的智慧。这种蘑菇已经发现了夜晚的光具有怎样的力量，并且知道如何让昆虫看到蘑菇，这些昆虫随后会把蘑菇的孢子带到森林里。这样一来，它就确保了自己的繁殖和地位，承担着关键的生态角色：枯木的降解者。真菌的能力不仅仅是繁殖。蘑菇强大的消化系统能降解纤维素、半纤维素和木质素，它能非常有效地清除土壤中长期残留的各种有毒化学物质。通过将复杂的分子分解成惰性小分子，火狐蘑菇在生物修复方面有着公认的强大作用。此外，蘑菇可以运营一个可食用和可生物降解的"生命网络"。蘑菇也有一个生物钟，无论在哪里，蘑菇的光总是在18:00—21:00之间最强。很明显，光照和日夜温差使蘑菇能够确定确切的时间。这与将海洋单细胞生物的基因植入芥菜基因组，使芥菜具有发光能力的提议形成了鲜明对比，后者就像让植物患癌症一样，而且稳定性不高。这项惊人的提议遇到了许多挑战，在得到全球关注之后，这个项目就被遗忘了。同时，很少有人知道"光的花蜜"以及发光蘑菇的独特作用。一旦我们接触到这些大自然的奇迹，我们可以设想许多可能的应用。这种对超乎我们想象的物种的观察，激发了我们的创造力。通过模仿大自然并改良相关的设计，我们能实现可持续发展。

动手能力
Capacity to Implement

　　没有电池怎么照明？火狐蘑菇不使用任何电池，因为它需要的一切都可以在自然界中找到。制订一个计划，用蘑菇和植物照亮你的街道和附近的房子。你周围的人对这个想法会有什么反应？准备充足的论据，向他们证明这是可行的。

教师与家长指南

故事灵感来自
This Fable Is Inspired by

詹妮弗·J·洛罗斯
Jennifer J. Loros

1971 年，詹妮弗·洛罗斯在卡布利洛学院和蒙特利半岛学院获得生物学学位。1984 年，她在加利福尼亚大学圣克鲁兹分校获得学士学位。随后，她在美国新罕布什尔州汉诺威的达特茅斯医学院获得了遗传学博士学位。她在那里开展生物化学方面的博士后研究。洛罗斯博士于 2000 年被任命为生物化学和细胞生物学教授，后来又被任命为分子和系统生物学教授。她研究生物发光，知道如何像蘑菇那样制造光。她想了解所有生物钟是如何受光和温度调节，并与现实世界保持同步的。

图书在版编目（CIP）数据

冈特生态童书.第八辑：全36册：汉英对照／
（比）冈特·鲍利著；（哥伦）凯瑟琳娜·巴赫绘；
何家振等译.—上海：上海远东出版社，2021
ISBN 978-7-5476-1773-1

Ⅰ.①冈… Ⅱ.①冈…②凯…③何… Ⅲ.①生态环境－环境保护－儿童读物—汉、英 Ⅳ.①X171.1-49

中国版本图书馆CIP数据核字（2021）第249940号

策　　划　　张　蓉
责任编辑　　祁东城
封面设计　　魏　来　李　廉

冈特生态童书
光之花蜜
［比］冈特·鲍利　著
［哥伦］凯瑟琳娜·巴赫　绘
贾龙智子　译

记得要和身边的小朋友分享环保知识哦！
八喜冰淇淋祝你成为环保小使者！

Energy 272

菠菜的电力

Spinach Power

Gunter Pauli

[比]冈特·鲍利 著
[哥伦]凯瑟琳娜·巴赫 绘
贾龙慧子 译

上海远东出版社

丛书编委会

主　任：贾　峰
副主任：何家振　闫世东　林　玉
委　员：李原原　祝真旭　牛玲娟　梁雅丽　任泽林
　　　　王　岢　陈　卫　郑循如　吴建民　彭　勇
　　　　王梦雨　戴　虹　翟致信　靳增江　孟　蝶

特别感谢以下热心人士对童书工作的支持：

匡志强　宋小华　解　东　厉　云　李　婧　陈　果
刘　丹　熊彩虹　罗淑怡　旷　婉　杨　荣　刘学振
何圣霖　廖清州　谭燕宁　韦小宏　李　杰　欧　亮
陈强林　王　征　张林霞　寿颖慧　罗　佳　傅　俊
胡海朋　白永喆　冯家宝

目录

菠菜的电力	4
你知道吗？	22
想一想	26
自己动手！	27
学科知识	28
情感智慧	29
艺术	29
思维拓展	30
动手能力	30
故事灵感来自	31

Contents

Spinach Power	4
Did You Know?	22
Think about It	26
Do It Yourself!	27
Academic Knowledge	28
Emotional Intelligence	29
The Arts	29
Systems: Making the Connections	30
Capacity to Implement	30
This Fable Is Inspired by	31

一棵菠菜坐在阳光下晒太阳，她看见一棵松树结满了营养充足的松果，想到自己只为人类健康作了微小的贡献，她有些难过。

　　"没必要垂头丧气呀，你为人类提供了可口健康的食物，不是吗？"松树说。

A spinach plant is sitting in the sun, and seeing a pine tree providing pine nuts full of nutrients, feels sad about her own contribution to people's health.

"No need to feel sad, you provide people with tasty and healthy food, don't you?" the pine tree says.

一棵菠菜坐在阳光下晒太阳……

A spinach plant is sitting in the sun ...

……吃掉你那健康的绿叶的人……

... people who eat your healthy green leaves ...

"是的,人类的确认为我是铁和钙的优质来源,但是尽管把我吃掉了,这些营养物质仍然得通过他们的肠胃才能进入血液和细胞。"

"你是说吃掉你那健康的绿叶的人并不能享受所有的好处?我不明白。"

"我的确充满营养,但这并不是事情的全部。"

"Yes, they do consider me a good source of iron and calcium, but once I am eaten, those good things must still get from their stomachs and intestines into their blood and cells."

"You mean that people who eat your healthy green leaves don't get to enjoy all the benefits? I don't understand."

"Well, it is true that I am full of nutrients, but no one tells the whole story."

"这可有点难以理解了。你能举个例子吗？"

"你知道西蓝花吗？"菠菜问道。

"当然了，西蓝花是另一种非常健康的绿色蔬菜，富含人们强健骨骼所需要的钙质。"

"This is a bit confusing. Perhaps you can give me an example?"

"You know broccoli?" Spinach asks.

"Oh yes, that other very healthy green vegetable, full of the calcium people need to build strong bones."

你知道西蓝花吗?

you know broccoli?

……更不理想……

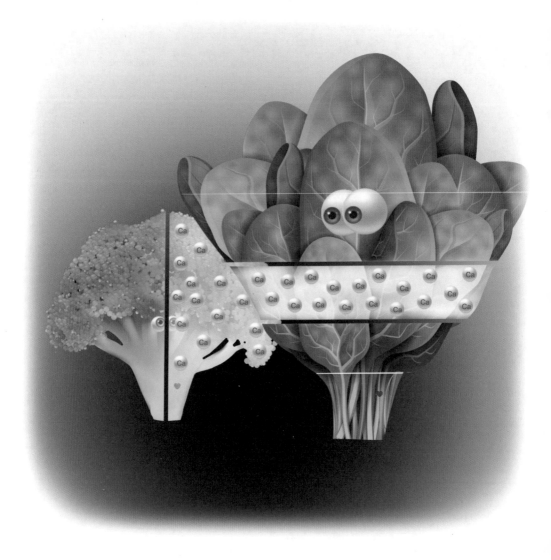

... even less impressive ...

"这个嘛，西蓝花中一半的钙质被身体吸收，而另一半流失了。"

"只有一半钙质去了该去的地方？有点可惜呀，但我想，那也聊胜于无吧。"

"我的表现更不理想，只有5%的钙质能留在他们的身体里，剩下的都流失了。"

"Well, half of the calcium in broccoli is absorbed into the body, while the other half goes out again."

"Only half gets to where it needs to be? Not very impressive, but better than nothing, I suppose."

"My performance is even less impressive, as only five percent of my calcium stays in their bodies. The rest gets flushed out."

"这真是太浪费了！流失的钙都去哪了？"

"冲进马桶里，然后永远流失了。"

"那么，如果你并不善于给孩子们提供健康的食物，你擅长什么？"

"我在发电方面表现不错。"

"What a lot of waste! And where does that end up?"
"Flushed down the toilet, and gone forever."
"So, if you're not the best at providing kids with healthy food, what are you good at?"
"I am great at making electricity."

冲进马桶里……

Flushed down the toilet …

……把我的暗蛋白质和硅混合……

... mixing my dark protein with silicon ...

"你在开玩笑吧!没人听说过这件事啊……"

"你不知道这件事,并不代表它不存在啊。人类把我的暗蛋白质和硅混合,用来发电。多亏了我,太阳能电池现在能发两倍的电了。"菠菜自夸道。

"You are kidding! No one in the world has ever heard of that…"

"Just because you don't know about it, doesn't mean it doesn't exist. By mixing my dark protein with silicon, people have created electricity. Thanks to me, solar cells can now make twice the power," Spinach boasts.

"怎么可能？！"

"你为什么不相信我呢？我清楚自己在说什么。我在谈论我的生活，而且我比你更了解我自己。"

"硅不是在很高的温度下制成的吗？在如此高的温度下，你的细胞无法存活啊。"松树说道。

"Impossible!"

"Why don't you believe me? I know what I am talking about. I am talking about my life – and I know myself better than you do."

"Is silicon not made at a very high temperature? None of your cells could ever survive," Pine Tree says.

……我更了解我自己……

...I know myself better...

当温度过高的时候……

When temperature is too high ...

"是时候上物理课了，我的朋友。当温度过高的时候，你必须降低压力，这点你记得吗？我的绿色细胞能够在真空中存活，这样一来，每个人都能利用和享受我的绿色电力。"

"我之前还以为生物学的世界与金属、矿物的世界是两个不一样的世界……"

"Time to go back to physics class, my friend. When temperature is too high then you must decrease pressure, remember? My green cells thrive in a vacuum, and that is how everyone can employ and enjoy my green power."

"And here I was thinking that the world of biology and the world of metals and minerals are two different worlds…"

"你的确意识到两个世界都存在于我们这个了不起的星球——地球，"菠菜说，"那么为什么不充分利用两者的结合呢？"

"确实，为什么不呢？让每个世界都尽其所能，为所有人提供食物和电力。"

……这仅仅是开始！……

"You do realise that both exist together on our awesome planet, Earth," Spinach says. "So why not make best use of combining the two?"

"Why not, indeed – with each world giving of its best, to provide food and power, for the benefit of all."

… AND IT HAS ONLY JUST BEGUN!…

……这仅仅是开始！……

... AND IT HAS ONLY JUST BEGUN! ...

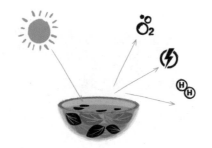

A simple extract from spinach leaves is able to produce electricity and hydrogen from water, using sunlight. The raw material is water, and outputs are electricity, hydrogen and oxygen.

菠菜叶中的简单提取物能够利用阳光，用水产出氢气并发电。水是原材料，输出的是电、氢气和氧气。

A bio-photo-electro-chemical process with spinach leaves converts sunlight into a flow of electrons. This is a new form of generation of clean fuels, from abundant sources like water and sun.

菠菜叶的生物光电化学过程将阳光转化为电子流。这是一种利用水和阳光等丰富资源生产清洁燃料的新方法。

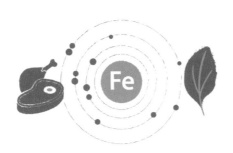

Spinach has high iron content, but this is low in bio-availability. Only 2% of the iron from spinach is actually absorbed. In comparison, as much as 35% of iron from meat is bio-absorbed.

菠菜的铁含量很高,但是生物利用度很低。菠菜中的铁只有2%被实际吸收了。相比而言,肉类中高达35%的铁能够被生物吸收。

Spinach has high levels of oxalic acid, which binds with iron, and which blocks the absorption of iron in the gut. That is why iron from animal-based protein is better absorbed by our bodies.

菠菜含有高浓度的草酸(乙二酸),这种酸能与铁结合,阻止肠道对铁的吸收。这就是为什么我们的身体更容易吸收动物蛋白中的铁。

Broccoli, Brussels sprouts, collards, kale, mustard greens, and Swiss chard, contain large amounts of highly absorbable calcium. Spinach is the exception. It does contain a lot of calcium but less is absorbed.

西蓝花、抱子甘蓝、散叶甘蓝、羽衣甘蓝、芥菜绿叶、瑞士甜菜都含有大量可吸收的钙质。菠菜是个例外。菠菜的确含有大量钙质，但是能够被人体吸收的部分很少。

Calcium is needed for healthy bones and teeth. It also plays an important role in blood clotting, nerve functioning, helping muscles to contract, and regulating normal heart rhythms.

健康的骨骼和牙齿需要钙质。钙在凝血、维持神经系统功能、帮助肌肉收缩和调节正常心率方面也扮演着重要角色。

About 99% of the body's calcium is stored in bones. The remaining 1% is in blood and muscles. If calcium levels drop too low in the blood, the parathyroid hormone will signal bones to release calcium in the bloodstream.

身体中大约99%的钙质储存在骨骼中。剩下的1%存在于血液和肌肉中。如果血液中的钙含量过低，人体就会释放甲状旁腺激素作为信号，让骨骼向血液中释放钙质。

Sardines, anchovies, sprat and mackerel are the richest sources of calcium as, once canned and cooked, the fish bones are edible. The Japanese eat Hamo eel, after mincing it to shorten the bones.

一旦罐装或烹饪过，在鱼骨可食用的情况下，沙丁鱼、凤尾鱼、西鲱和鲭鱼是最丰富的钙质来源。日本人吃海鳗前会把它们剁碎，这样它们的骨头就变短了。

Think about It　想一想

Are the worlds of biology and that of metals and minerals two different worlds?

生物的世界与金属和矿物的世界是两个不同的世界吗?

When it is too hot, can you survive?

你能在过热的环境里生存吗?

How good is a food that only delivers 5% of its nutrients to the body?

只能为身体提供其5%营养的食物是好食物吗?

Is all that is flushed down a toilet lost forever?

所有被冲进厕所的东西都永远消失了吗?

Do It Yourself! 自己动手！

List all the food items you include in your daily diet, and find out how much of the nutrients will ultimately get into your cells. Select those with the best absorption, and focus on eating more of these while reducing your intake of those that have poor absorption. This will reduce your overall intake of food, making it a sensible diet. Share the information you have gathered with your friends and family members, and ask them to share their point of view.

列出你日常饮食中的全部食物，然后找出有多少营养物质最终进入了你的细胞。选择那些营养吸收率最高的食物，专注于摄入更多的这类食物，同时减少摄入那些营养吸收率低的食物。这样做会减少你的食物总摄入量，并让你的饮食结构更合理。和家人、朋友分享你收集的信息，并请他们分享自己的观点。

TEACHER AND PARENT GUIDE

学科知识
Academic Knowledge

生物学	贝类富含铁；生物利用度指的是营养物质被人体吸收和利用的程度；西蓝花是十字花科蔬菜，含有多种抗氧化剂。
化学	为了通过光合作用产生电流，要添加铁基化合物；将太阳能转化为化学能；维生素C已经被证明能增强铁的吸收；植酸（肌醇六磷酸）可以减少过量的铁；西蓝花含有萝卜硫素和3-吲哚甲醇；人体需要维生素C来合成胶原蛋白，而西蓝花富含这两种物质；植物化学物质包括类胡萝卜素和多酚类，包括酚酸、类黄酮、芪类、木脂素类；松子富含维生素E、锌、单不饱和脂肪酸、镁和固醇；血红素具有可溶性，可被珠蛋白降解产物吸收。
物理	用真空来抵消高温。
工程学	太阳能电池板在制造过程中需要大量的能量并且具有毒性，这就要求太阳能电池板产业的转型，从单纯的无机工程转变为无机、有机相结合。
经济学	有些"健康食品"中的营养物质不能被人体吸收，生产这类食品十分低效，会带来经济损失。
伦理学	说谎和省略部分真相的区别；当某件事不为人知时，人们通常认为它是不可能的，甚至认为是不真实的。
历史	早在公元前400年，希腊医生希波克拉底就建议通过食用营养丰富的食物来预防和治疗疾病。
地理	中医中级别最高的医生是饮食医生；冲绳人是世界上最长寿的人群之一，他们相信并实践着正确的长寿饮食方式。
数学	计算食物中有多少营养物质能被身体吸收；通过随机对照试验（RCT）来评估营养物质对健康的影响；营养学非常适合引入数学概念，如排序、分类、统计、概率、估算、速率和比例。
生活方式	缺铁引起的红细胞数量减少会导致贫血症；铁元素不足会减少体内运输氧分的血红蛋白，导致身体虚弱、疲倦以及易怒。
社会学	植物疗法利用植物中的化合物治疗和预防疾病；允许每个物种进化到其最佳状态是很重要的。
心理学	营养心理学：你吃什么直接影响大脑的结构和功能，并最终影响心情；血清素是一种神经递质，在胃肠道中产生，帮助调节睡眠和食欲，调节情绪，抑制疼痛；研究表明，相比较于地中海饮食与传统日本饮食，西方传统饮食使人患抑郁症的风险提高25%—35%。
系统论	一个从水开始到水结束的封闭循环将太阳能转化为氢气中的化学能；人体能够自然调节对植物中铁的吸收来防止体内铁过多，动物性食物中的血红素铁容易被人体吸收，但难以调节，可能诱发炎症。

教师与家长指南

情感智慧
Emotional Intelligence

松 树

　　松树对菠菜的悲伤感同身受。他不明白为什么人类不能获得菠菜提供的所有益处，并且立刻询问原因。当他仍然不清楚时，他让菠菜举一个具体的例子。他很快分享了他对西蓝花的认识。松树想知道被浪费的营养物质去哪儿了。当他明白提供营养并不是菠菜的强项时，他试图通过询问菠菜的长处来安慰她。他自信地表示，菠菜的回答难以置信。当菠菜教他物理学，以及生物世界和金属世界可以合作的事实时，他的思维变得更开阔了。他意识到这有益于大家的共同利益。

菠 菜

　　菠菜了解松果的营养价值和自己的局限性，并为此感到伤心。她意识到人类并不完全了解她的营养价值，并对自己的表现感到失望。对于松树的问题，她的回答让松树感到困惑。当被要求举一个例子时，她将自己的表现与西蓝花作了比较，指出西蓝花的营养损失比自己少很多。她对此感到难过，但另一方面，她很高兴地宣布，她在发电方面做得很好。当菠菜被怀疑时，她坚持认为她很了解自己和自己的能力。然后，她详细解释了如何通过技术手段把生物和矿物的世界结合起来。这增强了她的信心，而且她为能够造福世界而感到自豪。

艺术
The Arts

　　我们怎样搭配食物才能使其中的营养物质被人体吸收？最好的方法是吃各类富含营养的食物。你要确保食物富含维生素C和铁。接下来加入一些健康的脂类。你要确保体内水分充足，因为没有水，血液就无法输送营养物质。你能用某种艺术形式来整合这些信息并分享给你的朋友和家人吗？

TEACHER AND PARENT GUIDE

思维拓展
Systems: Making the Connections

摄取食物是所有生物的本能。整个生命网络都建立在能量交换的基础上，依存于食物链的循环。这就引出了一个问题：什么算是食物？食物是任何可以被摄取和消化，并提供能量和维持生命的物质。早期人类以游猎、采集为生，四处游荡来寻找基本的生活条件：水、食物以及居所。几千年后，有些人类在水源附近形成了更稳定的聚落，他们在那里发展农业，以确保食物充足。关于食用以及药用植物的知识成为了他们生存能力的一部分。像驯养动物的知识一样，他们将关于植物的知识传授给了后代。饮食习惯取决于这些种群周边有哪些动植物。所以，饮食习惯的进化是由环境和生态系统的承载能力决定的。从那时起，各种植物都被用作食物、饲料和药物。传统可食用植物的不同部位都为人类提供了淀粉，香草和香料不仅增添了味道，也具有药用价值。可食用植物富含抗氧化剂、蛋白质、碳水化合物、矿物质等，这些化合物现在被提取出来，用于"救命药"中。过去几十年的研究表明，这些植物中的各种活性成分可以用来生产药物。可食用植物富含活性抗氧化成分，能对抗压力，进行自我保护。这种成分还对人体有促愈合和保护作用。从可食用植物中发现的化合物表明，古代的饮食习惯是以健康为导向的。这些从食物中提取的新一代的化合物以药物的形式存在，不能作为食物而每天大量食用。过去的人们使用草药来治疗。耐人寻味的是，有许多低等植物（如苔藓、地衣和蕨类植物）和许多被子植物（开花植物）已经被中国、朝鲜和印度等国家的古代文献记载了几个世纪。这表明了古人在植物药用方面具有非常广博的知识。因此，我们应该这样理解菠菜发电这件事：很多植物也在为我们的身体"发电"，没有它们的"发电"，我们的大脑、神经和心脏就无法工作。现在是时候关注我们身体的能源需求了。

动手能力
Capacity to Implement

你能用植物发电吗？我们需要把它们的根连起来。植物会在土壤中留下养分，养分分解的过程中会产生微小的电流。你能利用这些知识，用植物产生的电来点亮一个小LED灯吗？这可能有点难，但你应该尝试一下。

教师与家长指南

故事灵感来自
This Fable Is Inspired by

大卫·克里夫
David Cliffel

大卫·克里夫于1998年获得得克萨斯大学奥斯汀分校分析化学博士学位。1998—2000年，他在北卡罗来纳大学教堂山分校担任博士后研究助理。他的职业生涯始于田纳西州纳什维尔的范德比尔特大学，2000—2016年，他在那里担任教授，并在2016—2019年担任系主任。他被誉为化学界的科尼利尔斯·范德比尔特教授。他的研究小组利用从菠菜叶子中分离出来的光合蛋白分子开发了基于光系统I的生物混合太阳能电池。他的目标是通过使用小分子介质、纳米碳材料、导电聚合物和纳米颗粒，增加光系统I蛋白质分子和电极表面之间的直接电荷转移。

图书在版编目(CIP)数据

冈特生态童书.第八辑:全36册:汉英对照/
(比)冈特·鲍利著;(哥伦)凯瑟琳娜·巴赫绘;
何家振等译. —上海:上海远东出版社,2021
ISBN 978-7-5476-1773-1

Ⅰ.①冈… Ⅱ.①冈…②凯…③何… Ⅲ.①生态环
境-环境保护-儿童读物—汉、英 Ⅳ.①X171.1-49

中国版本图书馆CIP数据核字(2021)第249940号

策　　划　张　蓉
责任编辑　祁东城
封面设计　魏　来　李　廉

冈特生态童书
菠菜的电力
[比]冈特·鲍利　著
[哥伦]凯瑟琳娜·巴赫　绘
贾龙慧子　译

记得要和身边的小朋友分享环保知识哦!
八喜冰淇淋祝你成为环保小使者!

用牙齿走路
Walking Using Teeth

Gunter Pauli

[比]冈特·鲍利 著
[哥伦]凯瑟琳娜·巴赫 绘
贾龙智子 译

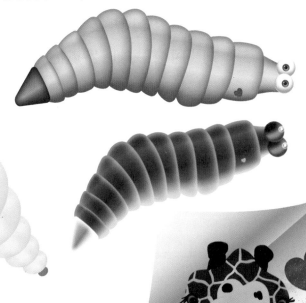

上海远东出版社

丛书编委会

主　任：贾　峰

副主任：何家振　闫世东　林　玉

委　员：李原原　祝真旭　牛玲娟　梁雅丽　任泽林
　　　　王　岢　陈　卫　郑循如　吴建民　彭　勇
　　　　王梦雨　戴　虹　翟致信　靳增江　孟　蝶

特别感谢以下热心人士对童书工作的支持：

匡志强　宋小华　解　东　厉　云　李　婧　陈　果
刘　丹　熊彩虹　罗淑怡　旷　婉　杨　荣　刘学振
何圣霖　廖清州　谭燕宁　韦小宏　李　杰　欧　亮
陈强林　王　征　张林霞　寿颖慧　罗　佳　傅　俊
胡海朋　白永喆　冯家宝

目录

用牙齿走路	4
你知道吗?	22
想一想	26
自己动手!	27
学科知识	28
情感智慧	29
艺术	29
思维拓展	30
动手能力	30
故事灵感来自	31

Contents

Walking Using Teeth	4
Did You Know?	22
Think about It	26
Do It Yourself!	27
Academic Knowledge	28
Emotional Intelligence	29
The Arts	29
Systems: Making the Connections	30
Capacity to Implement	30
This Fable Is Inspired by	31

几只幼虫在一堆泥土上碰面了。这些幼虫有的是从甲虫卵中孵化出来,另一些是从苍蝇和黄蜂卵中孵化出来的。

"我们都长得很像,这不是很有趣吗?任何人都很难判断我们中的哪一个会变成什么——甲虫、苍蝇、蝴蝶、蛾子、黄蜂,还是蚂蚁?"苍蝇幼虫说道。

Some larvae meet up on a pile of dirt. A few hatched from beetle eggs, others from those of flies and wasps.

"Isn't it interesting that we all look alike? It is hard for anyone to tell which of us will turn into beetles, or into flies, butterflies, moths, wasps, or ants," Fly Larva says.

几只幼虫在一堆泥土上碰面了。

Some larvae meet up on a pile of dirt.

或者蜜蜂……

Or bees ...

"或者蜜蜂。"蜜蜂幼虫补充道。"如果只是观察幼虫，没人会得到任何线索。除非有专家在我们孵化时监视我们。"

"你知道我们是动物中最成功的生命形式吗？"

"谁会知道昆虫是世界上最多样化的生物种群，而且我们昆虫都会经历这个简单的幼虫阶段？"

"Or bees," a bee larva adds. "No one has any clue, just looking at a larva. Not unless some expert monitors us as we hatch."

"Do you realise that we are the most successful form of animal life?"

"Who in the world would have known that insects are the most diverse group of organisms in the world, and that we all go through this simple larval phase?"

"没有人！我们呈圆柱形，皮肤柔软，没有腿和鼻子。我们沉迷于食物，生长迅速，且远离别人的视线。"

"那些注意到我们的人，认为我们是无助的生物。然而，我们是大自然超级成功的设计产物。没有生物比得上我们。"

"是的，我们巧妙的生存方法不引人注意。也许这样更好。我们不要自吹自擂，让每个人都知道我们认为自己有多好……"

"No one! We take the shape of a cylinder, have a soft skin, and skip legs and noses. We are obsessed with food, growing fast and staying out of sight."

"Those who do notice us, consider us helpless creatures. However, we are a super successful design of Nature. There is no match for us."

"Yes, our ingenious ways go unnoticed. Perhaps better so. Let's not brag about ourselves and let everyone know how good we think we are…"

我们沉迷于食物……

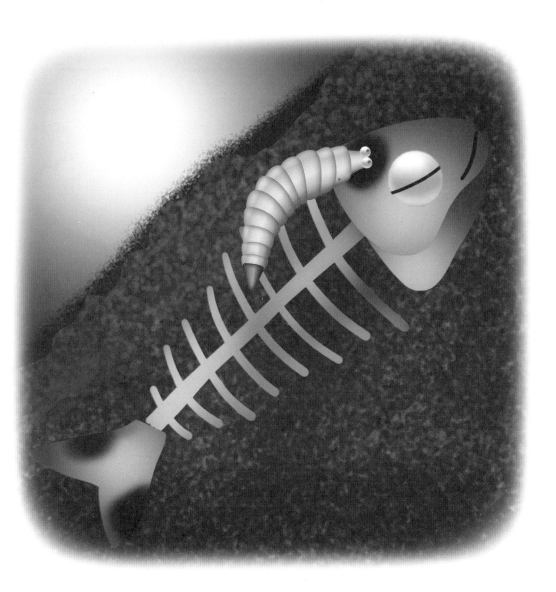

We are obsessed with food...

……只要我能吃得多……

... as long as I can just get to eat lots ...

"我不介意自己不为人知地活着,只要我能吃得多,长得快,茁壮成长。"蜜蜂幼虫说。

"而且我们吃得多好啊!我们可以在不到100小时的时间里,让自己的体形和体重增加数百倍。"

"因为我们的身体结构非常简单:没有腿,没有尾巴,没有鼻子。"

"I don't mind living unnoticed, as long as I can just get to eat lots, grow quickly and thrive in life," Bee Larva says.
"And what good eaters we are! We can increase our size and weight hundreds of times – in less than a hundred hours."
"Because we have such a very simple body: no legs, no tail, no nose."

"幸运的是，我们的皮肤非常柔软有弹性。即使我们长大10倍，同样的外皮也一样合身。"

"蜕皮需要消耗大量的能量。如果我们因长得太快而蜕皮，我们就会像蜘蛛那样，吃掉我们的旧皮。就这样，旧皮又得到利用。"

"Fortunately, our skin is very soft and flexible. The same jacket fits us, even when we grow ten times bigger."

"Moulting costs lots of energy. And if we do moult when growing too fast, we do what the spider does – eat our old skin. So, it's used again."

……我们的皮肤非常柔软有弹性。

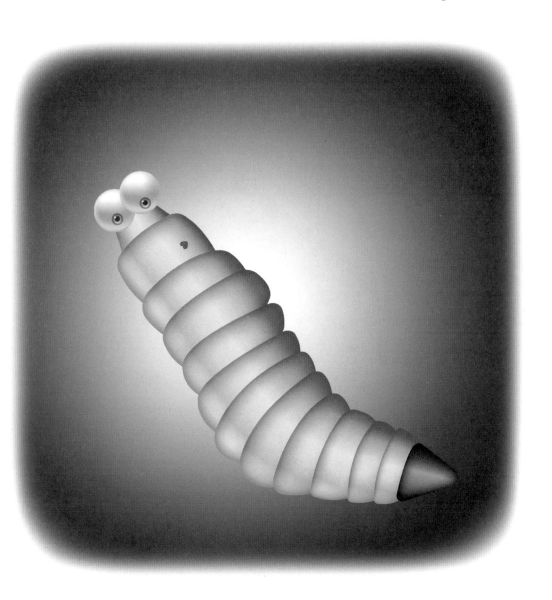

... our skin is very soft and flexible.

我们快速泵送氧气……

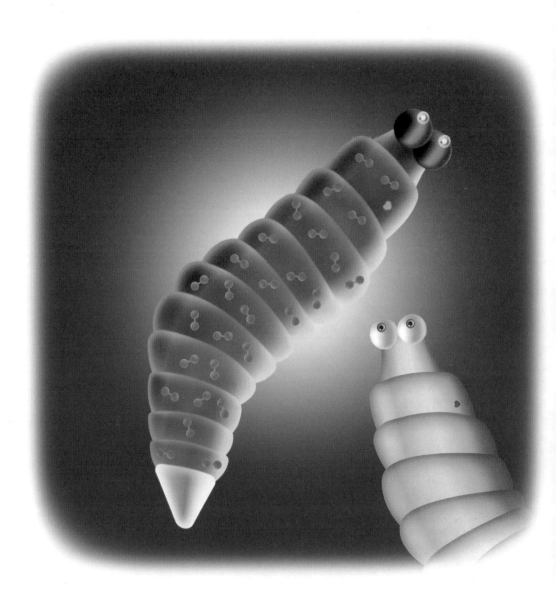

And we rapidly pump oxygen ...

"简单的设计是一门艺术：我们像管子一样，把身边的食物泵进体内。所有食物都能轻松有效地进入我们体内。我们的身体几乎没有摩擦力。"

"我们通过体内的空气快速泵送氧气，使氧气的流动速度比它在血液中快10 000倍。我们就是这样茁壮成长的。很酷吧？"

"Simple design is an art: we pump food around our body shaped like a tube. Everything moves through us easily and efficiently. We have no friction."

"And we rapidly pump oxygen through the air in our body, moving it 10,000 times faster than oxygen in blood. That's how we thrive. Cool, right?"

"别忘了,"苍蝇幼虫补充道,"我们可以藏在粪便、尸体、内脏甚至强酸性水里,并且仍能茁壮成长。没有捕食者会想在那里寻找我们,所以我们可以平静地生活。"

"很少有人意识到,躲在泥土和黑暗中有助于我们生存。"

"Don't forget," Fly Larva adds, "we can hide inside dung, a cadaver, guts or even very acidic water, and still thrive. No predator would ever want to look for us there, so we get to live in peace."

"Few realise that staying hidden in dirt and darkness helps us survive."

……我们可以藏在粪便、尸体里……

… we can hide inside dung, a cadaver …

……我们很快就会变得很胖……

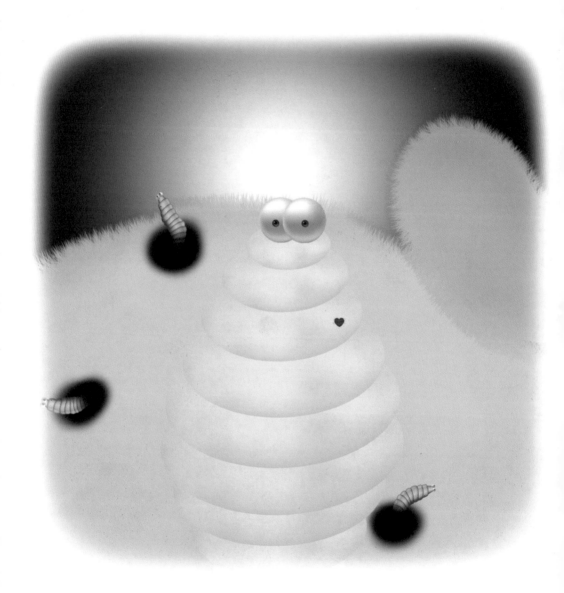

... we get very fat very fast ...

"我们唯一的问题是我们很快就会变得很胖。我们不能轻易地四处走动,因为我们没有腿,也没有鳍……"

"为了逃跑,我们可以进行最疯狂的跳跃,没有人能模仿。"

"Our only problem is that we get very fat very fast. And we can not get around easily because we don't have legs, or fins…"

"To escape, we can take the wildest leaps, ones no one can imitate."

"我甚至有两颗尖牙,让我能用门牙'行走'。"

"我们没有腿可以行走,也没有鳍可以游泳,我们借助牙齿,收缩和伸长身体,用这样的方式向前爬行。我们独特的出行方式就是规则,用腿和鳍行动才是例外。"

……这仅仅是开始!……

"And I even have a couple of fangs that allow me to 'walk' – on my front teeth."

"And not having legs to walk, or fins to swim, we crawl by contracting, expanding and moving forward, using our teeth. Our way of getting around is the rule – legs and fins are the exception."

… AND IT HAS ONLY JUST BEGUN!…

……这仅仅是开始！……

...AND IT HAS ONLY JUST BEGUN! ...

你知道吗？

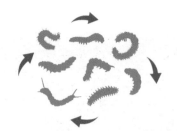

Larvae represent the most successful form of life. An estimated 45-60% of all living animal organisms pass through the larval stage. While the ultimate species differs a lot, all larvae look very much the same.

幼虫是最成功的生命形式。估计有45%—60%的动物会经过幼虫阶段。虽然各种完全成熟的物种有很大的不同，但所有的幼虫看起来都大同小异。

Beetles, flies, wasps, bees, ants, butterflies and moths all emerge from a cylindrical, soft body that feeds and breathes easily, grows very fast, reduces predation, and has a very high rate of survival.

甲虫、苍蝇、黄蜂、蜜蜂、蚂蚁、蝴蝶和蛾子都是从一个圆柱形的、柔软的身体开始成长的，这种身体便于进食和呼吸，生长非常快，不易被捕食，生存率非常高。

Drosophila larvae grow from hatchlings to pupation in only 96 hours. Honeybee larvae increase their body size 800 times in 96 hours; during that time they are fed 1,900 times by specialised nursing bees.

果蝇幼虫从孵化到化蛹只需96小时。蜜蜂幼虫在96小时内身体增大800倍；在此期间，它们被专门的育幼蜂喂食1 900次。

Specialised nursing bees examine their larvae several times a minute, and feed them up to 140 times in that time. Queens can be fed 1,600 times in 96 hours. Having no legs allow them to burrow very efficiently.

专门的育幼蜂每分钟要检查它们的幼虫好几次，并在这段时间里喂它们多达140次。蜂王在96小时内被喂食1 600次。由于没有腿，它们可以非常高效地钻进蜂巢。

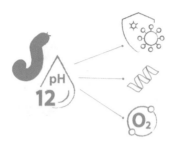

Larvae are able to produce a very high pH (up to 12). This allows very fast digestion of protein, creates a defence against parasites and viruses, and circulates oxygen more rapidly.

幼虫能够产生非常高的pH值（高达12）。这帮助幼虫快速消化蛋白质，抵御寄生虫和病毒，并加快氧气循环。

Diffusion of oxygen in water is 10,000 times slower than in air. Water is also more viscous than air. Thus, the larvae have an extraordinary breathing system (trachea) that provides energy very efficiently.

氧气在水中的扩散速度是空气中的万分之一。水的黏滞度也比空气强。因此，幼虫有一个特别的呼吸系统（气管），可以很高效地提供能量。

Blood nutrients such as sugars and lipids can be 50 times higher in concentration in insects than in vertebrates. This suggests a very high efficiency in circulating nutrients through the body.

昆虫"血液"中的糖和脂质等营养物质的浓度可能是脊椎动物的50倍。这表明营养物质在昆虫体内的循环效率非常高。

The simple, worm-like shape of larvae allows them to grow fast and stay hidden from the external world, ensuring that once they are established, their survival rates are high.

幼虫形状简单,像蠕虫一样,这可以让它们快速生长,并对外界保持隐蔽。这意味着一旦卵孵化成幼虫,它们的存活率就很高。

Think about It 想一想

Is moving around without legs the standard for animals?

不用腿行走,对动物而言是常态吗?

Would you like to be the same as everyone else?

你愿意和其他人一样吗?

Larvae are the most successful form of life?

幼虫是最成功的生命形式吗?

What about getting 100 times fatter in 100 hours?

在100小时内长胖100倍是什么感觉?

Do It Yourself!   自己动手!

Find out which animals gain the most weight in the shortest period of time. We must find those that can be fed to gain weight quickly, and become a reliable, rich source of protein. Make a list of the champions of protein production, and then calculate how much these will need to produce to help eradicate world hunger. Once you have your findings, share them with your friends and family members. Be prepared to be surprised, as they may have found some additional champions.

找出在最短的时间内体重增加最多的动物。我们必须找到那些能够快速增加体重的动物,作为可靠、丰富的蛋白质来源。列一张蛋白质生产冠军的名单,然后计算需要养殖多少这些动物才能在全世界消除饥饿。一旦你有了自己的发现,与你的朋友、家人分享。准备接受惊喜,因为他们可能已经找到了其他冠军。

TEACHER AND PARENT GUIDE

学科知识
Academic Knowledge

生物学	昆虫、两栖动物和刺胞动物会经历幼虫期；蝌蚪是一种幼虫；幼虫生活在一个独特的环境中，有躲避天敌的庇护所，减少了与成熟个体的资源竞争；与家禽、牛、猪相比，昆虫转化食物的效率更高；黑水虻是一种腐生昆虫；从幼虫变为成虫。
化学	幼虫产生热量，提高肠道酶的活性；黑水虻幼虫蛋白质含量为33%，高于棕榈仁（14%），脂肪含量为38%，低于棕榈仁（45%）；从幼虫身上提取的油的抗氧化活性强于棕榈仁油；幼虫富含钙、铁、钾、镁、磷、锌和维生素，包括维生素B_1、维生素B_2、维生素B_3、维生素B_{12}。
物理	幼虫身体的摩擦力较小，并且能使体内充满氧气。
工程学	原本需要花钱来处理的废物，被转化为种植作物的基料；需要净化的污水被转化为饲料。
经济学	由于幼虫疯狂消耗各种有机质，因此幼虫是可持续废物管理的关键；幼虫蛋白质和脂肪含量高，是动物饲料和生物柴油生产中有用的添加剂；消耗100千克饲料和2年时间，可产出10千克牛肉，而消耗等量的饲料和3周的时间就可产出20千克蟋蟀；饲料中重要的蛋白质来源是大豆和鱼粉，大豆与粮食竞争土地，而鱼粉则由于海洋过度捕捞，其供应越来越紧张。
伦理学	我们应该用化学物质杀死昆虫，把它们从用来喂动物的植物中消灭掉，还是吃那些以植物废料为食的昆虫？这些植物废料没人要，甚至连牛都不要。
历史	欧洲历史上，蝉的幼虫曾是一些地方的特色菜；北美洲的一些原住民猎杀无翼摩门蟋蟀。
地理	非洲一些民族，会在春雨降临时享用有翅白蚁；蜻蜓在一些地区是美味佳肴；黑水虻生活在热带和亚热带地区。
数学	计算并比较用幼虫或家畜生产蛋白质的生态足迹。
生活方式	我们需要改变生活方式，因为我们不再有能力像过去那样生产饲料和饲养动物来满足我们的食欲；人类必须改变生活方式，我们必须寻找更可持续的蛋白质来源。
社会学	苍蝇幼虫饲养系统可以适用于大规模的市政堆肥，以及小规模的个体养殖场，甚至城市家庭；顽固的饮食偏好。
心理学	恐惧心理对消费者食用昆虫意愿的影响；虾和龙虾是甲壳类动物，因此是节肢动物，就像蟑螂，但我们没有表现出厌恶。
系统论	由于麦糟的保质期不到48小时，因此大多数酿酒副产品最终都会被填埋，这可能会污染周围的环境，而这些副产品可以喂给幼虫；垃圾填埋场中的每吨麦糟释放出513千克二氧化碳当量的温室气体，但利用幼虫可以防止这种情况的发生。

教师与家长指南

情感智慧
Emotional Intelligence

苍蝇幼虫

苍蝇幼虫是乐观的,而且很高兴成为非凡的幼虫生命形式的一员。他很自信,知道对于动物来说,幼虫发育阶段是地球上最成功的生命形式。他知道他们成功的生存方式并不为人所知,因此幼虫可以快乐地躲起来。他主张保持谦虚和低调,但又吹嘘幼虫的胃口。他详细解释了他们简单的身体设计,以及柔软而有弹性的皮肤,这些都是幼虫成功的原因。让他感到自豪的是,幼虫能在其他生物不愿意生活的环境中悄悄地生存并茁壮成长。苍蝇幼虫意识到他们不能行走或游泳,但这并不影响他下定决心去获取食物。

蜜蜂幼虫

蜜蜂幼虫证实了在幼虫期很难区分不同昆虫的事实。蜜蜂幼虫自信地提到,许多动物都要经历幼虫期,这证明了幼虫期是保护后代的最有效形式。他满足于不被注意、肮脏和黑暗的生活状态,因为这能保障他茁壮成长。蜜蜂幼虫赞美幼虫的简单设计和不浪费能源的成长方式。蜜蜂幼虫知道自己的身体是如何工作的,并随意地分享幼虫通过体内空气而不是血液泵送氧气这一信息。蜜蜂幼虫认为生活在肮脏和黑暗的环境中是有益的。蜜蜂幼虫很机智,说他们的无足运动方式是常态,而不是例外。

艺术
The Arts

通过一系列图画向人们展示,世界上有许多没有腿、翅膀或鳍的动物,通过伸展、收缩身体和使用牙齿向前移动。这些信息可能会让很多人感到惊讶,甚至感到好笑。想办法用你的艺术创造笑声。你甚至可以模仿这些动物的动作。

TEACHER AND PARENT GUIDE

思维拓展
Systems: Making the Connections

所有人都要吃饭。对动物饲料而言，昆虫是鱼粉或豆粕的理想替代品。昆虫的蛋白质和脂肪含量高，各类氨基酸和脂肪酸含量均衡，富含人体所需的微量元素。昆虫可以有效地将有机废物转化为高质量的蛋白质，这些废物原本会被丢弃在垃圾场，造成环境污染。苍蝇幼虫还可用于改善环境卫生和人体健康状况。这些举措将有助于改善贫困社区的环境。例如，用1吨幼虫饲料来代替0.5吨鱼粉和0.5吨豆粕，可以减少对土地的需求，提高能源使用效率，进而缓解全球土地压力。如果用有机资源喂养昆虫，昆虫饲料生产的可持续性是最高的，因为这些有机资源目前不适合用作牲畜饲料。对于小规模养殖户来说，牲畜养殖最主要的成本是饲料成本，饲料成本高达全部成本的70%。豆粕和鱼粉，特别是鱼粉价格正在迅速上涨。因此，中低收入国家的农民需要既有效又负担得起的替代品。一项针对肯尼亚小规模家禽养殖户的调查显示，他们意识到昆虫饲料给他们带来的机遇。小农户可以在当地养殖苍蝇幼虫，他们既可以将这些幼虫加入自己配制的饲料中，也可以出售。这将为农民提供机会，积极参与新兴的昆虫农业产业链。通过这种方式，他们可以减少对国际饲料生产商的依赖，同时通过生产昆虫饲料获得收入。这有助于改善小农的生计和粮食安全。苍蝇幼虫的生产成本涉及基础设施投资，如场地和容器，相关费用很低。生产昆虫饲料的另一个好处是，收获苍蝇幼虫后的剩余物是一种肥料，这提供了额外的经济和生态效益。总之，昆虫饲料不仅有竞争力，而且对环境也有好处，提高了生态系统的承载能力，特别是在自给自足的农民感到难以生存的地区。

动手能力
Capacity to Implement

去网上、博物馆或科技馆观察一下各类幼虫。仔细看，你会发现有些是灰色的，有些是黑色的，有些是白色的。它们中的大多数在受到干扰时不会移动。研究它们的形态和结构。拍几张照片，然后确定幼虫的类型，以及这些幼虫成熟后会变成什么样。这能提高你的观察力，比如，能区分家蝇和黑水虻。试着成为幼虫专家，并准备解决世界粮食问题！

教师与家长指南

故事灵感来自
This Fable Is Inspired by

埃尔斯杰·皮特塞
Elsje Pieterse

　　埃尔斯杰·皮特塞出生在纳米比亚，在一座农场长大。她于 1992 年在南非比勒陀利亚大学获得农业和动物科学学士学位。她从一开始就担任非洲最成功的动物蛋白公司 AgriProtein 的科学总监。2006 年，她在南非斯泰伦博斯大学获得了动物科学博士学位。随后，她开始在斯泰伦博斯大学任教，讲授可持续的替代蛋白质来源，以及利用昆虫进行营养循环。皮特塞博士研究用耗水更少、造成水污染更少的饲料来生产食物时的效率，尤其是在家禽养殖领域。她培养了 30 多名昆虫养殖方向的研究生，造就了新一代的食品专家。她创造了每公顷土地蛋白质产量的纪录。

图书在版编目（CIP）数据

冈特生态童书.第八辑:全36册:汉英对照/
(比)冈特·鲍利著;(哥伦)凯瑟琳娜·巴赫绘;
何家振等译.—上海:上海远东出版社,2021
ISBN 978-7-5476-1773-1

Ⅰ.①冈… Ⅱ.①冈…②凯…③何… Ⅲ.①生态环境－环境保护－儿童读物—汉、英 Ⅳ.①X171.1-49

中国版本图书馆CIP数据核字(2021)第249940号

策　　划	张 蓉
责任编辑	祁东城
封面设计	魏　来　李　廉

冈特生态童书
用牙齿走路
[比]冈特·鲍利　著
[哥伦]凯瑟琳娜·巴赫　绘
贾龙智子　译

记得要和身边的小朋友分享环保知识哦！
八喜冰淇淋祝你成为环保小使者！

Energy 270

像石头一样下坠

Drop like a Rock

Gunter Pauli

［比］冈特·鲍利 著
［哥伦］凯瑟琳娜·巴赫 绘
贾龙慧子 译

上海远东出版社

丛书编委会

主　任：贾　峰

副主任：何家振　闫世东　林　玉

委　员：李原原　祝真旭　牛玲娟　梁雅丽　任泽林
　　　　王　岢　陈　卫　郑循如　吴建民　彭　勇
　　　　王梦雨　戴　虹　翟致信　靳增江　孟　蝶

特别感谢以下热心人士对童书工作的支持：

匡志强　宋小华　解　东　厉　云　李　婧　陈　果
刘　丹　熊彩虹　罗淑怡　旷　婉　杨　荣　刘学振
何圣霖　廖清州　谭燕宁　韦小宏　李　杰　欧　亮
陈强林　王　征　张林霞　寿颖慧　罗　佳　傅　俊
胡海朋　白永喆　冯家宝

目录

像石头一样下坠	4
你知道吗？	22
想一想	26
自己动手！	27
学科知识	28
情感智慧	29
艺术	29
思维拓展	30
动手能力	30
故事灵感来自	31

Contents

Drop like a Rock	4
Did You Know?	22
Think about It	26
Do It Yourself!	27
Academic Knowledge	28
Emotional Intelligence	29
The Arts	29
Systems: Making the Connections	30
Capacity to Implement	30
This Fable Is Inspired by	31

一只无聊的小老鼠冒险爬上一颗树,那里有一只猫头鹰在休息。要知道,那可是以老鼠为食的猛禽呀。小老鼠的生活非常安逸,她的宝宝们食物充足,快速地成长着,因此她有时间淘气,去打扰这只猫头鹰。

"你好,猫头鹰先生,你最近过得怎么样?"老鼠太太问道。

A bored mouse ventures into a tree, where an owl, a bird of prey that eats mice, rests. All is well with her, her little ones are growing fast with all the food available, so she has time to be naughty and disturb the owl.

"Hello, Mr Owl. How are you doing?" Mrs Mouse asks.

一只无聊的小老鼠冒险爬上一颗树……

A bored mouse ventures into a tree ...

你有了宝宝,要照顾他们……

Having babies and taking care of them ...

"哦，看看谁在这。嗯，让我看看，你已经当妈妈啦！你有了宝宝，要照顾他们，这真让人感动。"

"我想知道你有没有看到过在这附近飞翔的游隼。"

"当然，那些鸟一直在周围飞来飞去。对于我们这么聪明的猫头鹰来说，一只游隼可不是对手。我们带来智慧，你知道的。"

"Oh, look who is here. Growing up to be a mommy, I see. Having babies and taking care of them. I am impressed."

"I wonder if you have seen the falcon flying around."

"Sure, those birds are around all the time. A falcon is no match for a wise owl like me. We bring wisdom, you know."

"游隼看得比你远呀。"

"只是在白天而已。我可是夜视冠军。"

"游隼飞得比你高。"

"她可不飞,她只是借着暖气流漂浮在高空中罢了。"

"The falcon can see better than you."
"Only during the day. I am the champion of sight at night."
"The falcon flies higher than you."
"She does not fly, she is just kept aloft by hot air."

游隼看得比你远呀。

The falcon can see better than you.

但我有更强的升力……

But I have more lift ...

"游隼飞得可快了。"

"但我有更强的升力。"猫头鹰回答。

"但是你在晚上鬼鬼祟祟的,你并不捕猎,你只是伏击你的猎物。"

"The falcon is really fast."
"But I have more lift," Owl responds.
"You are sneaky at night, though. You do not hunt, you ambush your prey."

"游隼使用蛮力,而我开创了沉默的艺术……"

"沉默?我宁愿称其为偷偷摸摸。"

"嘿!你为什么这么喜欢游隼呢?"猫头鹰问道。

"我并不喜欢或讨厌你们中的任何一个,我只是想知道谁是最强的。保持谦逊并且知道有别人比自己更优秀总是好事,尤其当一个人认为自己最棒的时候。"

"That falcon uses brute force, whereas I have developed the art of silence…"

"Silence? I would call that stealth."

"Look, why are you so in favour of the falcon?" Owl asks.

"I am not in favour or against either of you, I just want to know who is the best. It is always good to be humble and know there are others who are better, especially when one thinks one is the best."

游隼使用蛮力……

That falcon uses brute force ...

……可以咬断兔子的脖子。

... can break the neck of a rabbit.

"游隼是最强的?我的双爪很强壮,一旦我抓到猎物,它就完蛋了。我能一口气把它咽下去。"

"游隼嘴尖部位的牙齿很强壮,可以咬断兔子的脖子。"

"游隼可以用牙齿咬死兔子?这我可不知道。"

"A falcon, the best? My talons are so strong, once I get a hold of my prey, it's over and done with. I gulp it up down in one go."

"The falcon has such a strong tooth at the tip of its beak that it can break the neck of a rabbit."

"Falcons can kill rabbits with a tooth? I did not know that."

"一只从空中急速下降的游隼只需要击中一只鹤的翅膀，就可以抓住他饱餐一顿。"

"嗯，是的，我同意。游隼俯冲的速度很快。"

"不，不，她不仅仅是快，她可是全世界最快的鸟类。"老鼠说道。

"A falcon dropping through the sky only has to hit the wing of a crane to catch him as a meal."

"Hmm, yes, agreed. The falcon is fast, going down."

"No, no, she is not just fast, she is the fastest bird in the whole wide world," Mouse says.

……击中一只鹤的翅膀，就可以抓住他……

... hit the wing of a crane to catch him ...

……游隼不需要拍打翅膀……

... falcon doesn't need to flap her wings ...

"但并不是靠她自己的力量,记得吗?那是自由落体运动。我拍打一次翅膀的效果相当于游隼拍打五次。这下如何?"

"嗯,猫头鹰先生,游隼甚至不需要拍打一次翅膀就可以让自己的速度达到400千米/时。"

"But not by her own force, remember? It is a free fall. Where a falcon needs to flap her wings five times, I only need to do so once. How about that?"

"Well, Mr Owl, the falcon doesn't need to flap her wings even once to reach four hundred kilometres an hour…"

"那是因为她利用重力下降，就像一块石头那样。"

"游隼知道如何利用一些你不会利用的东西，她利用暖气流上升，几乎不用拍打翅膀。"

"好吧，你赢了。她在某些方面更优秀，但我更聪明，可以安静而优雅地飞翔。"

……这仅仅是开始！……

"That's because she uses the force of gravity to drop down like a rock."

"The falcon knows how to use something else you don't, she uses the warming air for lift,… hardly flapping."

"Alright, you win. She is better at some things, but I am the wise one, who flies silently and gracefully!"

… AND IT HAS ONLY JUST BEGUN!…

……这仅仅是开始！……

... AND IT HAS ONLY JUST BEGUN! ...

Did You Know?

你知道吗？

In 1940, Britain issued the "Destruction of Peregrine Falcons Order", where peregrines were to be killed under wartime emergency regulations, as these were a threat to the pigeons that carried messages for the army.

1940年，英国颁布了《消灭游隼令》，规定根据战时紧急条例必须杀死游隼，因为游隼对给军队送信的鸽子构成了威胁。

The toxins of organochlorine pesticides used in agriculture, most notably DDT, concentrates up the food chain. These toxins thinned the raptors' eggshells, and increased adult mortality.

农业中使用的有机氯农药，尤其是DDT，其毒素随食物链富集。这些毒素使猛禽的蛋壳变薄，也会增加成鸟的死亡率。

驯鹰术

De Arte Venandi cum Avibus, or *The Art of Falconry*, was written in the 13th century, by Frederick II of Hohenstaufen, the Holy Roman Emperor and King of Sicily and Jerusalem.

《驯鹰术》在13世纪由神圣罗马帝国皇帝、西西里和耶路撒冷国王，霍亨斯陶芬王朝的腓特烈二世所著。

In *The Book of Saint Albans* written in 1486, Dame Juliana Berners warns that destroying falcon's eggs is punishable by a year in prison; and poaching a falcon from the wild risked having your eyes gouged out.

在写于1486年的《圣奥尔本斯书》中，朱莉安娜·博纳斯爵士警告称：破坏游隼的蛋将被判处一年监禁；如在野外偷猎一只游隼，眼睛将会被剜出来。

Falcons prey on birds as large as a crane, as small as a hummingbird, and as elusive as a swift. They also hunt bats. They occasionally seize prey from other raptors.

游隼可以捕猎各种各样的鸟类，大到鹤，小到蜂鸟，以及像飞燕一样灵活的鸟类。它们也捕食蝙蝠。它们偶尔还会抢夺其他猛禽的猎物。

Owls swallow their prey whole, like snakes do, and then regurgitate the undigested parts in a pellet. One barn owl family will eat 3,000 rodents in a breeding season and can devour 25 kilograms of gophers.

猫头鹰能像蛇一样，将猎物整个吞下去，然后未消化的部分像个小球一样，继续被猫头鹰反刍。在繁殖季，一窝猫头鹰能吃掉3 000只啮齿动物，或25千克囊鼠。

In Greek mythology, the Little Owl was a companion to Athena, the Greek goddess of wisdom. That is why owls symbolise learning and knowledge. Seeing an owl during battle was a sign of a coming victory.

在希腊神话中，小猫头鹰是希腊智慧女神雅典娜的伙伴。这就是为什么猫头鹰象征着学习与知识。在战争中看到猫头鹰标志着胜利即将到来。

Owls have been associated with witches and evil. While this may sound like Halloween fun, many cultures still have superstitions about owls and in some places owls are killed based on such mistaken beliefs.

猫头鹰被认为和女巫、邪恶事物有所关联。虽然这听起来像个万圣节玩笑，但许多文化仍然对猫头鹰抱有迷信，在某些地方，猫头鹰因为这种错误的信念而被杀害。

Think about It   想一想

If you are the best, is there someone who is better?

如果你是最优秀的那个人，还会有人比你更优秀吗？

What do you do when you are bored?

当你无聊的时候你会做什么？

When you lose, do you lose with grace?

当你失败的时候，你会优雅地接受失败吗？

Do you like to argue for and against, or argue for the better?

你喜欢为辩论而辩论，还是为了更好的结论而辩论？

Do It Yourself!

 自己动手！

The owl and the falcon are two birds that compete for prey, each with its great performance but a unique style. Identify other instances of animals that compete for prey in Nature. See which comparisons you and your friends can find. Comparing your findings will train you in building up arguments, and ensuring that you have a sharp logic.

猫头鹰和游隼是两种互相争夺猎物的鸟类，它们都表现出色但风格迥异。找出其他自然界中不同动物争夺猎物的例子。看看你和你的朋友能找到哪些相似的例子。比较你们的发现会训练你建立论点，并确保你有清晰的逻辑。

TEACHER AND PARENT GUIDE

学科知识
Academic Knowledge

生物学	游隼是一种猛禽；猫头鹰的眼睛是柱状而非球状，占其体重的3%；猫头鹰的视杆细胞数量是视锥细胞的30倍；游隼可以看到1.6千米外的兔子；视锥细胞在明亮环境下工作，视杆细胞在昏暗环境下工作。
化学	鸟的翅膀由β-角蛋白构成，它的喙和爪子也是这种物质构成的；β-角蛋白和构成哺乳动物的指甲、爪子和毛发的α-角蛋白不一样。
物理	上升的热气柱为鸟类飞行提供升力；升力随空气密度和翅膀表面积的增加而增加；猫头鹰羽毛的形状能改变空气湍流，从而降低噪音；柔软的猫头鹰羽毛能吸收大多数猎物产生的高频声音；游隼通过极速俯冲精准地抓捕多动的猎物，与速度较慢的低空攻击相比，这么做成功率更高。
工程学	游隼和导弹的转向，其背后的数学规律是相似的；着陆的飞机发出的噪音不是来自发动机，而是来自它周围的气流。
经济学	一般的经济学家与环境经济学家就经济增长、社会效益、生态可持续性、人类义务、公平问题进行的辩论很重要，这种辩论是为了设计一套稳定的经济体系，在不伤害大自然的前提下，使所有人都受益。
伦理学	优雅地接受他人赢得辩论是一种艺术；新一代父母有自己的尽责方式，承认这种代际差异是一种艺术。
历史	有记录的最早的游隼狩猎是在8 000年前的伊朗；猫头鹰代表智慧女神雅典娜，因而被视作智慧的象征；在17世纪，不同社会阶级的人能养什么鹰是有规定的；在16世纪，阿兹特克人训练鹰。
地理	猫头鹰和游隼生活在南极洲以外的世界各地；在沙漠、森林、草原甚至北极苔原中都能看到猫头鹰；游隼生活在开阔的空间里，在海岸附近成长，从苔原到沙漠都能看到它们。
数学	游隼俯冲时的最大速度取决于体重、俯冲角度和俯冲时间；游隼竖直俯冲时速度能达到89—112米/秒；升力与空气流速的平方成正比。
生活方式	辩论能力取决于批判、倾听、研究、信息处理、创造、沟通和说服的能力。
社会学	一个倡导辩论的社会具有成为倡导者的力量，帮助我们在以共同利益为先的前提下，用这些技能改善我们的学校、社区、国家和整个世界。
心理学	"战斗或逃跑"的压力；谦逊，知道总会有人比自己更优秀；在辩论中训练论证能力；主张、结论、数据、论据和主张背后的原因；辩论时要注意声音的清晰度、响度、音高、速率、准确度和发音，以及外表、手势、肢体语言、眼神交流和面部表情。
系统论	进化为每个物种带来了惊人的变化，每个物种都建立了自己的生态位，评价其完美程度不需要与其他生物比较，而是要赞美每个物种，赞美它们做最好的自己。

教师与家长指南

情感智慧
Emotional Intelligence

老 鼠

老鼠非常自信,大胆地接近猫头鹰,并和他说话。她谈到了猫头鹰的竞争对手:游隼。她上来就说游隼表现得更好。基于游隼比猫头鹰看得更远,飞得更高更快这些事实,她提出了一系列强有力的论点。当被问及为何偏爱游隼时,老鼠说服猫头鹰展现自己谦卑的一面,并准备承认有人比自己表现得更好。当猫头鹰再次开始争论时,老鼠继续提供更多的令人惊讶的事实和信息,以至于猫头鹰让步了。老鼠以此为切入点,展开更多的论点,这让她赢得了辩论。

猫头鹰

猫头鹰和老鼠进行了一场随意的、非正式的交谈。当老鼠开始用一长串的事实来说明在每一点上游隼都比猫头鹰优秀的时候,猫头鹰保持镇静,用平静的声音、清晰的论点逐一回复。当每个论点都被激烈的语言破坏时,他直接问老鼠为什么支持游隼。当老鼠再次坚持说游隼更优秀时,他提出了一些有利于自己的强有力的论点,但很快就被老鼠反驳了。于是,他认为这场辩论已经没有必要再继续下去了,他说老鼠赢了。然而,胜利只是形式上的,从他的结束语可以看出,他在沉默中优雅地飞翔,保持了他作为一只伟大的、聪明的鸟的姿态。

艺术
The Arts

猫头鹰有着严肃的面部表情,圆圆的脸上有一双有压迫力的眼睛,构成了一幅有趣的画面。试着捕捉这个画面。拿起一支铅笔,用几根线条在纸上再现它引人注目的样子。不要画脸上其他部位,只画它的脸和眼睛。比较你和你朋友的画,讨论怎样改进。在描绘猫头鹰时,一个小小的变化会产生巨大的不同!

TEACHER AND PARENT GUIDE

思维拓展
Systems: Making the Connections

　　今天的大自然,包括其中的生态系统和现存物种,是长期进化的结果。当我们仔细研究每一个物种的表现时,我们对创造和进化的奇迹感到敬畏。一只鸟怎么可能不花力气就从1千米的高空精确、快速地下降,然后再利用暖气流的力量上升到空中呢?由完全相同的材质构成的羽毛,怎么可能有着这么多种不同的用处呢?有些羽毛能利用物理定律,并与天然胶水巧妙结合,从而安静地飞行。鸟类的眼睛构造也很神奇,猫头鹰的眼睛是柱状而非球状。更不用说能量效率了,猫头鹰拍打一次翅膀的效果,相当于游隼拍打5次翅膀。每种生物都有独特而有力的特征。为支持或反对一种生物而争论,或赞美一种生物比另一种生物优秀,没有任何直接的意义。这种辩论的间接意义则在于能够很好地训练思维,让我们陈述存在的事实,并使用它们进行比较和说明。这则寓言中的老鼠进行了一场重要的辩论,有助于在双方辩友的头脑中形成差异和自信。猫头鹰保持其优势地位,倾听和回应老鼠,尽管老鼠为了争论而争论,坚持不懈地用强有力的语言来确认自己的位置。当辩论进行到某一环节,猫头鹰决定不再进一步讨论细节。他宁愿把形式上的胜利留给那只弱小的老鼠。他知道猫头鹰已经在世界各地成功地生存了许多年。这就是这个寓言留给我们的信息或启示:我们都有各自的特点,都能在生活中找到自己的位置,找到一个让我们快乐和健康的地方,以改善我们的生活。

动手能力
Capacity to Implement

　　在这个巨大的世界里,我们都是渺小的。我们要做的练习是弄清楚一只小老鼠如何能够与一只猫头鹰进行宏大的辩论,并在辩论中获胜。分析老鼠做了什么,怎么做的。由此,你可以学到一些辩论的方法,建立论点,组织你的思路,以强化每一个人的角色、地位。老鼠不仅仅是一个讨人厌的东西,事实上,她还帮助猫头鹰明确了自己对生态系统的独特贡献。最后,不管一个人是大是小,是聪明还是精力充沛,我们都有自己要扮演的角色。即使是微小的事物也有其独特的作用,我们永远不要觉得自己是微不足道的。我们都是同一个世界的一部分。

教师与家长指南

故事灵感来自
This Fable Is Inspired by

路易斯－哈德里安·泡利
Louis-Hadrien Pauli

路易斯－哈德里安是一名热爱自然的小学生，他乐于发现每个物种最好的一面。然而，除了学习大自然的奇迹之外，他还喜欢辩论、比较和挑战。对于一个学生来说，好奇是一个很好的特质，在好奇之外，他还常常为了支持和反对某一事物而辩论，不断提出新的想法，并让自己的思维更清晰。路易斯－哈德里安从很小的时候就开始练习，提出一个主张或结论，收集数据（真实的和想象的），并通过论点背后的原因构建论据，这可能是真的，也可能是假的。他在辩论时重视说话的清晰度、音量、音高、速率、发音，并结合外表、手势、肢体语言、眼神交流和面部表情，所有这些都以一个灿烂的微笑收尾——即使他输了辩论，他仍然很享受这个过程。

图书在版编目（CIP）数据

冈特生态童书.第八辑：全36册：汉英对照 /
（比）冈特·鲍利著；（哥伦）凯瑟琳娜·巴赫绘；
何家振等译. —上海：上海远东出版社，2021
ISBN 978-7-5476-1773-1

Ⅰ.①冈… Ⅱ.①冈…②凯…③何… Ⅲ.①生态环境－环境保护－儿童读物—汉、英 Ⅳ.①X171.1-49

中国版本图书馆CIP数据核字（2021）第249940号

策　　划	张　蓉
责任编辑	祁东城
封面设计	魏　来　李　廉

冈特生态童书

像石头一样下坠
［比］冈特·鲍利　著
［哥伦］凯瑟琳娜·巴赫　绘
贾龙慧子　译

记得要和身边的小朋友分享环保知识哦！
八喜冰淇淋祝你成为环保小使者！

Energy 269

脑中指南针

A Compass in My Brain

Gunter Pauli

［比］冈特·鲍利 著
［哥伦］凯瑟琳娜·巴赫 绘
贾龙智子 译

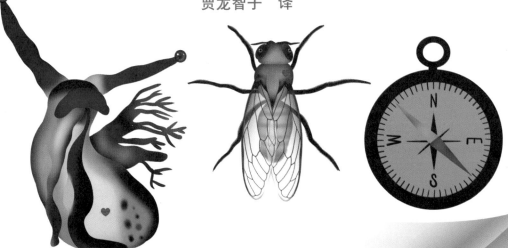

丛书编委会

主　任：贾　峰
副主任：何家振　闫世东　林　玉
委　员：李原原　祝真旭　牛玲娟　梁雅丽　任泽林
　　　　王　岢　陈　卫　郑循如　吴建民　彭　勇
　　　　王梦雨　戴　虹　翟致信　靳增江　孟　蝶

特别感谢以下热心人士对童书工作的支持：

匡志强　宋小华　解　东　厉　云　李　婧　陈　果
刘　丹　熊彩虹　罗淑怡　旷　婉　杨　荣　刘学振
何圣霖　廖清州　谭燕宁　韦小宏　李　杰　欧　亮
陈强林　王　征　张林霞　寿颖慧　罗　佳　傅　俊
胡海朋　白永喆　冯家宝

目录

脑中指南针	4
你知道吗?	22
想一想	26
自己动手!	27
学科知识	28
情感智慧	29
艺术	29
思维拓展	30
动手能力	30
故事灵感来自	31

Contents

A Compass in My Brain	4
Did You Know?	22
Think about It	26
Do It Yourself!	27
Academic Knowledge	28
Emotional Intelligence	29
The Arts	29
Systems: Making the Connections	30
Capacity to Implement	30
This Fable Is Inspired by	31

有只果蝇正在利用地球磁场来寻找向北飞的路线。一只海蛞蝓从海底抬起头来，观察着果蝇的飞行，问道：

"你真的知道北极在哪儿吗？"

A fruit fly is using the Earth's magnetic field to find his way north. A sea slug looks up from the ocean floor, and observing his flight, asks,

"Do you really know where the North Pole is?"

有只果蝇正在利用地球磁场……

A fruit fly is using the Earth's magnetic field …

一股强大的水流就能很容易地把我卷走。

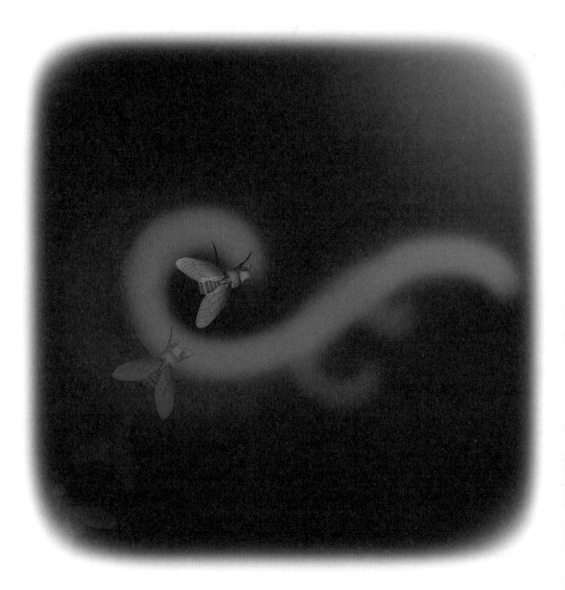

A strong current could easily sweep me away.

"既然你生活在海底,在泥地里挖洞,北极在哪儿对你来说重要吗?你显然不需要指南针。"

"在我生活的这个地方,一股强大的水流就能很容易地把我卷走,或者海浪会把我翻了个身。当我终于停下来的时候,我一点也不知道自己在哪里,也看不到朋友和家人。"

"Why is the North Pole important to you if you are living on the sea floor, burrowing in the mud? Surely you do not need a compass."

"Here where I live, a strong current could easily sweep me away, or the waves could tumble me head over tail. And when I finally come to rest, I won't have a clue where I am, with no friends or family in sight."

"是的,水的力量是很强大的。但就为了在这么小的地方找路,你真的需要一个指南针吗?"

"哦,我不需要指南针,因为我的整个身体就是指南针!"

"对不起。这听起来太离谱了!你被海浪冲走并撞到石头上了吗?"

"True, the water can be very powerful. But do you really need a compass just to find your way around such a small area?"

"Oh, I don't need a compass, as my whole body is one!"

"Excuse me. That sounds outrageous! Has a wave hit you against a rock?"

我的整个身体就是指南针!

My whole body is one!

我全身都有含金属的细胞……

I have cells with metals all over me...

"听着，我不需要用指南针来找食物。我周围的食物有很多。我身体里的指南针指引我回家。"

"你是怎么做到的？你的身体里有金属吗？"

"事实上，的确如此！我全身都有含金属的细胞，我的导航系统工作得很好。你不明白，我根本一点也不担心自己怎么导航。我对你的导航方式感到好奇。"

"Look, I do not need a compass to find my food. There is plenty of that around. My body compass directs me back to my home."

"And how do you do that? Does your body have any metals in it?"

"As a matter of fact, it does! I have cells with metals all over me, and my navigation system works great. You do not get that I am not at all worried about myself. I am curious about you."

"你并不迁徙,那为什么我的导航方式对你很重要呢?"

"你说得对,我不迁徙。不过,我的内置指南针无论如何都能帮我找到我的家,即使它从不遥远。"

果蝇解释道:"好吧,如果你真的想知道我是如何导航的,我可以告诉你,我的大脑里有一个很小的传感器。"

"You do not migrate, so why is it important to you?"
"You are right, I do not migrate. But my built-in compass helps me find my home anyway, even if it is never far away."
"Well, if you are really interested to know how I navigate, I have a tiny sensor in my brain," Fruit Fly explains.

我的大脑里有一个很小的传感器。

I have a tiny sensor in my brain.

……感觉到磁力有多强……

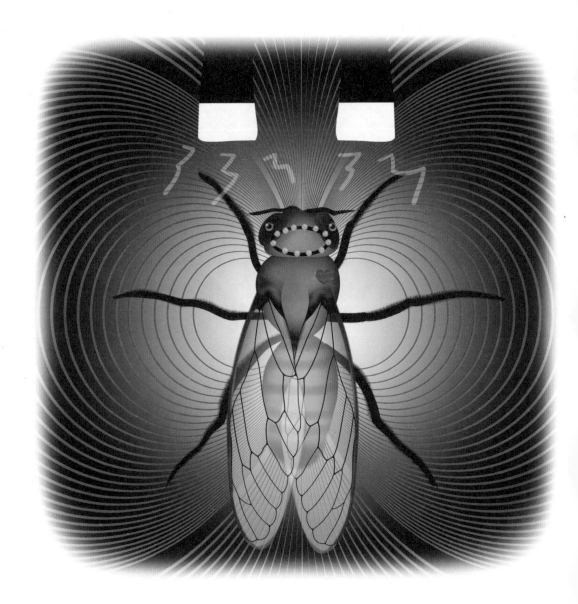

… senses how strong magnetic forces are …

"像指南针一样工作的传感器?"

"比那更好。我的大脑里有一圈看起来像指南针的细胞。"

"一圈细胞?内置在你小小的脑子里?"海蛞蝓问道。

"是的,它像指南针一样工作,告诉我方向。它能感觉到磁力有多强,以及它们来自哪个方向。"

"A sensor that works like a compass?"
"Better than that. My brain has a circle of cells that looks like a compass."
"A circle of cells? Built into your tiny brain?" Sea Slug asks.
"Yes, it's working like a compass, telling me the direction. It senses how strong magnetic forces are, and which direction they are coming from."

"这听起来很厉害,也很聪明。我得说,对苍蝇来说太聪明了。"

"对不起!我不只是一只苍蝇,我是一只果蝇。"

"但是你看起来就像一只普通的苍蝇。长着大大的红眼睛的苍蝇。"

"现在你太粗鲁了。告诉我,你知不知道我们和人类有很多共同点?"

"That sounds powerful, and very clever. Too clever for a fly, I'd say."
"Excuse me! I am not just any old fly, I am a fruit fly."
"But you look just like a common fly. One with big, red eyes."
"Now you are being rude. Tell me, did you know we have a lot in common with people?"

我是一隻果蠅。

I am a fruit fly.

人类并不总是最聪明的。

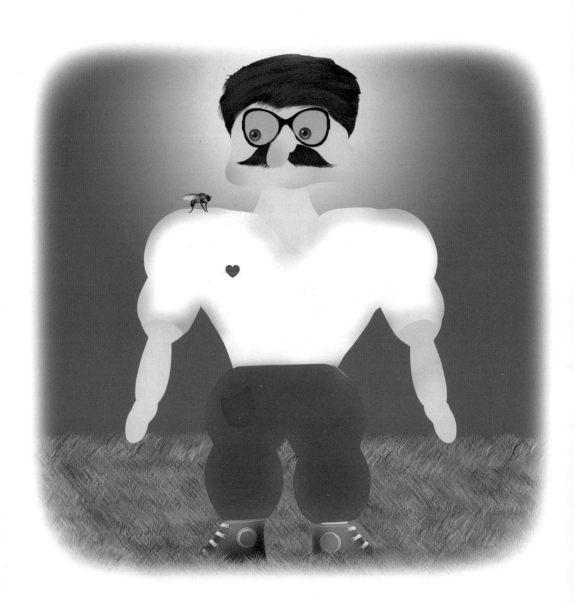

People are not always the smartest.

"嗯，人类并不总是最聪明的。"海蛞蝓评论道。

"哦，他们会学习，就像我一样。以我的体形来说，我有一个非常大且忙碌的大脑，帮助我学习、寻找食物、建立记忆、保卫领土。它使我成为导航方面的冠军。"

"因为你的大脑里有一个指南针！"

"Well, people are not always the smartest," Sea Slug remarks.

"Oh, they will learn. Like me. I have a very big and busy brain for my size, helping me to learn, find food, build a memory, and defend my territory. And it makes me a champion at navigating."

"Because you have a compass built into your brain!"

"是的，我用指南针找到森林里最甜的发酵水果，从而能够享用免费的饮料。这给了我很多能量。"

"嗯，我得说，你真了不起。你不仅聪明，而且懂得如何享受生活。对一只被大多数人讨厌的小昆虫来说，这还算不错。"

……这仅仅是开始！……

"Yes, and I use my compass to find the sweetest fermented fruit in the forest, to enjoy a free drink. That gives me lots of energy."

"Well, I must say, you are impressive. You are not only clever, but you also know how to enjoy life. Not bad, for a tiny insect, that is disliked by most people."

… AND IT HAS ONLY JUST BEGUN!…

……这仅仅是开始！……

… AND IT HAS ONLY JUST BEGUN! …

Did You Know?
你知道吗？

The fruit fly (*Drosophila melanogaster*) has a very rapid life cycle. One pair can produce hundreds of eggs within 10 to 12 days. Their short life span makes it easy to study genetic evolution over generations.

果蝇（黑腹果蝇）的生命周期非常短。一对果蝇在10—12天内可以产下数百枚卵。它们的寿命很短，这使对果蝇的多世代遗传进化研究变得很容易。

The fruit fly, only a few millimetres long, has 14,000 genes, and human being has 24,000. A surprising 75% of the genes that cause diseases in humans are also found in the fruit fly.

只有几毫米长的果蝇拥有14 000个基因，人类有24 000个。令人惊讶的是，导致人类疾病的基因中有75%可以在果蝇的基因中找到。

Fruit flies can simulate diseases that affect humans. Fruit flies eat a lot of sugar and have diabetes symptoms. Their brains have more than 100,000 neurons.

果蝇可以模拟影响人类的疾病。果蝇吃很多糖，有糖尿病症状。它们的大脑有超过100 000个神经元。

Some animals that navigate rely on visual cues and a sense of the direction their bodies are facing. Other animals navigate using polarised light from the sun or sensitivity to the Earth's magnetic field.

有些动物依靠视觉线索和对身体朝向的方向感来导航。其他动物利用太阳发出的偏振光或对地球磁场的感知来导航。

In the fruit fly, neurons are arranged in a circle, showing changes in direction by lighting up in a clockwise or counter-clockwise direction. They keep track of the fruit fly's direction, based only on body movements.

在果蝇体内，神经元呈环形排列，通过顺时针或逆时针方向依次"点亮"来显示方向的变化。这些神经元只根据身体的运动来确定果蝇的方向。

The sea slug (*Tritonia diomeda*) aligns itself precisely at an angle of 87.6° east to the Earth's north-south magnetic axis. It uses the Earth's magnetic field as an orientation cue.

海蛞蝓能精确地朝向地球南北磁轴以东87.6°角。它以地球磁场为方向指示。

There are thousands of sea slug species in the world, with over five hundred in the Great Barrier Reef of Australia alone. They have feathery gills that allow them to breathe.

世界上有数千种海蛞蝓，而在澳大利亚大堡礁就有500多种。它们通过羽毛般的鳃来呼吸。

Sea slugs have the bright colours of reef-dwelling species, indicating that they are under constant threat from predators. The bright colours serve as a warning that it has toxic stinging cells and an offensive taste.

海蛞蝓有着礁栖物种特有的鲜艳颜色，表明它们经常受到捕食者的威胁。鲜艳的颜色起到了警示作用——它有含毒的刺细胞和强刺激性味道。

Think about It

想一想

Can you figure out where the North Pole is, without a compass?

不用指南针的情况下你能弄明白北极在哪里吗?

How does it feel knowing that the fruit fly has half the number of genes we have?

当你知道果蝇的基因数量是我们的一半的时候,你感觉如何?

Are fruit flies smart?

果蝇聪明吗?

Do we have a brain just to think, or also to enjoy life?

我们的大脑仅仅是用来思考的,还是也用来享受生活的?

Do It Yourself!

Identify the cues that allow you to find magnetic north, when you are at home, in the classroom, or walking in a park. List at least three ways that will help you determine the four directions: north, south, east and west. Do not use a compass and do not rely on your smartphone. Now share the ways of orienting yourself with your friends and family members.

当你在家、教室或公园里散步时，找出能让你找到磁北的线索。列出至少三种可以帮助你确定东南西北的方法。不要使用指南针，也不要依赖智能手机。现在，与你的朋友、家人分享你的导航方法吧。

TEACHER AND PARENT GUIDE

学科知识
Academic Knowledge

生物学	海蛞蝓是食肉动物，以海绵、珊瑚、鱼卵、螃蟹、虾、龙虾和蛤蜊为食；果蝇在几天内经过幼虫和蛹的阶段，变为成虫；海蛞蝓能将头部与身体分开，再生身体；有些海蛞蝓能吸收藻类的叶绿体，进而利用太阳能；果蝇眼睛细胞的特异性分化、形成和形态发生；果蝇的昼夜节律也发生在每个人类细胞中。
化 学	水果发酵产生的异戊酸乙酯吸引了果蝇。
物 理	角度路径整合；第一个指南针是由磁石制成的；十几种细胞以六边形阵列的样式排列，这优化了视野的覆盖范围；海水是电导体，因此与地球磁场相互作用。
工程学	陀螺罗盘，一种不受铁磁金属影响的非磁性罗盘；GPS接收器起罗盘的作用，可确定纬度、经度和海拔高度；传感器。
经济学	果蝇是世界上危害性最强的果蔬害虫之一，造成了巨大的经济损失。
伦理学	苍蝇应该被认为是一种较低等的生命形式吗？
历 史	早期人类生活在洞穴里，有果蝇相伴；指南针发明于中国，磁罗盘最初用于风水和算命。
地 理	磁北极不断地向西北方向移动；果蝇起源于南非；哥伦比亚太平洋沿岸的丽海龟有能力确定它们的确切出生地点，并在15年后返回。
数 学	地球磁场不完全对称；磁感应方程；用计算机模拟地球发电机需要计算一组非线性偏微分方程的数值解。
生活方式	我们有着以北极为参照确定自己的方向，并且总是沿着南北轴线移动的文化；然而，我们失去了在没有仪器的情况下确定方向的能力。
社会学	对第六感的认识，与我们再次找到回家的路的能力有关，当你拥有这种能力时，会产生一种独立感，当你没有这种能力时，会产生一种依赖感。
心理学	需要确定性，知道迷失方向时自己在哪里，以及如何找到参照点；在没有任何人工工具的情况下（比如在黑暗中），有些人生来就有第六感，能帮助自己导航。
系统论	许多生物有感知地球磁场并与之相互作用的能力，但我们并不认为这就是所谓的第六感。

教师与家长指南

情感智慧
Emotional Intelligence

海蛞蝓

　　海蛞蝓向果蝇说明了尽管他生活在海里,但他需要知道北方在哪里的原因。起初,他质疑果蝇,随后与果蝇直接交流。这激发了他的兴趣,进而与果蝇更深入地交流。他过于自信,认为对于一只苍蝇来说,果蝇太聪明了。他不尊重果蝇。他拒绝承认果蝇的独特性,除了果蝇的红眼睛。海蛞蝓认为与人类有共同基因并不是很好,因为他认为人类不太聪明。当果蝇指出自己有能力享用发酵饮料时,海蛞蝓改变了对果蝇的看法,向她表示尊重,但海蛞蝓仍然提到人们不喜欢果蝇的事实。

果　蝇

　　果蝇想知道为什么即便海蛞蝓只是在海底的一个小范围内生存活动却需要一个指南针。当海蛞蝓指出他的整个身体就是一个指南针时,果蝇完全不相信。当海蛞蝓坚持这是事实时,果蝇产生了好奇心,并想知道这是如何运作的。当海蛞蝓问起果蝇的导航方式时,她回应了海蛞蝓对其导航系统的兴趣。果蝇说她脑中的传感器可能看起来像指南针,但性能更好。果蝇坚持认为,果蝇与人类有很多共同之处,而且即使他们的大脑很小,也能成功运作。果蝇最后引起海蛞蝓注意的论点是,果蝇的大脑不仅有很多实用功能,还可以帮助果蝇找到甜美的发酵果汁。

艺术
The Arts

　　科学家这样描述果蝇:红色大眼睛的背后有个指南针。果蝇的眼睛是六边形的。让我们来玩玩六边形的物体,用不同的艺术形式来表现它,并给它涂上果蝇眼睛特有的红色。

TEACHER AND PARENT GUIDE

思维拓展
Systems: Making the Connections

关于地磁场的起源，有一种说法是，地球外核中液态金属的流动产生电流，地球绕轴自转导致这些电流产生在行星周围延伸的磁场。地磁场对维持地球上的生命极其重要。如果没有地磁场，我们将暴露在来自太阳的大量辐射下，我们的大气层也会被破坏。包括鸟类和海龟在内的动物可以探测到地球磁场，并利用这些磁场进行导航。这是一种感应能力，允许动物探测磁场以感知特定的方向、位置和海拔高度。这种感应能力存在于细菌、类人猿、软体动物和所有主要脊椎动物中。细菌的每个细胞都能充当一个磁偶极子，因此细菌能辨别自己的方位并沿着地球磁场线移动。磁感应能力使线虫、软体动物、脊椎动物、蜜蜂、蚂蚁和白蚁能在巢穴内部和周围导航。巴西无刺蜂是一个特殊的例子，其复杂的感知能力使它能够利用触角上成千上万的毛发状粒子来区分海拔高度、位置和方向的差异。在过去的50年里，科学家们发现了这种第六感的更多特征，但我们仍然缺乏对动物磁感应的感受器的全面了解。这个系统是如此敏感、如此微小，却又如此完善，即使它的磁性物质含量可能低于百万分之一，而且可能只有一立方毫米大小。许多动物有能力感知地球磁场的微小波动，以此计算出自己在地图上的坐标，误差达到几千米甚至更小。这些动物的感知系统必须能够辨别周围磁场的微小差异，才能绘制出足够详细的磁场图。许多动物都有这种感知地球磁场微小波动的能力，科学家至今难以解释这种能力。不幸的是，我们人类似乎已经失去了这种能力，我们应该对此表现出谦卑的态度。对果蝇及它极其发达的方向感的研究，应该会启发我们更多地了解地球上不同动物如何使用不同的导航系统为自己指引方向。

动手能力
Capacity to Implement

说出20种具有复杂磁感应系统的现存物种。研究细菌、线虫、鸟类和果蝇是如何导航的。比较这些物种和人类的导航方式。问自己一个问题：哪种导航系统能在没有外部电源输入或卫星辅助的情况下运转？在泛用性和独立性方面进行评分，并选择你希望引入社区的导航系统。

教师与家长指南

故事灵感来自
This Fable Is Inspired by

伊莎贝尔·吉拉多
Ysabel Giraldo

伊莎贝尔·吉拉多于 2005 年在美国俄克拉荷马大学动物学系获得理学学士学位。2014 年,她在美国波士顿大学获得生物学博士学位。她对昆虫的行为着迷,并试图了解昆虫的神经系统及其神经机制是如何进化的。她专注于研究昆虫导航,以及昆虫如何利用环境线索,特别是天体线索,如太阳的位置或天空中光的偏振模式。她还对与昆虫导航有关的各种神经元感兴趣。她在美国加利福尼亚大学河滨分校昆虫学系建立了自己的实验室。

图书在版编目（CIP）数据

冈特生态童书.第八辑：全36册：汉英对照 /
（比）冈特·鲍利著；（哥伦）凯瑟琳娜·巴赫绘；
何家振等译.—上海：上海远东出版社，2021
ISBN 978-7-5476-1773-1

Ⅰ.①冈… Ⅱ.①冈…②凯…③何… Ⅲ.①生态环
境—环境保护—儿童读物—汉、英 Ⅳ.①X171.1-49

中国版本图书馆CIP数据核字（2021）第249940号

策　　划　张　蓉
责任编辑　祁东城
封面设计　魏　来　李　廉

冈特生态童书
脑中指南针
[比]冈特·鲍利　著
[哥伦]凯瑟琳娜·巴赫　绘
贾龙智子　译

记得要和身边的小朋友分享环保知识哦！
八喜冰淇淋祝你成为环保小使者！

Housing 268
蜈蚣、豪猪还是蜂鸟？
Centipede, Porcupine or Hummingbird?
Gunter Pauli

[比] 冈特·鲍利 著
[哥伦] 凯瑟琳娜·巴赫 绘
李原原 译

上海远东出版社

丛书编委会

主　任：贾　峰
副主任：何家振　闫世东　林　玉
委　员：李原原　祝真旭　牛玲娟　梁雅丽　任泽林
　　　　王　岢　陈　卫　郑循如　吴建民　彭　勇
　　　　王梦雨　戴　虹　翟致信　靳增江　孟　蝶

特别感谢以下热心人士对童书工作的支持：

匡志强　宋小华　解　东　厉　云　李　婧　陈　果
刘　丹　熊彩虹　罗淑怡　旷　婉　杨　荣　刘学振
何圣霖　廖清州　谭燕宁　韦小宏　李　杰　欧　亮
陈强林　王　征　张林霞　寿颖慧　罗　佳　傅　俊
胡海朋　白永喆　冯家宝

目录

蜈蚣、豪猪还是蜂鸟?	4
你知道吗?	22
想一想	26
自己动手!	27
学科知识	28
情感智慧	29
艺术	29
思维拓展	30
动手能力	30
故事灵感来自	31

Contents

Centipede, Porcupine or Hummingbird?	4
Did You Know?	22
Think about It	26
Do It Yourself!	27
Academic Knowledge	28
Emotional Intelligence	29
The Arts	29
Systems: Making the Connections	30
Capacity to Implement	30
This Fable Is Inspired by	31

一只海鼠正使劲儿挖泥，寻找螃蟹作为美餐。一条鲭鱼游过，问她：

"你今天找到足够的食物了吗？"

"嗯，你找到了我，你就有了食物，"海鼠这么说道，然后平静地问，"你现在就吃我吗？"

A sea mouse is digging deep into the mud, looking for a meal of crabs. A mackerel passes by and asks,
"Did you find enough food today?"
"Well, you found me, so you have your food," the sea mouse says, and then calmly asks, "Are you eating me right away?"

一只海鼠正使劲儿挖泥。

A sea mouse is digging deep into the mud.

我确实喜欢吃海鼠……

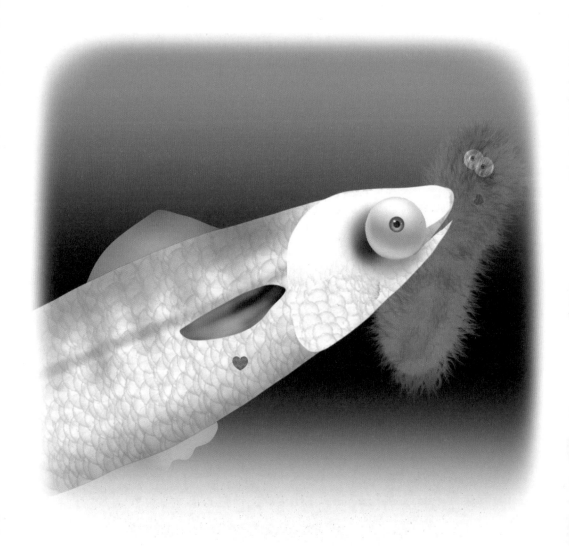

I do enjoy eating sea mice...

"我确实喜欢吃海鼠,因为海鼠有丰富的蛋白质,但我的肚子已经饱了。我只是很好奇,你明明是条小虫,为什么人们还叫你老鼠?"

"哈哈,一定是因为我的牙齿很坚固,像老鼠一样。我能嚼碎蟹壳。我闪亮的毛发——那不是真正的毛发——一定让人们把我当成了老鼠。"

"I do enjoy eating sea mice, so rich in protein, but my belly is already full. I just wondered why you are called a mouse when you are a worm?"

"Ha-ha, it must be because I have strong teeth, just like a mouse. I can crunch crab shells. My shiny hair, that isn't really hair, must have made people mistake me for a mouse."

"哦，如果他们走近看，看到你有那么多条小腿，他们会更糊涂的，他们会称你为海里的蜈蚣！"

"完全正确！但你知道，我们与你提到的任何动物都非常不同，即使我们确实有一些共同点。"海鼠说。

"当然，你还有鬃毛，所以你看起来像一只豪猪！"

"Well, if they did look closer, and saw your many tiny legs, they would have been more confused, and would've called you the centipede of the sea!"

"Exactly! But you know, we are very different from any of the animals you mention – even if we do have something in common," Sea Mouse says.

"Of course you also have bristles, so you look like a porcupine!"

……海里的蜈蚣!

... the centipede of the sea!

……这些神奇的豪猪。

... these wonderful porcupines.

"是的,我们借用了这些神奇的豪猪的构想。我的鬃毛很独特,是中空的,就像北极熊的毛一样。"

"当太阳照在你身上时,你就会呈现出彩虹般美丽的颜色——就像一只蜂鸟。"

"Yes, we've borrowed the idea for these wonderful porcupines. My bristles are unique, and hollow – just like a polar bear's hair."

"And when the sun shines on you, you display all the beautiful colours of the rainbow – just like a hummingbird."

"谢谢你注意到我很漂亮，而且赏心悦目。"

"这一定是有目的的。你能给点提示吗？"鲭鱼问道。

"为什么每件事都要有目的？有时候我们只是为了美丽而美丽。"

"Thank you for noticing that I am beautiful, and pleasing to the eye."

"It must serve a purpose. Can you shed some light on it?" Mackerel asks.

"Why does everything have to have a purpose? Sometimes we can just be beautiful for beauty's own sake."

……就像一只蜂鸟。

... just like a hummingbird.

只是为了呈现彩虹色，好用来炫耀吗？

Just to show off your iridescence?

"告诉我,为什么你的毛发是中空的,并且比人类的头发还细?只是为了呈现彩虹色,好用来炫耀吗?"

"我自己也是直到最近才知道。但发现自己往往需要时间。所以我们要一直保持好奇心,极富创造力地去发现我们从未知道甚至想象过的东西。"

"So tell me, why do you have hollow hair, thinner than human hair? Just to show off your iridescence?"

"Until recently, I myself had no clue. But it often takes time to discover something about yourself. So we always have to be inquisitive, and very creative in finding out things we never knew or even imagined."

"确实,不过在自然界中,所有不需要的东西都会消失。你知道我为什么两边都有条纹吗?"

"当你们成群结队游泳时,它可以帮助你测量自己和其他鱼之间的距离。你们都游得很快,可不想经常撞到一起!"

"That's true, but in Nature everything that is not needed disappears. You know why I have stripes on both sides?"

"It helps you measure the distance between yourself and the others when you swim in schools. You all swim fast, and don't want to bump into each other all the time!"

你知道我为什么两边都有条纹吗?

You know why I have stripes on both sides?

你很聪明，我的朋友。

You are clever, my friend.

"你很聪明,我的朋友。我真是服了。"鲭鱼承认。

"别忘了你以我们为食。我知道捕食者是如何行动的。"

"现在告诉我,为什么你的毛发是空心的?"

"You are clever, my friend. I am impressed." Mackerel admits.

"Don't forget that you feed on us. I know how predators function."

"Now tell me, why are your hairs hollow tubes?"

"我们一直不知道它有什么用,直到人们发现它是铸造微小纳米线的完美模具。想象一下,由一只来自大海的'老鼠'制造的纳米线!你不以我为荣吗?"

……这仅仅是开始!……

"We never knew what it was good for, until people discovered that it makes a perfect mould for casting tiny nanowires. Just imagine – nanowire produced by a 'mouse' from the sea! Are you not proud of me?"

... AND IT HAS ONLY JUST BEGUN!...

……这仅仅是开始！……

... AND IT HAS ONLY JUST BEGUN! ...

Did You Know?
你知道吗？

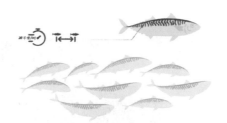

Mackerel swim in schools, at a speed of up to 20 km per hour. They need to coordinate their movements, and the hypothesis is that the stripes on the fish's side helps it measure distance and speed.

鲭鱼成群游动，速度可达20千米/时。它们需要协调彼此的动作，有一种假说认为，鲭鱼身上的条纹有助于它测量距离和速度。

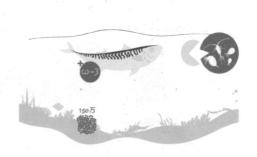

Mackerel are a rich source of omega-3 fatty acids. The female can lay up to 1.5 million eggs at the time, which float to the surface where they feed on zooplankton.

鲭鱼富含 ω–3 脂肪酸。雌鲭鱼一次可以产150万个卵，它们会浮到水面，以浮游动物为食。

Mackerel have a brilliant blue-green colour, broken by black curved lines. The green on the back gives way to iridescent blue, fading to white and enhanced by subtle pink. The colours fade rapidly after death.

鲭鱼呈明亮的蓝绿色，中间有黑色的曲线。背面的绿色渐变为五彩斑斓的蓝，再渐变为白色，淡淡的粉红色更丰富了整体的色彩。鲭鱼死后会迅速褪色。

The broken lines of black on the back make it difficult for a predator to see the fish when attacking from above. The white belly breaks up the fish's outline against the light when a predator approaches from below.

鲭鱼背上的黑色虚线使得捕食者从上面攻击时很难看到它。当捕食者从下面靠近时，白色的腹部在光线下使鱼的轮廓变得模糊。

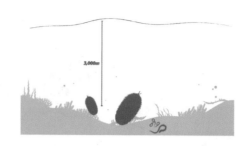

The sea mouse is a marine worm, related to the earthworm. Its body has about 40 segments. Its Latin name *Aphrodita* is a reference to the goddess Aphrodite. They can be found at a depth up to 3,000 metres.

海鼠是一种海洋蠕虫，与蚯蚓有亲缘关系。它的身体大约有40节。它的拉丁语名字 *Aphrodita* 源于女神阿芙罗狄忒。可以在水下3 000米深的地方找到它们。

The structural colouration caused by light triggers colour through hexagonal crystal cylinders, which perform more efficiently than man-made fibre optics.

光通过六棱柱状晶体，产生结构色效应，其性能比人造光纤强很多。

It is hard to tell which end of the sea mouse is its head, and which its tail. Its head is hidden, but the two tiny horns protruding in front indicate the head. There is one eye on either side of a single antenna.

很难分辨海鼠的哪一端是头，哪一端是尾巴。它的头是隐藏的，但前面突出两个小角的那端是它的头。在一个触角的两侧各有一只眼睛。

The sea mouse eats other worms, but also young crabs, especially hermit crabs. They swallow their prey, which can be three times its size, whole and head first.

海鼠吃其他蠕虫，也吃小螃蟹，尤其是寄居蟹。海鼠从头开始整个吞下猎物，其尺寸可能是海鼠的3倍。

When will we be able to use a hollow hair as a mould?

我们什么时候才能用中空的头发做模具?

Are you prepared to talk to your predator?

你准备好和捕食你的家伙谈谈了吗?

Would you eat a fish that has 1.5 million eggs?

你会吃一条肚子里有150万个卵的鱼吗?

How does it feel when someone you don't know says, "I'm impressed"?

当一个你不认识的人对你说"我很佩服",你会有什么感觉?

We are surrounded by iridescence, but do we realise how important it is in bringing beauty into our daily lives? Iridescence can be seen in bird feathers like the peacock and also in beetles, flies, butterflies, seashell nacre, fish scales, and even snakes, like the boa. We also see it in soap bubbles and gemstones like the opal. Compile a list of 100 natural and man-made materials that have iridescence. Collect pictures of these examples, and share this often-neglected source of beauty with friends and family members.

我们被五彩缤纷的事物包围着，但我们是否意识到让美进入我们的日常生活是多么重要？像孔雀这样的鸟类有彩虹色的羽毛，甲虫、苍蝇、蝴蝶、贝壳珍珠层、鱼鳞，甚至像蟒蛇这样的蛇身上也有彩虹色。我们还可以在肥皂泡和蛋白石等宝石中看到彩虹色。列出100种具有彩虹色的天然和人造材料。收集这些材料的照片，并与朋友、家人分享这些经常被忽视的美。

学科知识
Academic Knowledge

生物学	海鼠是一种海洋蠕虫；海鼠是一种活跃的捕食者，以小螃蟹和其他蠕虫为食；海鼠有刚毛和鳞片，但没有毛发；鲭鱼是一种远洋觅食鱼类，有约30个品种；雌鲭鱼一次产30万—150万个卵，这些卵在开放海域自由漂浮；鲭鱼没有鱼鳔，所以必须不停地游泳以避免下沉。
化 学	刚毛由甲壳质构成；鲭鱼富含ω-3脂肪酸；阿拉斯加鲭鱼汞含量低，其他鲭鱼汞含量高；鲭鱼富含辅酶和组氨酸，这是它快速腐烂的原因。
物 理	海鼠利用颜色变化作为防御机制；海鼠鬃毛的折射能力可与蛋白石相媲美；彩虹色；灵感源于北极熊的仿生隔热材料。
工程学	海鼠能发光并操纵光；北极熊的毛发存在空腔结构，这些空腔的形状和间距形成了北极熊独特的白色外表，使北极熊的毛发具有显著的保温性、耐水性和弹性，人们希望隔热材料也能有这些特性；海鼠的刺毛被用来制造又长又便宜的纳米线；中空的毛发被用作培养纳米线的模具，方法是在一端放置一个金电极，从另一端将铜离子或镍离子射入空腔。
经济学	为了在3 000米的深度产生强烈的光，海鼠有能力以最有效的方式利用到达这个深度的极少量光；每年有500万吨鲭鱼被捕捞，但过度捕捞导致其数量锐减；秋季捕获的鲭鱼的脂肪含量为30%，春季最低，仅为3%。
伦理学	人类怎么能将鱼类资源消耗到无鱼可捕的地步呢？
历 史	1758年，在卡尔·林奈的《自然系统》第10版中，鲭鱼首次被描述。
地 理	在北大西洋3 000米深的地方发现了海鼠；直齿真鲨跟随迁徙的鲭鱼群离开马达加斯加；在法国，鲭鱼的传统处理方式是用盐腌制，这样就能在全国各地销售。
数 学	海鼠刺毛内的六棱柱结构传递光的效率比人造光纤高得多；光速约为300 000千米/秒，可见光的波长约为380—740纳米，眼睛和脑将不同的波长"解读"为不同的颜色；紫光波长约为380—450纳米，红光波长约为630—740纳米。
生活方式	用小钱做大事。
社会学	鲭鱼因它的条纹和小鳞片被称为"海中的老虎"，这些小鳞片使它的表面非常光滑；在生态、进化、保护和管理之间的相互作用中，对人类引起的变化的行为反应至关重要。
心理学	通过钓鱼寻求刺激。
系统论	由于北海的鲭鱼被过度捕捞，生态平衡遭到破坏，许多渔业工作岗位也大量流失。

教师与家长指南

情感智慧
Emotional Intelligence

鲭鱼 鲭鱼是捕食者,但令人惊讶的是,他问海鼠是否找到了足够的食物。他很快解除了海鼠的担忧,然后问海鼠,她为什么被称为老鼠。鲭鱼取笑海鼠,说她可能被误认为是海里的蜈蚣,长得还像豪猪,甚至是蜂鸟。鲭鱼想知道这些奇怪的特征到底有什么用。鲭鱼看海鼠没有正面回答他,就更具体地询问为什么她的毛发是中空的。鲭鱼认为,在自然界中没有用处的东西最终会消失。鲭鱼承认,他没想到海鼠能如此准确地回答自己的问题。

海鼠 海鼠冷静地面对捕食者鲭鱼。她挑衅地问对方,是马上吃掉自己还是再等一等。当谈到海鼠是一种蠕虫时,她确认自己的优势之一就是坚固的牙齿可以咬碎螃蟹的壳。她知道自己和其他动物有共同之处,但这些动物也都有自己的独特之处。她想知道为什么所有的事情都必须有目的,为什么不能为了美而美。当鲭鱼执着地提问时,她承认,通常需要花些时间才能发现事物的真相。当对方问她鲭鱼的条纹有什么作用时,她给出准确的回答。海鼠承认,为了生存,她必须充分了解捕食者。最后她解释了自己的中空毛发的用途。

艺术
The Arts

学习如何呈现彩虹色!往塑料容器中倒入一半温水。加几滴透明的指甲油,让它在水面形成一层非常薄的膜。现在小心地将一张黑色砂纸放入容器中,当指甲油膜稳定下来后,慢慢将砂纸取出,这样指甲油就会均匀地附着在砂纸上。晾干后,你就有了一张彩虹纸。现在你可以把它折成你喜欢的样子,你的彩虹纸会像蜂鸟的羽毛一样色彩缤纷。

TEACHER AND PARENT GUIDE

思维拓展
Systems: Making the Connections

对于工程师来说，北极熊的毛发是制造锁住热量的合成材料的理想模板，这种材料由天然和可再生原料制成。中国的材料科学家们现在已经开发出了这样一种隔热材料，它复制了北极熊毛发的结构，同时将其扩展为由许多毛发组成的材料，用于建筑和航空航天领域。与人类或其他哺乳动物的毛发不同，北极熊的毛发是中空的。在显微镜下观察，每根毛发的中心都有一条长长的圆柱形通道。这些空腔的形状和间距不仅使毛发呈白色，也使毛发具有较强的保温性、耐水性和弹性，这些都是隔热材料的理想特性。另一种毛发中空的动物是海鼠。现在，海鼠的毛发被用作培养纳米线的模具，方法是在一端放置一个金电极，从另一端将铜离子或镍离子发射到空腔中。我们应该运用我们的创造性思维，去探索自然现象并应用到生产中。在过去的20年里，社会的进步依赖于计算机和通信系统的集成来提供这些高度智能化的移动服务。在过去的10年中，出现了第三种集成：更加多功能的小型设备。一部手机集成了照相机、气压计、夜视仪等设备的功能。只有当所有组件都非常小的时候，集成才是可行的。即使是极限运动专用相机GoPro也太大了，无法塞进手机。这正是需要工程创新的领域，因为电子线路不仅非常昂贵和沉重，而且限制了新型设备进一步集成到我们的手机中。这就是用海鼠中空的毛发作为模具的天才之处。海鼠的毛发为生态系统创造了令人难以置信的价值，现在也为当地的经济发展增添了巨大的动力。这也将对海鼠的研究带入一个全新的阶段，因为它使曾经非常昂贵的纳米线的生产变得可行。最后，我们从中了解到，自然界中看似奇怪的生物，如毛发中空的北极熊和海鼠，对现代生活的质量有显著的影响。现在我们只希望这些创新产品也能使用仿生材料，以实现可持续生产。

动手能力
Capacity to Implement

让我们来制造肥皂泡。肥皂泡有天然的彩虹色，但我们的配方可能更有优势，因为它更环保。你需要水、甘油、洗涤剂和糖。根据你的喜好和创意，把所有的原料混合好，然后用它吹泡泡。展现彩虹吧！

教师与家长指南

故事灵感来自
This Fable Is Inspired by

帕维尔·西科尔斯基
Pawel Sikorski

帕维尔·西科尔斯基出生于波兰。他于 1998 年获得弗罗茨瓦夫理工大学材料科学硕士学位。2002 年，他在英国布里斯托大学获得高分子物理学博士学位。在这所大学做了一年助理研究员后，他在挪威科技大学生物技术系从事博士后研究，并于 2011 年成为该大学的教授。他的研究小组致力于生物材料和生物纳米技术，特别是纳米制造的应用。这就是他的团队专注于使用海鼠的毛发作为模具生产纳米线的原因。

图书在版编目(CIP)数据

冈特生态童书.第八辑:全36册:汉英对照/
(比)冈特•鲍利著;(哥伦)凯瑟琳娜•巴赫绘;
何家振等译.—上海:上海远东出版社,2021
ISBN 978-7-5476-1773-1

Ⅰ.①冈… Ⅱ.①冈…②凯…③何… Ⅲ.①生态环境-环境保护-儿童读物—汉、英 Ⅳ.①X171.1-49

中国版本图书馆CIP数据核字(2021)第249940号

策　　划	张　蓉
责任编辑	祁东城
封面设计	魏　来　李　廉

冈特生态童书

蜈蚣、豪猪还是蜂鸟?

[比]冈特•鲍利　著
[哥伦]凯瑟琳娜•巴赫　绘
李原原　译

记得要和身边的小朋友分享环保知识哦!
八喜冰淇淋祝你成为环保小使者!

Housing 267

月亮水泥

Cement on the Moon

Gunter Pauli

[比]冈特·鲍利 著
[哥伦]凯瑟琳娜·巴赫 绘
李原原 译

上海远东出版社

丛书编委会

主　任：贾　峰
副主任：何家振　闫世东　林　玉
委　员：李原原　祝真旭　牛玲娟　梁雅丽　任泽林
　　　　王　岢　陈　卫　郑循如　吴建民　彭　勇
　　　　王梦雨　戴　虹　翟致信　靳增江　孟　蝶

特别感谢以下热心人士对童书工作的支持：

匡志强　宋小华　解　东　厉　云　李　婧　陈　果
刘　丹　熊彩虹　罗淑怡　旷　婉　杨　荣　刘学振
何圣霖　廖清州　谭燕宁　韦小宏　李　杰　欧　亮
陈强林　王　征　张林霞　寿颖慧　罗　佳　傅　俊
胡海朋　白永喆　冯家宝

目录

月亮水泥	4
你知道吗？	22
想一想	26
自己动手！	27
学科知识	28
情感智慧	29
艺术	29
思维拓展	30
动手能力	30
故事灵感来自	31

Contents

Cement on the Moon	4
Did You Know?	22
Think about It	26
Do It Yourself!	27
Academic Knowledge	28
Emotional Intelligence	29
The Arts	29
Systems: Making the Connections	30
Capacity to Implement	30
This Fable Is Inspired by	31

剑麻和竹子正在讨论未来。剑麻因其纤维而闻名，竹子则作为强韧而美观的建筑材料而闻名。两者都生长迅速，而且生命周期很长。

"你原产于墨西哥，后来在世界各地被广泛种植，这是真的吗？"竹子问。

A sisal plant and a bamboo are discussing the future. The sisal is celebrated for its fibres and the bamboo for its great strength and beauty for building. Both are fast growers, and last long.

"Is it true that you came from Mexico, and have been replanted everywhere in the world?" the bamboo asks.

剑麻和竹子……

A sisal plant and a bamboo ...

……现在海洋中充满了塑料微粒……

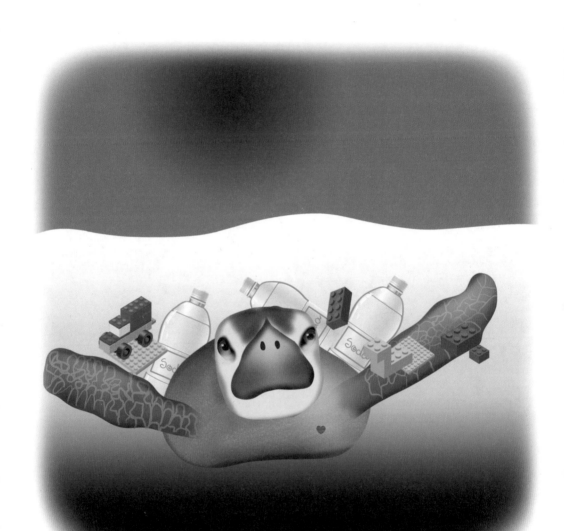

… now our oceans are full of tiny bits of plastics …

"是的,用我们做成的绳子最结实。人人都想要我们。"

"用石油制成的塑料出现后,所有人都抛弃了你们?"

"现在海洋中充满了塑料微粒,假如最终是你我的纤维进入海洋,那就会被当作食物。"

"Yes, we make the toughest ropes ever. Everyone wanted us."

"Then plastics made from petroleum came, and everyone abandoned you?"

"And now our oceans are full of tiny bits of plastics – whereas if your and my fibres should end up in the ocean, it would be considered food."

"我也被抛弃了。我曾是全世界首选的建筑材料，直到出现了混凝土和水泥。"

"那些灰暗的东西怎么就占领世界了呢？你们竹子不仅可以承受数百年的压力，而且非常柔韧，可以把碳固定几十年……并且……"

"I was abandoned too. I used to be the world's first choice in building material, until concrete and cement came along."

"How could that dusty stuff take over the world? Bamboo is able to resist stress for hundreds of years. And you are flexible, and you can fix carbon for decades… and …"

我曾是全世界首选的……

I used to be the world's first choice ...

……混凝土和水泥有什么区别？

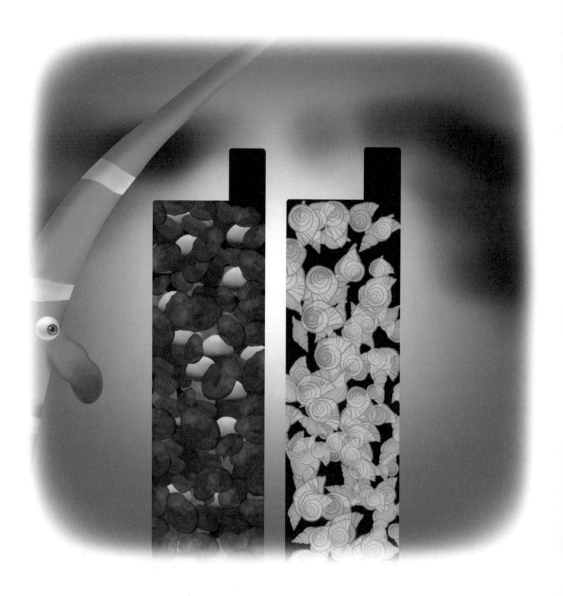

... difference between concrete and cement?

"我知道!告诉我,混凝土和水泥有什么区别?"竹子问道。

"我也刚刚知道:大多数混凝土是由牡蛎等海洋贝类的壳烧制而成;水泥则由岩石烧制而成。"

"烧贝壳和石头就可以生产建筑材料?"

"I know! Tell me, what's the difference between concrete and cement?" Bamboo asks.

"I've just figured this out: Most concrete is made by burning oysters and seashells. And cement is made from burning rocks."

"Burning seashells, and putting rocks on fire makes a building material?"

"燃烧过程中,二氧化碳会从这些坚硬的固体中被释放出来,排放到空气中。"

"二氧化碳?就是它导致了气候变化!竹子吸收它,水泥释放它。这不合理啊。人类知道吗?"

"为了制造混凝土,人们将水泥、河水,以及来自采石场、河流和矿井的沙子混合。"

"Burning takes carbon dioxide out of these hard solids and puts it in the air."

"Carbon dioxide? But that causes climate change! Bamboo captures it, and cement releases it. That does not make much sense. Do people know?"

"And to make concrete, people mix cement with water taken from rivers, and sand from quarries, rivers and mines."

就是它导致了气候变化!

But that causes climate change!

这就是为什么人类把钢筋放入水泥中……

That is why people put steel bars in cement …

"现在，这块易碎又缺少韧性的水泥板有什么用呢？"竹子问道。"当你拉它的时候，它就会裂开。"

"是的，不过当你挤压它们的时候，它们就能很好地承受压力。"

"这就是为什么人类把钢筋放入水泥中，使其更坚固，这样他们就能建造数百米高的建筑。"

"Now what is the use of this brittle and inflexible sheet of cement?" Bamboo asks. "When you pull on it, it breaks apart."

"Yes, but when you press it together, it holds so well."

"That is why people put steel bars in cement, to make it stronger, so they can build hundreds of metres high."

"为什么没有人想到把我的纤维放进去呢？只需要少量的剑麻或竹子，就可以增强水泥板的韧性。而我们也将重获人生的意义。"

"哦，人类决定加入的是矿物质和玻璃纤维。"

"And why doesn't anyone put some of my fibres in? Just a small percentage of sisal or even bamboo will give it great flexibility. And we will have a purpose in life again."

"Oh, people have decided to mix in minerals and glass fibres."

……没有人想到把我的纤维放进去呢?

... doesn't anyone put some of my fibres in?

……在月球上制造水泥……

... making cement on the Moon ...

"他们不知道我们剑麻和竹子有多好,是吗?你听说过工程师们正在设想在月球上制造水泥吗?"

"瞧,一到月球,我们就肯定出局了!在那儿我们无法生存。月球上可能有很多灰尘,但那里没有电力,没有氧气,也没有水来制造混凝土。"

"They do not know how good we are, do they? Have you heard that engineers are now imagining ways of making cement on the Moon?"

"Look, we are definitely out of the game when it comes to the Moon! We can not live there. There may be a lot of dust on the Moon, but there is no power, no oxygen, and no water to make concrete."

"但是对于水,有一个非常实用的解决方案!你能想象谁能在月球上提供最大的水源吗?"

"嗯,那一定是人类!别告诉我他们正在考虑用月亮上的沙子和尿液来制造混凝土!"

……这仅仅是开始!……

"But there is a very practical solution for the water part! Can you imagine who could provide the biggest source of water on the Moon?"

"Well, that must be people! Don't tell me they are now thinking of making concrete from moon sand and urine!"

... AND IT HAS ONLY JUST BEGUN!...

……这仅仅是开始！……

... AND IT HAS ONLY JUST BEGUN! ...

你知道吗？

Sisal is used in Morocco and Central America as a skin treatment. In the Bahamas, the sisal flower bud is cooked with salt, to treat jaundice. The plant is also used for treating indigestion.

剑麻在摩洛哥和中美洲被用于治疗皮肤病。在巴哈马群岛，人们用盐煮剑麻花蕾，用来治疗黄疸。剑麻也被用于治疗消化不良。

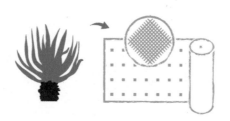

Sisal is used in paper, cloth, dartboards, geotextiles, filters, carpets, mattresses, as a strengthening agent to replace glass fibre and asbestos in composite materials.

剑麻被用于生产纸、布料、镖靶、地工织物、过滤器、地毯、床垫，还可作为复合材料中替代玻璃纤维和石棉的增强剂。

Sisal is drought resistant, grown in poor soil in drought-prone tropical regions. In China, sisal has been planted on relatively rich soils. Demand for increased food production pushes sisal out of these areas.

剑麻抗旱，生长在易干旱的热带地区贫瘠的土壤中。在中国，剑麻曾被种植在相对肥沃的土壤上。不断增加的粮食需求使得这些地区的剑麻逐渐消失。

Using agave sugar from sisal for liquor production is a longstanding tradition in central Mexico. Tequila is distilled from sisal with lower fibre content. The income from drinks and fibre makes sisal competitive.

在墨西哥中部，使用剑麻中的龙舌兰糖酿酒是一种悠久的传统。龙舌兰酒由纤维含量较低的剑麻蒸馏而成。来自饮料和纤维的收入使剑麻具有竞争力。

Seashells burnt at high heat (850° C) create calcium oxide, and when mixed with water produces calcium hydroxide that can be used to build houses, or as the basis for fresco painting, that will last centuries.

贝壳在高温下（850℃）燃烧会生成氧化钙，当与水混合时会生成氢氧化钙。氢氧化钙可用来建造房屋，或者作为壁画的基底，可以保存几个世纪。

Historic buildings, in the North Atlantic, Norway, Denmark, British Isles, and the Faroe Islands, dating from the Middle Ages until the 20th century, were constructed with seashells. It was a key source for paint.

北大西洋、挪威、丹麦、不列颠群岛和法罗群岛的中世纪至20世纪的历史建筑都用到了贝壳。贝壳是生产涂料的重要原料。

Lime kept under water becomes lime putty, or liquid rock. Once it dries, with exposure to air in a building or a fresco, it absorbs carbon from the atmosphere and turns back into a hard rock.

石灰放在水里会变成石灰膏，也就是液体岩石。一旦它变干，暴露在建筑或壁画周围的空气中，它就会从大气中吸收碳，变回坚硬的岩石。

Concrete made from lunar regolith, known as lunarcrete, was first proposed in 1985, to cut costs of construction on the Moon. Urea from urine could make lunar concrete more malleable before it hardens.

1985年，人们首次提出用月球表土制造混凝土，以降低在月球上建造建筑物的成本。尿液中的尿素可以使月球混凝土在硬化前更具延展性。

Think about It
想一想

Making concrete on the moon with urea?

可以用尿素在月球上制作混凝土吗?

Making fibres for carpets turns a drought-resistant plant competitive?

生产地毯纤维可以使抗旱植物更具竞争力吗?

Why use unhealthy asbestos when bamboo and sisal fibre is abundantly available?

竹子和剑麻纤维这么丰富,为什么还要使用不健康的石棉呢?

Burning seashells to build a wall or to paint fresco?

烧贝壳是用于建墙还是画壁画?

Study the use of fibres in building materials. What is the functional use of glass fibre and asbestos fibre in cement? Both have questionable health effects, and neither can be recycled once mixed as cement. It is time to build up the arguments that permit us to replace these materials with the natural ones that are abundant and renewable, while fixing carbon. List at least ten benefits, to prove that the industry does not have any reason to continue with the building traditions of the past. Test these arguments with friends and family, and have them strengthened.

研究纤维在建筑材料中的用途。玻璃纤维和石棉纤维在水泥中的作用是什么？这两种材料对健康都有可疑的影响，而且一旦与水泥混合就不能循环利用。是时候证明我们应当用丰富、可再生、可固碳的天然材料取代像玻璃纤维这样的材料了。列出至少十个好处，以证明建筑行业没有任何理由再因循守旧了。与朋友、家人一起验证这些论据，并加以补充。

TEACHER AND PARENT GUIDE

学科知识
Academic Knowledge

生物学	龙舌兰是只结一次果的草本植物。
化学	剑麻制品被聚丙烯取代；建筑材料的矿物成分主要为无水硅酸三钙、硅酸二钙、铝酸三钙、铝酸铁四钙、硫酸钙；将贝壳中的碳酸钙转化为氧化钙，再与水混合就会生成氢氧化钙；生石灰与水发生剧烈反应，释放巨大的热量；贝壳的无机相中，碳酸钙（方解石、文石、球霰石和一些非晶体）大约占95%—99.9%；碳酸钙在825℃左右分解为氧化钙和二氧化碳。
物理	剑麻因其耐久性、强度、拉伸能力、耐盐水性和对某些染料的亲和性而受到青睐。
工程学	纤维填充提高了塑料作为一种结构材料的强度和刚度，取代了在大多数国家被禁止的石棉。
经济学	聚丙烯已经渗透到剑麻的核心市场——农用麻绳和绳索，这实际上限制了剑麻的价格，迫使剑麻原料价格整体下降；替代剑麻来制作绳索或麻绳的聚丙烯树脂，仅占世界聚丙烯树脂产量的1.7%；社会和大自然承担的外部成本，并未反映在合成材料和矿物材料的价格上。
伦理学	鱼类和贝类摄入大量的塑料微粒，这些微粒可能会进入人体；外部成本由社会承担，污染环境换来的利润只被少数人享有；一旦发现、记录和计算出相应的后果，原本无意中造成的损害就会被认定为附带损害。
历史	早在一万年前，玛雅人就用剑麻纤维制作渔网；烧贝壳的做法起源于一万年前的中东地区；1986年，阿波罗16号采集的40克月球表土样品被用于生产月球混凝土。
地理	剑麻原产于墨西哥，巴西是最大的生产国。
数学	经过50次从零下27℃到室温的温度变化后，月球混凝土仍能承受17兆帕的压缩压强。
生活方式	生活中发生的坏事，也许最终会变成好事。
社会学	竹子象征着高尚、正直、坚韧和刚强，中国文学经常描写竹子。
心理学	生活中平凡的事情可能会产生意想不到的结果。
系统论	合理利用自然界中的植物可以改善人类生活，减少污染；农民收入的减少导致城市化，许多农民搬到棚户区生活，导致贫困和营养不良，这是全社会要承担的代价。

教师与家长指南

情感智慧
Emotional Intelligence

竹子

竹子非常清楚正在发生的危机,也知道自然产品正被人造产品取代的事实。竹子认为自己正被易造成污染的物质替代。竹子质疑并试图寻找真相。当竹子听说建筑材料是通过燃烧贝壳和岩石制成的,还释放大量二氧化碳时,他感到非常困惑。竹子想知道人类是否真的那么无知。竹子质疑水泥板对环境的影响,但认可水泥板建造摩天大楼的能力。竹子声称,如果在地球上没有机会作改变的话,就应该去月球试试。竹子的见解是,如果一个人想在月球探险中获得成功,那就有必要使用可在当地获取的东西。

剑麻

剑麻指出海洋正受到污染,并表明如果换成剑麻,那剑麻将是海洋生物的食物。剑麻对那些灰暗、笨重、排放二氧化碳的东西占领了市场的事实感到困惑。剑麻也开始寻找解决方案,并质疑为什么要用对人体健康有害的产品来加固水泥。剑麻不仅想为自己找到解决方案,还与竹子共享解决方案,因为两者有加强水泥韧性所必需的纤维。然后,剑麻开始想象看似不可能的事情:在月球上制造水泥。即使竹子声称天然产品不适合那里的环境,剑麻仍保持积极的态度和开放的思维。

艺术
The Arts

想过画一幅壁画吗?这是一个很大的挑战,但也许我们可以进行小小的尝试。买一些石灰膏,这种材料具有惊人的特性,随着时间的推移,它会变得更硬,因为它会继续从大气中吸收碳,将其还原为固体石灰岩。这会让你的壁画保存很长时间。你可以试着在自己房间的墙上画画。

TEACHER AND PARENT GUIDE

思维拓展
Systems: Making the Connections

地球正在生产无数可生物降解和可再生的产品，这些产品在技术和环境方面都具有非凡的性能。可惜的是，它们正逐渐失去人们的青睐。像竹子和剑麻这样的产品被抛弃，取而代之的是聚合物和强化水泥，因为它们更便宜且随时都可以买到，但对环境的负面影响也非常明显。这意味着从长远来看，它们并不便宜，因为社会承担的成本增加了，这些产品进一步破坏了我们所有人都依赖的生态系统。这是一个困境，因为从天然到合成、从生物到矿物的转变与严重的经济恶化彼此交织。在农村，工作岗位大幅减少，人们在那里无法创造更多的价值，只能选择离开。这导致了去农村化和城市化，大量贫困农民现在生活在棚户区，加剧了贫困和营养不良问题。经济模型只计算产品的成本并专注于供应链管理，在供应链管理中，每个人只关注自己的产品，而忽视了对大环境造成的不必要副作用和意外后果。当我们知道有必要减少碳排放，有充分的证据表明石棉是不健康的，为什么没有一种共同的责任感来督促我们改变现在的做法，利用当地和全球可用的资源，确保全社会的共同利益被置于中心地位？这不是一种创新，而是基本原则的回归。现在我们需要一个根本的转变，那就是利用现有的可再生资源。随着世界人口持续增长，生态系统正承受着越来越大的压力，现在是时候了解一些看似很小的决定对数百万人的日常生活和生计产生的深远影响，并了解这是如何影响我们的生态系统和大气的。除非我们学会预见和避免这些意外后果，否则我们将继续承受附带的损害。

动手能力
Capacity to Implement

现在看一看用来建造住房的建筑材料。把用到的主要材料列成一张清单。有多少是矿物材料？有多少是消耗大量能源（或排放温室气体）的？有多少是人工合成的？检查哪些材料是在人们知道它对环境和健康有害之前制造和销售的。现在，用创新的眼光观察，哪些建筑材料是现成的、天然的、可再生的，并且可以通过加工实现现有材料相同功能的建筑材料。列出你希望使用的建筑材料。我们这么做不是为了批判，而是为了做得更好。

教师与家长指南

故事灵感来自
This Fable Is Inspired by

安娜－莉娜·科约尼克森
Anna-Lena Kjøniksen

安娜－莉娜于 1993 年获得挪威奥斯陆大学物理化学学士学位。1996 年，她获得了高分子与胶体科学硕士学位。1999 年，她在同一所大学获得了高分子科学博士学位。随后，她成为了一名专利审查员，这是她的第一份工作。在转入奥斯特福德大学学院之前，安娜－莉娜曾在奥斯陆大学药学院担任研究员多年。她正在领导一个研究项目，她与欧洲航天局合作，开发一种用于月球建设的地质聚合物混凝土，这种混凝土是在真空环境以及月球白天的平均温度下，利用月球表面土壤加工而成的。

图书在版编目（CIP）数据

冈特生态童书.第八辑:全36册:汉英对照/
(比)冈特·鲍利著;(哥伦)凯瑟琳娜·巴赫绘;
何家振等译.—上海:上海远东出版社,2021
ISBN 978-7-5476-1773-1

Ⅰ.①冈… Ⅱ.①冈…②凯…③何… Ⅲ.①生态环
境-环境保护-儿童读物—汉、英 Ⅳ.①X171.1-49

中国版本图书馆CIP数据核字(2021)第249940号

策　　划	张　蓉
责任编辑	祁东城
封面设计	魏　来　李　廉

冈特生态童书
月亮水泥
[比]冈特·鲍利　著
[哥伦]凯瑟琳娜·巴赫　绘
李原原　译

记得要和身边的小朋友分享环保知识哦！
八喜冰淇淋祝你成为环保小使者！

Housing 266

森林里的水母

Jellyfish in the Forest

Gunter Pauli

[比] 冈特·鲍利 著
[哥伦] 凯瑟琳娜·巴赫 绘
李原原 译

上海远东出版社

丛书编委会

主　任：贾　峰
副主任：何家振　闫世东　林　玉
委　员：李原原　祝真旭　牛玲娟　梁雅丽　任泽林
　　　　王　岢　陈　卫　郑循如　吴建民　彭　勇
　　　　王梦雨　戴　虹　翟致信　靳增江　孟　蝶

特别感谢以下热心人士对童书工作的支持：

匡志强　宋小华　解　东　厉　云　李　婧　陈　果
刘　丹　熊彩虹　罗淑怡　旷　婉　杨　荣　刘学振
何圣霖　廖清州　谭燕宁　韦小宏　李　杰　欧　亮
陈强林　王　征　张林霞　寿颖慧　罗　佳　傅　俊
胡海朋　白永喆　冯家宝

目录

森林里的水母	4
你知道吗？	22
想一想	26
自己动手！	27
学科知识	28
情感智慧	29
艺术	29
思维拓展	30
动手能力	30
故事灵感来自	31

Contents

Jellyfish in the Forest	4
Did You Know?	22
Think about It	26
Do It Yourself!	27
Academic Knowledge	28
Emotional Intelligence	29
The Arts	29
Systems: Making the Connections	30
Capacity to Implement	30
This Fable Is Inspired by	31

开着漂亮黄色花朵的金缕梅注意到树上有一个雪球。她很困惑，于是仔细看了看，然后说：
　　"你的伪装真聪明！现在要不是夏天，我还以为你是雪呢。"

A witch hazel tree, with its beautiful yellow flowers, notices a snowball in a tree. Being confused, she takes a closer look, and then says,
"What clever camouflage you are using! If it weren't summer, I would have sworn you were snow."

……注意到树上有一个雪球。

... notices a snowball in a tree.

我不是外星人。

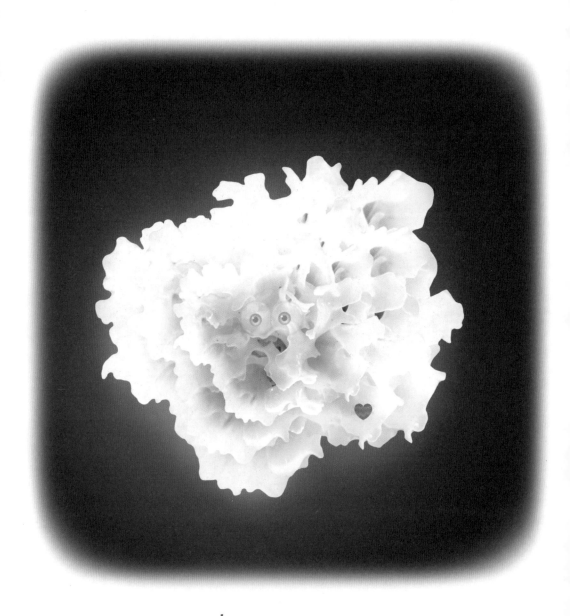

I am not an extra-terrestrial.

"真的吗？我知道有些人觉得我很奇怪，甚至认为我是从遥远星球掉下来的。但是别担心，我不是外星人。你看到的是一种菌菇，一朵银耳。"

"你自称是菌菇，但现在我近距离看你，觉得你更像是一块洗澡用的海绵，甚至是一只水母。从来没见过你这样的真菌！"

"Really? I know some find me very strange, and even think I fell to Earth from a star far away. But don't worry. I am not an extra-terrestrial. You are looking at a mushroom, a Tremella, no less."

"You may claim to be a mushroom, but now that I see you close-up, you look more like a soapy bath sponge, or even a jellyfish. Never a fungus!"

"哦，那你想要看什么样的？有茎有帽的菌菇？"

"对啊，那才正常呀。"

"所以你认为我不正常？"

"Oh, so what did you expect? A mushroom with a stem and a hat?"

"Yes, that would be normal."

"So you don't consider me normal?"

有茎有帽的菌菇？

A mushroom with a stem and a hat?

……我们是一家人。

... we are all family.

"不,不,不是说你不正常,但你看起来确实和我认识的其他菌菇不一样。"

"不是看起来不一样,我是真的不一样。即便如此,我们仍是一家人。我们不摄取食物到体内,而是在体外消化食物。与众不同难道不正是大自然生物多样性的体现吗?"

"No, no, I did not mean to say that you are abnormal, but you certainly look very different to other mushrooms I know."

"I don't only look different, I am different. And yet, we are all family. We don't ingest our food but digest it outside of our body. Is being different not what biodiversity in Nature all about?"

"我想你是对的……但告诉我，是什么让你如此特别？"

"既然你问了，我可以陪你聊几个小时。不过还是让我们长话短说：我喜欢水。"

"你喜欢水吗？地球上的所有物种都需要水。"

"Guess you are right... But tell me what makes you so special?"

"Now that you ask, I can entertain you for hours. But let's keep it short and simple: I love water."

"You love water? Well, everyone on Earth needs water."

我喜欢水。

I love water.

……有弹性、柔软,不会出现皱纹……

... elastic, soft, and wrinkles don't show ...

"在提出你的意见之前,请听我把话说完!我们吸收的水分比任何物种都多,也能更好地储存水分。"

"就像没有足够的水不行一样,喝太多也不好!"

"当你像我一样体内有很多水的时候,你就会有弹性、柔软,皮肤上不会出现皱纹。"银耳回答。

"Please hear me out before you offer me your opinion! We can absorb more water than anyone else, and store it better too."

"Just like not having enough water can not be good, having too much can't be any good either!"

"When you have a lot of water, like me, you are elastic, soft, and wrinkles don't show on your skin," Tremella replies.

"这样你就能做出理想的美容产品。你知道人们不喜欢自己显老,如果你能保证帮他们消除皱纹……"

"我什么都不承诺,我只会履行!毕竟,没有什么比说到做到更重要。"

"我们不要在这里玩文字游戏。我只是一棵植物:一颗有叶子、花和树皮的树。但我能使人变美丽。"

"Then you would make the ideal beauty product. You know that people do not like to look old, and if you can promise to take their wrinkles away..."

"I do not promise anything, I deliver! After all, there is nothing more important than to say what you do, and do what you say."

"Let's not play with words here. I'm only a plant; one with leaves, flowers and bark. But what I have to offer, is beauty."

……人们不喜欢自己显老……

... people do not like to look old ...

我就是著名的金缕梅！

I am the famous Witch Hazel!

"你打算怎么做到这点?"

"好吧,我就是著名的金缕梅!"

"从没听说过使人变美丽的女巫……"

"我也没听说过,也不知道为什么人们叫我女巫。但我可以在皮肤上施'魔法'以淡化疤痕。这已经被证明了。"

"And how do you propose doing that?"
"Well, I am the famous Witch Hazel!"
"Never heard of anyone called a witch that brings beauty…"
"Nor have I, and I have no idea why I am called a witch. But I can apply magic to the skin and take scars away. That has been proven."

"好吧，如果你能帮助别人变得漂亮，别人怎么叫你都没关系。你应该对世界上的所有女人施魔法。"

"为什么只有女人？"

……这仅仅是开始！……

"Well, if you can help make people beautiful, it doesn't matter what they call you. You should apply your magic to every woman in the world."

"And why only the women?"

... AND IT HAS ONLY JUST BEGUN!...

……这仅仅是开始！……

... AND IT HAS ONLY JUST BEGUN! ...

Did You Know?
你知道吗？

The Native Americans used witch hazel for its medicinal properties. The early settlers in America learned about the remedies from the native cultures, and its use became widely established in the settlements.

美洲原住民因为金缕梅有药用价值而使用金缕梅。早期的美洲殖民者从当地文化中了解到这种疗法，于是金缕梅在殖民地也得到了广泛应用。

The word "witch" in "witch hazel" is not related to the practice of magic, but derived from the old English word wice, which means flexible or pliable. Witch hazel trees are popular ornamental plants.

金缕梅（witch hazel）中的 witch 一词与魔法无关，它来源于古英语单词 wice，意思是灵活的、柔韧的。金缕梅是很受欢迎的观赏植物。

The Tremella is a parasitic mushroom living off other mushrooms growing on wood. There are over 100 different species found worldwide.

银耳是一种寄生菌，寄生在木头上的其他菌菇身上。世界上有超过100种不同的银耳。

The Tremella fusiformis is considered a super food. Using it in tonics has been part of Traditional Chinese Medicine (TCM) dating back to 200 AD.

银耳被认为是一种超级食物。中医在补药中使用银耳，这可以追溯到公元200年。

The mushroom is commonly known as "the snow fungus" or even "the beauty mushroom". Its gelatinous fruiting bodies give it the appearance of a jellyfish, or a bath sponge.

银耳通常被称为"雪耳",甚至是"美人菇"。它的胶质子实体使它看起来像水母或浴用海绵。

Tremella has been clinically used in TCM to replenish fluids in the body, to treat chest congestion, asthma, and constipation, to balance blood sugar levels and cholesterol (reduces LDL), and to treat inflammation.

银耳在中医临床被用于补充体液,治疗胸闷、哮喘、便秘,能平衡血糖和胆固醇水平(降低低密度脂蛋白水平),还能治疗炎症。

In ancient China, Yang Guifei (719-756, Tang Dynasty) was considered one of the most beautiful women in Chinese history. Yang Guifei attributed the secret of her lasting beauty to the Tremella mushroom.

在中国古代，杨贵妃（公元719—756年，唐朝）被认为是中国历史上最美丽的女性之一。杨贵妃将她持久美丽的秘诀归功于银耳。

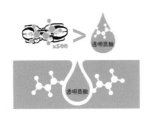

Tremella's water holding capacity is nearly 500 times its own weight. This is greater than hyaluronic acid, a common ingredient used in skincare products to retain the skin's moisture.

银耳能储存接近自身重量500倍的液体。这比透明质酸更有效。透明质酸是护肤产品中用来保持皮肤水分的常见成分。

Think about It
想一想

Whatever is not normal is abnormal? — 任何非常规的事物都是反常的吗?

Are beauty products for men and women? — 美容产品是男女都需要的吗?

Can you play with words? — 你会玩文字游戏吗?

If you look different, can you still be part of the family? — 如果你看起来与家族成员不同，你还是这个家族中的一员吗?

Do It Yourself!
自己动手！

Do you have a dry skin? Have you been bitten by a mosquito and felt the itch afterwards? What remedy did you take? Or did you leave it to the skin to recover without any help? Now if you used anything, check what it was and find out if you, your friends and family are using one of the natural remedies well known in TCM, and in the over-the-counter drug market. Report on who has gone with tradition, and who has put their trust in modern chemistry.

你皮肤干燥吗？被蚊子咬了之后你有没有觉得痒？你用了什么药？还是在没有任何帮助的情况下让皮肤自行恢复？如果你用了什么药，查一查它是什么，看看你和你的朋友、家人是否在使用一种著名的中药或者非处方药。统计一下，谁遵循传统，谁信任现代化学。

TEACHER AND PARENT GUIDE

学科知识
Academic Knowledge

生物学	金缕梅的花和果实一起出现；菌菇不摄入食物，而是从体外吸收营养，因此没有胃；银耳是一种寄生菌，一种吃其他真菌的真菌；世界上有100多种银耳；菌根真菌的菌丝体与土壤保水能力、水力传导度和渗透性呈正相关。
化 学	金缕梅多酚包含类黄酮、鞣酸；药物中的酒精会让皮肤变干；金缕梅与芦荟等保湿成分结合；银耳富含维生素D；银耳引起细胞凋亡；β-葡聚糖是在银耳中发现的一种多糖，是免疫刺激剂。
物 理	金缕梅种子荚爆开，种子可以喷射10米远；维生素D是一种脂溶性维生素；银耳所含的多糖化合物具有很强的保水性；透明质酸（HA）是一种由皮肤分泌的有保湿作用的糖，其分子比银耳所含的保湿分子更大，因此银耳更能保湿；保水性是由质地、结构和有机质含量决定的。
工程学	水蒸气蒸馏过程；做斑贴试验以避免不良反应。
经济学	金缕梅是一种非处方药，价格具有竞争力；传统药物市场的规模与西药市场相同，估计有5 000亿美元。
伦理学	谁来定义什么是正常的？为什么除此之外的一切都被视作不正常，而不是创新，也不是进化中生物多样性的证明？
历 史	2 000年来，中国和日本的药师将银耳用作一种滋阴补药；1753年，林奈为银耳命名，并将其归为藻类；1894年，清朝时期，中国开始培植银耳；在历史上，女巫比施暴者更容易受到攻击。
地 理	野生银耳分布在亚热带、寒带、温带和热带地区；金缕梅生长在北美洲、中国和日本。
数 学	流行病学家过去抱怨数据太少，现在数据却太多；数学和医学一起为临床试验服务。
生活方式	金缕梅叶和树皮提取物被用作收敛剂，可以收紧皮肤、止血，治疗昆虫叮咬、螫伤；维生素D可以促进钙的吸收，从而帮助我们保持骨骼和牙齿强健；皮肤水肿或水潴留。
社会学	日本人称银耳为"白树水母"；银耳被用于东南亚的汤和甜点中；需要找到平衡——太多或太少对健康都不好；承诺（而非法律协议）的价值；大众文化中的巫术被认为是一种迷信；女巫是女性受压迫的象征；女巫揭示了对女性权力的矛盾感受。
心理学	当我们发现新的生命形式时，应该保持开放的心态；由谁来定义什么是正常的？我们对女巫的解读不仅反映了她们，也反映了我们。
系统论	银耳吃其他真菌来获取营养，而不是像许多众所周知的菌菇那样吃木头；堆肥能保持土壤的水分。

教师与家长指南

情感智慧
Emotional Intelligence

金缕梅

当看到银耳时,金缕梅无法掩饰自己的惊奇。她自然公开地表达她对菌菇的印象。她一板一眼,指出常规的菌菇应该有的样子。金缕梅反驳银耳的说法,质疑他偏离常规,坚称他看起来不像菌菇。她强烈地表达自己的观点,并指出所有物种都需要水。随后,金缕梅在谈话中寻求平衡,这使得谈话内容更有哲理,表达方式更温和,甚至她认可银耳给皮肤补水的功能。她拒绝通过文字游戏来展示自己的智慧,并表明自己是一种没有被菌菇认出来的著名植物。金缕梅承认,被人称为女巫令她感到困惑,不过她确实可以给女性甚至男性施"魔法"。

银　耳

银耳非常清楚自己长得奇怪。银耳反问金缕梅觉得自己应该长什么样,菌菇的正常外观又应该是什么样。银耳坚持自己的与众不同,因为这就是生物多样性的体现。银耳不太愉快,因此他只简短地告诉金缕梅他的长处。当被金缕梅打断时,银耳请求对方先听他说明一下自己吸收和储存水分的能力。当金缕梅继续争论时,银耳并没有回应,而是直接描述充足的水分对美容的好处。银耳坚信结果比承诺重要。当金缕梅声称自己同样有美容功效时,银耳表明自己的困惑:女巫也能使人变美丽?银耳很务实:名字不重要,结果才重要。

艺术
The Arts

艺术的蜕变:画四幅银耳,一幅看起来像外星生物,一幅像浴用海绵,一幅像雪球,最后一幅像水母。你能画出那些微妙的差别,使你的每一幅画都与其他几幅截然不同,即使它们描绘的都是同一事物?

TEACHER AND PARENT GUIDE

思维拓展
Systems: Making the Connections

现代西医和传统医学都提高了人们的生活质量。西医必须听取完整的诊断，然后才能决定采用的药物。西医有一个明确的原则：因果关系要有明确的证据，副作用也要非常有限，因此才可以推荐使用相应的治疗方法。通过双盲测试、使用安慰剂和第三方独立审查，人们会得出肯定的结论，这是一种非常有说服力的证明药物有效的方式。传统医学在中国、印度以及其他很多文化中都得到了成功实践。传统医学的方法和体系不同，因此不一定能用西医的标准来衡量。首先，很多传统药物并不是用来治疗疾病的，而是用来加强免疫系统。这种明确侧重于预防而非治疗的做法是一个根本区别。当我们的目标不是治愈而是避免疾病发生时，我们为何要去证明叶子、树皮、根或菌菇的提取物能治愈疾病呢？证明药物的预防效果比证明治疗效果更困难。第二个主要区别是，传统医学不是在单一和孤立的分子基础上运作的，而是使用一些混合物，这些混合物与人体发生了一系列复杂的相互作用，这与简单的因果关系不同。第三个主要区别是，西医追求标准化、可复制、控制成本和追求利润。传统医学的高度个性化形式无法提供符合高水平工业化社会和用来进行统一治疗的理论依据。根本原因在于，每个人都是不同的，都有一个复杂的、难以标准化的新陈代谢和免疫系统，难以得出统一的疗效预测。千百年来，金缕梅一直为北美洲的原住民服务，但现代医学却认为金缕梅"无效"。事实是，这些传统疗法可能仅仅在现代西方医学的框架下被视为没有价值，而在其他研究框架下还是很有价值的。毕竟，现代医学不能宣称自己是唯一有用的医学，我们也应当关注其他医学。在大自然中，促进健康和复原力是生态系统服务的一部分，它是免费向所有人提供的。

动手能力
Capacity to Implement

找点金缕梅种子。它是一种在世界各地广泛销售的观赏植物。如果你住在中国、北美洲或日本，一定要买当地特有的品种。观察金缕梅的生长，看看它的叶子和花朵萌发的过程，检查树皮的厚度。研究一下如何从这种植物中提取活性成分，并制成缓解蚊虫叮咬的药物。

教师与家长指南

故事灵感来自
This Fable Is Inspired by

爱丽丝·W·陈
Alice W. Chen

爱丽丝·陈出生于中国南京，先后就读于中国台湾大学和美国堪萨斯大学劳伦斯分校。她在澳大利亚悉尼大学和英国伦敦帝国理工学院获得博士学位。她曾与因发现青霉素而获诺贝尔奖的恩斯特·伯利斯·柴恩博士共事。陈博士是一名训练有素的科学家，她花了多年时间研究菌菇。她特别关注灵芝在免疫学研究中的应用。它在中国被称为灵芝，在日本被称为万年茸，经常被捧为"长生不老的菌菇"。爱丽丝发表了许多关于灵芝在生物医学应用中的潜在效用的论文。她是国际公认的真菌学家，同时也是一名全职艺术家。她经常说她的科学背景帮助她成为艺术家。例如，在画一朵花时，她首先会仔细观察花朵，仔细检查它的所有生物学结构，以彻底理解它，作为她的灵感。爱丽丝在罗彻斯特大学纪念美术馆教亚洲毛笔画已经 20 年了。她发表了大量关于种植药用菌菇的文章，并将她的发现用于创建菌菇种植公司。

图书在版编目(CIP)数据

冈特生态童书.第八辑:全36册:汉英对照/
(比)冈特·鲍利著;(哥伦)凯瑟琳娜·巴赫绘;
何家振等译.—上海:上海远东出版社,2021
ISBN 978-7-5476-1773-1

Ⅰ.①冈… Ⅱ.①冈… ②凯… ③何… Ⅲ.①生态环境—环境保护—儿童读物—汉、英 Ⅳ.①X171.1-49

中国版本图书馆CIP数据核字(2021)第249940号

策　　划	张　蓉
责任编辑	祁东城
封面设计	魏　来　李　廉

冈特生态童书
森林里的水母
［比］冈特·鲍利　著
［哥伦］凯瑟琳娜·巴赫　绘
李原原　译

记得要和身边的小朋友分享环保知识哦！
八喜冰淇淋祝你成为环保小使者！

鱼肚里的小蠕虫

Worms in My Fish

Gunter Pauli

［比］冈特·鲍利 著
［哥伦］凯瑟琳娜·巴赫 绘
李原原 译

上海远东出版社

丛书编委会

主　任：贾　峰
副主任：何家振　闫世东　林　玉
委　员：李原原　祝真旭　牛玲娟　梁雅丽　任泽林
　　　　王　岢　陈　卫　郑循如　吴建民　彭　勇
　　　　王梦雨　戴　虹　翟致信　靳增江　孟　蝶

特别感谢以下热心人士对童书工作的支持：

匡志强　宋小华　解　东　厉　云　李　婧　陈　果
刘　丹　熊彩虹　罗淑怡　旷　婉　杨　荣　刘学振
何圣霖　廖清州　谭燕宁　韦小宏　李　杰　欧　亮
陈强林　王　征　张林霞　寿颖慧　罗　佳　傅　俊
胡海朋　白永喆　冯家宝

目录

鱼肚里的小蠕虫	4
你知道吗？	22
想一想	26
自己动手！	27
学科知识	28
情感智慧	29
艺术	29
思维拓展	30
动手能力	30
故事灵感来自	31

Contents

Worms in My Fish	4
Did You Know?	22
Think about It	26
Do It Yourself!	27
Academic Knowledge	28
Emotional Intelligence	29
The Arts	29
Systems: Making the Connections	30
Capacity to Implement	30
This Fable Is Inspired by	31

一条鲸蠕虫很高兴海里有了越来越多的鲸。他遇到一只乌贼，评论道：

"现在周围的鲸比以往任何时候都多。看来人类终于懂得保护和照顾这些神奇的哺乳动物了。"

"你可能会为更多的鲸而高兴，但我认为人类不会为更多的蠕虫而高兴。"

A whale worm is happy that there are more and more whales in the oceans. He meets a squid and remarks:

"There are more whales around than ever before. It seems that people are finally protecting and caring for these amazing mammals."

"You may be happy that there are more whales, but I don't think that people are happy that there are more of you worms."

一条鲸蠕虫遇到一只乌贼……

A whale worm meets a squid ...

你就会让它生病……

you make it sick...

"要我说,自己活也让别人活。"鲸蠕虫回答道。

"这说起来容易做起来难。每当你进入一个生物体内,你就会让它生病。我见过,当你寄生在那些可怜的鲑鱼的肚子里之后,它们是怎么流血的。"

"嗯,我们每个人都有自己的角色要扮演,而且似乎总是会对其他事物产生一些负面影响。"

"Live and let live, I say," Whale Worm replies.
"That is easier said than done. Whenever you get into a body you make it sick. I saw how those poor salmon bled after you infested their bellies."
"Well, we each have our role to play, and there always seems to be some negative effect on something else."

"比如，当人们吃完寿司后，由于寿司厨师没有确保鱼肉里的你们都被消灭干净，导致他们不得不跑去厕所的时候……"

"我很抱歉给大家带来不适，但他们应该知道先把鱼冷冻起来，或者煎一下，这样我们就变得完全无害了。"

"直到最近，人们才注意到你的存在。但现在你比以往任何时候都多——每条鱼似乎都被感染了！我不喜欢这样。"

"Like when people eat sushi and have to run to the toilet, when the sushi chef has not made sure the fish is without any of you …"

"I am sorry to have caused anyone discomfort, but they know they should first freeze the fish, or fry the steak, to render us completely harmless."

"Until recently, hardly anyone noticed that you were around. But now there are more of you than ever before – with every fish seemingly infested! I don't like it."

……当人们吃寿司……

... when people eat sushi ...

鲸越来越多……

There are more and more whales ...

"但我们越来越多是个好消息！"

"得了吧！好消息？"

"是的。我们之所以这么多，是因为鲸越来越多。随着鲸的数量回升，我们的幼虫也多了起来。"

"请解释一下你是如何把鲸和你们的幼虫联系起来的。"

"But more of us is good news!"

"Come on! Good news?"

"Yes. We are only this abundant because there are more and more whales. As the whale population rebounds, we can have many more larvae."

"Please explain how you connect whales with your larvae."

"我的卵被磷虾吃掉了,对吧?"

"那是鲸最喜欢的食物!"

"没那么快,乌贼太太。是你先吃了我们。"

"不不,我没有吃你们,我讨厌虫子!我更喜欢富含铁的磷虾。"

"My eggs are eaten by krill, right?"
"Which is the whales favourite food!"
"Not so fast, Mrs Squid. You eat us first."
"No, I don't, I hate worms! I prefer krill, which is rich in iron."

我更喜欢富含铁的磷虾。

I prefer krill, which is rich in iron.

这就是生命的循环……

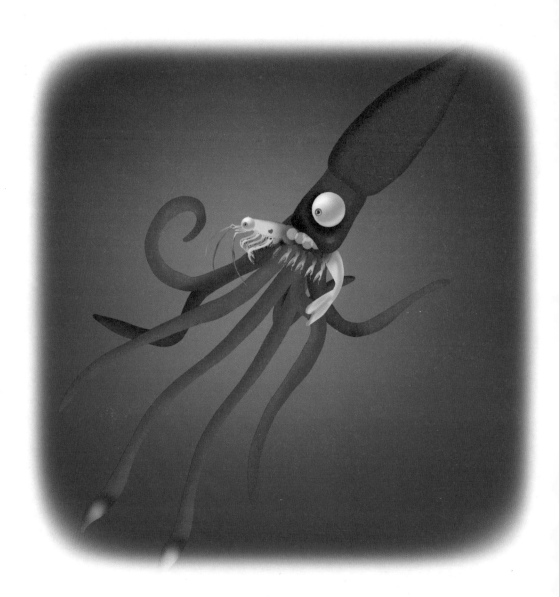

This is the cycle of life...

"对，这就是问题的关键：你吃的磷虾里面有我们的卵。"

"真的吗？我从来没听说过。这真让人倒胃口。"

"这就是生命的循环，而你就是这个生命循环的一部分。"

"Well, that is just it: you eat the krill with our worm eggs inside them."

"Really? I've never heard of that. It's not appetising at all."

"Well, this is the cycle of life, and you form part of this life cycle."

"这么说，你们是寄生虫？"乌贼夫人问道。

"你以磷虾为食，它们在不知不觉中吃了我的卵，所以你也在不知不觉中吃了我的卵。"

"真的吗？那你们是在哪里让卵孵化成虫的呢？"

"So, you are a parasite?" Mrs Squid asks.

"And you prey on krill, that unknowingly eat my eggs, and so do you."

"Really? And where do you hatch from your eggs to become worms again?"

……寄生虫?

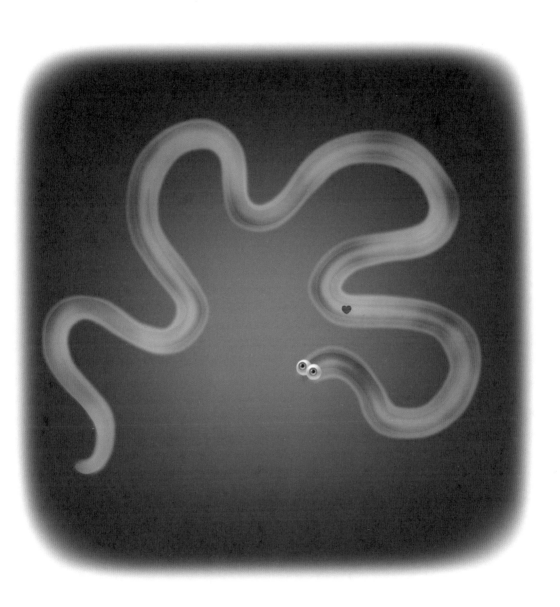

... a parasite?

在海豚或鲸温暖的身体里。

Inside the warm body of a dolphin or a whale.

"在海豚或鲸温暖的身体里。"鲸蠕虫回答道。

"那你的卵排出后,又怎么从鲸的身体里转移到磷虾的身体里呢?"

"通过鲸排出的大量粪便。"

"Inside the warm body of a dolphin or a whale," Whale Worm replies.

"And when you lay eggs, how do your eggs get from the whale to the krill?"

"With the massive flow of poo the whale releases."

"磷虾喜欢吃鲸的粪便！现在我明白了，生命的循环是如何让所有物种都受益的，也明白你所说的'自己活也让别人活'是什么意思了，但我真不希望你是我生活中的一部分……"

"好吧，不管你喜不喜欢我，让我们都继续生活下去，也让其他物种繁荣昌盛。"

……这仅仅是开始！……

"And the krill love to eat the whale poo! Now I see how there are cycles promoting life for everyone. And what you mean by 'live and let live', but I do wish that you were not part of my life…"

"Well, whether you like me or not, we both keep life going – and we make others thrive."

... AND IT HAS ONLY JUST BEGUN!...

……这仅仅是开始！……

... AND IT HAS ONLY JUST BEGUN! ...

你知道吗?

In 1978 scientists reported finding less than one whale worm, on average, per 100 fish. By 2015, they were finding more than one Anisakis simplex worm, on average, per individual fish.

1978 年，科学家报告说，平均每 100 头鲸身上有不到一条鲸蠕虫。到 2015 年，他们发现平均每头鲸身上有不止一条简单异尖线虫。

The world population of blue whales decreased by 99% in the Southern Hemisphere (from 320,000 to 1,000). The natural fertilisation of the oceans' ecosystem also dramatically decreased, and now finally rebounds.

南半球的蓝鲸数量减少了 99%（从 32 万头减少到 1 000 头）。海洋生态系统的自然肥力也急剧下降，现在终于恢复了。

Faecal pellets and the moulted exoskeletons of dead organisms sink to the ocean floor, providing carbon and nitrogen, where bacteria bind the particles into flakes described as "marine snow" by scientists.

粪粒和死亡生物蜕下的外骨骼沉入海底，提供碳和氮，细菌将这些颗粒结合成被科学家们称为"海洋雪"的薄片。

Each whale eats up to several tons of krill per day, excreting most of the iron in their faeces, which contains roughly 10 million times the concentration of iron of the surrounding water.

一头鲸每天要吃掉好几吨磷虾，并将大部分铁通过粪便排出体外，粪便中铁的浓度大约是周围海水的1000万倍。

Whale and seal faeces make many more tons of nitrogen available per year than the input of all rivers combined, and approximately the same as input from the land.

鲸和海豹的粪便每年产生的氮比所有河流产生的总和还要多，约等于所有陆地产生的氮量。

With the sun's energy, nitrogen, phosphorus, iron and other nutrients in the water, phytoplankton multiplies rapidly due to the whale faeces, forming the base of all major marine ecosystems.

在太阳能以及氮、磷、铁等营养物质的共同作用下，浮游植物因鲸的粪便而迅速繁殖，形成了地球上主要海洋生态系统的基础。

Marine larvae, copepods, krill, shrimps, mussels, clams, scallops, anchovies and other grazing fish feed on plankton, and on every life form that feeds on them. As the whale population recovers, so does marine life.

海洋幼虫、桡足类、磷虾、虾、贻贝、蛤蜊、扇贝、凤尾鱼和其他植食性鱼类都以浮游生物和所有以浮游生物为食的生物为食。随着鲸的数量回升，众多海洋生物也在恢复。

The biological downward pump brings biological waste to the bottom of the sea, while whales create an upward pump by diving up to one mile deep, and on surfacing expelling their faeces near the surface.

向下的生物泵将生物废物带到海底，而鲸则潜入1000多米深的地方，然后浮出水面，将它们的粪便排到接近海面的区域，从而创造了一个向上的生物泵。

Think about It
想一想

Is excrement good for the environment?

粪便对环境有好处吗?

Is the number of whales important for our planetary health?

鲸的数量对我们星球的健康重要吗?

Worms in fish: a good sign or a bad sign?

鱼肚里有蠕虫,这是好兆头还是坏兆头?

Should we harvest krill to make dog food?

我们应该捕捞磷虾做狗粮吗?

We need to compare the impact that whale excrement has with the impact of human excrement. Draw up a chart showing the annual numbers of whales, and the subsequent loss of nutrition in the oceans during the massive killing of whales, and now the gain in nutrition as the population rebounds. Now draw up a chart showing the annual human population, and the corresponding increase in human excrement that goes into the oceans untreated every year. What are your conclusions? What do you suggest should be done to redress the imbalance?

我们需要比较鲸的粪便和人类粪便的影响。绘制一张图表，显示每年鲸的数量，大规模捕鲸期间海洋营养物质的损失，以及现在随着鲸的数量回升而增加的营养物质。再绘制一张图表，显示每年的人口以及相应的未经处理进入海洋的人类排泄物的增长量。你的结论是什么？你认为应该采取什么措施来纠正这种不平衡？

TEACHER AND PARENT GUIDE

学科知识
Academic Knowledge

生物学	浮游植物喂养海洋幼虫、桡足类、磷虾、虾、贻贝、蛤蜊、扇贝、凤尾鱼和其他植食性鱼类；磷虾在南极积聚和储存铁。
化 学	在鲸的粪便中发现的生殖激素和肾上腺皮质激素可以用来判断鲸何时性成熟，以及雌鲸是否怀孕或哺乳；太阳的能量以及海水中的氮、磷、铁为浮游植物提供养分。
物 理	冷冻和煮沸能消灭寄生虫；海洋"降雪"是食物触及深海动物和海底群落的主要途径；植物漂浮在深水中以获取养分。
工程学	与鱼类有关的食品卫生需求；城市垃圾在排入大海之前需要进行处理；化粪池的设计；使用旱厕。
经济学	鲸、磷虾和浮游植物提供的免费生态系统服务是公共资源的一部分。
伦理学	人类对自己对生态系统造成的深远影响是多么无知，如捕杀鲸，或将未经处理的人类排泄物排入海洋；将对废物的管理委托给他人，而不进行源头处理；看起来不好的方面，可能实际上是有好处的；如果你只狭隘地关注消极的方面，你就会忽视每种生物扮演的重要角色；透过显而易见的表象去看事物。
历 史	1951年，蕾切尔·卡森在《我们周围的海洋》一书中写道，海洋物质的向下漂移是"有史以来最大的降雪"；20世纪30年代，捕鲸活动达到顶峰。
地 理	近海的营养物质供应不足，只有风暴才能把凉爽、营养丰富的海水带到海面；世界人口集中在沿海地区。
数 学	南极铁循环的复杂模型；比较分析鲸的数量下降带来的影响及其对营养循环的影响，以及人口增长的影响和过量的人类粪便的影响；大数定律：大量试验所得结果的平均值应该接近期望值，并且随着试验次数的增加，平均值趋近于期望值。
生活方式	吃生鱼导致的寄生虫感染；吃生鱼的文化。
社会学	不愿妥善处理人类排泄物的行为让人联想到边际感知效应（冲一次厕所，感觉不会有多大危害），而预计2050年人口将达到90亿，这意味着将会产生巨大的整体影响。
心理学	拒绝处理自己的废物；我们宁愿不知道让我们感到不舒服的事实。
系统论	卵在海水中孵化，幼虫被甲壳类动物吃掉，甲壳类动物接着被鱼或乌贼吃掉，鱼或乌贼又被鲸、海豹和海豚等海洋哺乳动物吃掉，卵通过宿主的粪便排出；一鲸落，万物生。

教师与家长指南

情感智慧
Emotional Intelligence

鲸蠕虫

鲸蠕虫高兴地看到鲸的数量正在回升。当乌贼批评他给海洋生物带来的负面影响时，他给出了一个富有哲理的回答。当乌贼继续批评鲸蠕虫时，他为自己辩护，说他只是在扮演大自然赋予他的角色。接下来，他为给其他动物带来的不适道歉，并提出了杀死寄生虫的方法（冷冻和高温烹饪）。然后，他热情地分享了关于生命循环的信息，并一步一步地解释这个循环，还说明了自己在海洋中通过食物循环进行繁殖的过程。尽管乌贼厌恶鲸蠕虫，鲸蠕虫仍保持积极的生活态度。

乌贼

乌贼对鲸蠕虫提供的信息表示怀疑。她告诉鲸蠕虫，她不喜欢他的存在。她指出鲸蠕虫对其他物种造成了伤害。当鲸蠕虫道歉并提出杀死寄生虫的方法时，她直接指出，现在所有的鱼都感染了寄生虫。她很困惑，不明白为什么鲸蠕虫越多越好，她大胆地要求对方给出详细解释。她以好胜的方式回应鲸蠕虫的观点，但不介意自己的思路顺着鲸蠕虫的逻辑走。即使得到的信息不符合自己的愿望，她也能继续耐心地听取，包括揭示她自己也是鲸蠕虫生命循环的一部分。最后，她接受了事实，但她不希望鲸蠕虫是自己生活中的一部分。

艺术
The Arts

现在，你的挑战是，同时表达极端微小生命形式的美丽和重要性，以及地球上最大的哺乳动物给生态系统带来的巨大影响。使用你喜欢的艺术材料，找到一种富有创意的方式来表达大与小、细微与庞大之间的和谐关系，展示两者在生命和生命循环中是多么重要。

TEACHER AND PARENT GUIDE

思维拓展
Systems: Making the Connections

生命之网中，一切都是相互联系的。鲸的粪便滋养着整个海洋食物网。在海洋中，植物将食物网固定在光照充足的浅水区。然而，在较深的水域，植物必须漂浮在有阳光照射的区域。因此，食物网就是从这里开始的——微小的单细胞藻类，被称为浮游植物，靠油滴漂浮在水中，利用太阳能、氮、磷和铁，迅速繁殖。这构成了海洋生态系统的重要基础，浮游植物供养了海洋幼虫、桡足类、磷虾、虾、贻贝、蛤蜊、扇贝、凤尾鱼，以及所有以它们为食的物种。数以亿计的动物粪便颗粒、蜕下的外骨骼和死去的有机体慢慢地下沉到海参和海中各类虫子所在的海底，为它们提供高碳高氮的食物。这种遍布海洋的"降雪"是深海动物获得营养的主要途径。在营养物供应不足的近海，风暴和洋流上涌会搅动并驱使密度高、营养丰富的海水上升到温暖的海洋表面。在南极，表层海水缺乏铁，这限制了浮游植物的生长，从而限制了整个系统的营养循环。磷虾在这方面起着关键作用，因为磷虾以浮游植物为食，并富集了大约四分之一的铁。正是因为这个原因，南极磷虾的过度捕捞对整个生态系统具有潜在的破坏性。鲸的数量的回升对海洋来说是个好消息，因为它刺激磷虾的增长，而人类随意向海洋排放排泄物则是一个坏消息，因为这会带来多种病菌和寄生虫。人类排泄物中的激素和抗生素增加，导致疾病的增加和病菌抗药性的增强。随着越来越多的人类排泄物进入水井、湖泊、小溪、河流和海洋，人类健康和环境也受到越来越大的影响。同样，我们正在研究鲸蠕虫的侵扰。不仅要了解创造和维持健康的食物循环的动力，我们也需要研究人类行为造成的问题，并以积极的方式解决这些问题。

动手能力
Capacity to Implement

你能做些什么来维持海洋的食物循环？让我们从厕所开始。旱厕能避免将不健康和有毒的废物排入自然水体。然而，在城市里，这可能不太实际。另一种方法是在家里安装飞机上用的那种厕所。这种厕所可以减少对环境的不利影响。确认一下你所在的地区允许做什么，以及如何解决这个问题。与朋友、家人讨论你的发现，并在学校分享你的见解。

教师与家长指南

故事灵感来自
This Fable Is Inspired by

切尔西·伍德
Chelsea Wood

切尔西·伍德于 2006 年获得美国新罕布什尔州达特茅斯学院生态学和进化生物学学士学位。随后，她在科学期刊《生态与环境前沿》担任助理编辑，并于 2013 年在斯坦福大学生物系获得博士学位。2013—2014 年，作为科罗拉多大学博尔德分校的博士后研究员，她开始研究生态系统管理，以测试生物多样性对寄生虫传播的影响。在密歇根大学研究员协会和生态与进化生物学系工作了两年之后，她在西雅图的华盛顿大学创建了自己的实验室。在这里，她和她的团队研究海洋和淡水环境中的寄生虫生态学。

图书在版编目(CIP)数据

冈特生态童书.第八辑:全36册:汉英对照/
(比)冈特·鲍利著;(哥伦)凯瑟琳娜·巴赫绘;
何家振等译.—上海:上海远东出版社,2021
ISBN 978-7-5476-1773-1

Ⅰ.①冈… Ⅱ.①冈…②凯…③何… Ⅲ.①生态环
境-环境保护-儿童读物—汉、英 Ⅳ.①X171.1-49

中国版本图书馆CIP数据核字(2021)第249940号

策　　划	张　蓉
责任编辑	祁东城
封面设计	魏　来　李　廉

冈特生态童书
鱼肚里的小蠕虫
[比]冈特·鲍利　著
[哥伦]凯瑟琳娜·巴赫　绘
李原原　译

记得要和身边的小朋友分享环保知识哦!
八喜冰淇淋祝你成为环保小使者!

Housing 264

我的舌头脏吗?

Is My Tongue Dirty?

Gunter Pauli

[比] 冈特·鲍利 著
[哥伦] 凯瑟琳娜·巴赫 绘
李原原 译

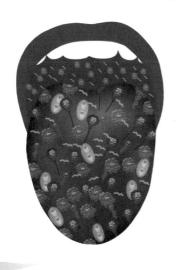

上海远东出版社

丛书编委会

主　任：贾　峰
副主任：何家振　闫世东　林　玉
委　员：李原原　祝真旭　牛玲娟　梁雅丽　任泽林
　　　　王　岢　陈　卫　郑循如　吴建民　彭　勇
　　　　王梦雨　戴　虹　翟致信　靳增江　孟　蝶

特别感谢以下热心人士对童书工作的支持：

匡志强　宋小华　解　东　厉　云　李　婧　陈　果
刘　丹　熊彩虹　罗淑怡　旷　婉　杨　荣　刘学振
何圣霖　廖清州　谭燕宁　韦小宏　李　杰　欧　亮
陈强林　王　征　张林霞　寿颖慧　罗　佳　傅　俊
胡海朋　白永喆　冯家宝

目录

我的舌头脏吗？	4
你知道吗？	22
想一想	26
自己动手！	27
学科知识	28
情感智慧	29
艺术	29
思维拓展	30
动手能力	30
故事灵感来自	31

Contents

Is My Tongue Dirty?	4
Did You Know?	22
Think about It	26
Do It Yourself!	27
Academic Knowledge	28
Emotional Intelligence	29
The Arts	29
Systems: Making the Connections	30
Capacity to Implement	30
This Fable Is Inspired by	31

有些细菌不仅活跃在人的舌头上,还侵入口腔的每一个角落。一个个头较大的细菌抱怨道:

"不知道是谁干的,这嘴里的气味越来越难闻,我在这儿都待不下去了。"

Some bacteria families have not only populated someone's tongue, but also invaded every possible corner of the person's mouth. One of the bigger kinds of bacteria complains:
"I don't know who is responsible, but the smell in this mouth is getting so bad that I can not stand living here any longer."

细菌活跃在人的舌头上……

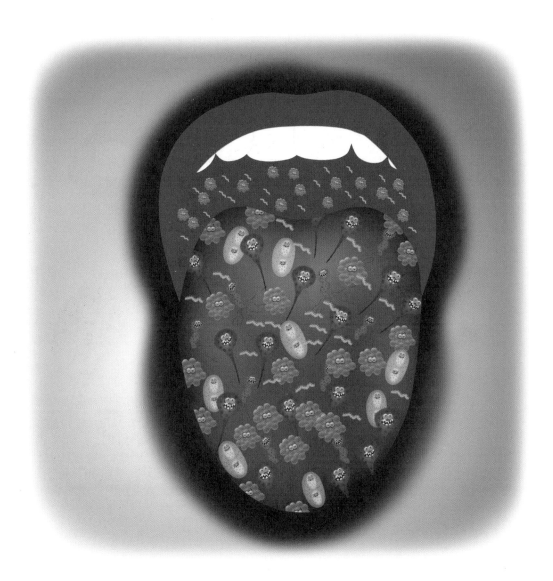

bacteria families populated someone's tongue ...

……再加上烂白菜、大蒜和鱼的气味。

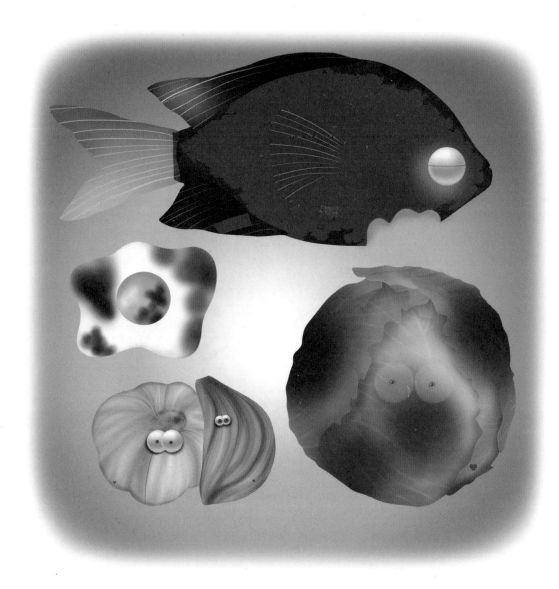

... with the smell of rotten cabbage, garlic and fish.

一个小一点的细菌回应道:"看看是谁在说话!难道你不是制造臭鸡蛋味的罪魁祸首吗?"

"这是一种臭鸡蛋,再加上烂白菜、大蒜和鱼的气味。没有气味比这更难闻的了。"

One of the smaller kinds of bacteria replies, "Look who is talking! Are you not the one responsible for the smell of rotten eggs?"

"This is a rotten egg smell, combined with the smell of rotten cabbage, garlic and fish. The smell can hardly get worse."

"我记得我刚到这根舌头上定居的时候,这个人还是个新生儿,那时的舌头多干净。但现在看来,是我们细菌的数量太多了。"小细菌说。

"当然,婴儿出生时有张超级干净的嘴。但是现在有1 000多种不同的细菌入侵了这张嘴,我甚至不知道还有多少真菌。说到数量过多……"

"I remember when I first settled on this tongue, when this person was a newborn baby. It was so pure and clean. But now it seems that there are far too many of us here," Small Bacterium says.

"Of course, babies are born with a super clean and pure mouths. But now over a thousand different families of bacteria have invaded it, and I do not even know how many fungi. Talk of overpopulation…"

超级干净的嘴……

Super clean and pure mouths ...

……特别是那些生活在牙龈里的……

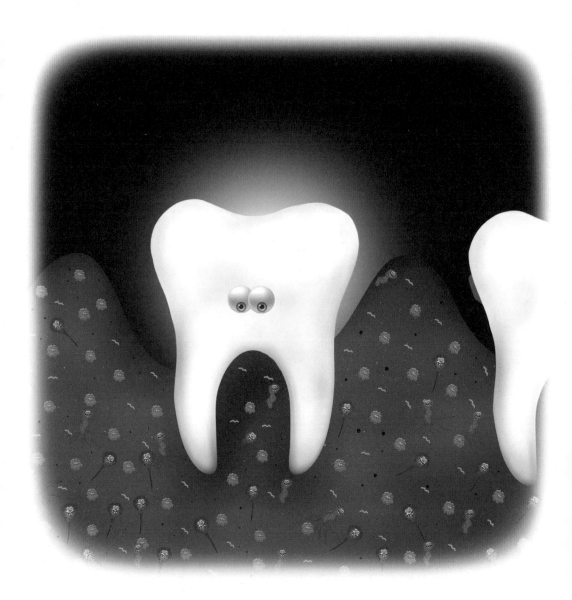

... especially those living in the gums ...

"是的，我相信生活在这张嘴中的细菌和真菌比地球上的人还多。想象一下我们留下的那些污垢。难怪人们有口臭。"

"你的数据大错特错，朋友。其实每克牙菌斑就含1 000亿个细菌。我们之中有些细菌比其他细菌都要难闻，特别是那些生活在牙龈里的，他们会释放恶臭。"

"Yes, I believe there are more bacteria and fungi living right here in this mouth than there are people on Earth. Just imagine all the dirt we all are leaving behind. No wonder people have bad breath."

"Your numbers are way off, my friend. There are 100 billion bacteria per every gram of plaque alone. And some of us smell worse than others; especially those living in the gums, where they give off their gasses."

"因此，现在人们使用那些可怕的化学漱口水。这种水能杀死所有的细菌，无论有益菌还是有害菌。"

"这就是人类的问题！他们想用一种简单的方法，解决所有的问题，尽管这是完全不可能的。"

"And for that people are nowadays using those awful chemical mouth rinses. Those that kill off all of us bacteria – the good and the bad."

"That is the problem with people! They want easy solutions, where one remedy fixes all problems. Even though that's simply not possible."

可怕的化学漱口水……

Awful chemical mouth rinses ...

只要出现一点不适,他们就服用抗生素。

They take antibiotics for every little complaint.

"他们吃的含糖食物，使身体处于压力之下，导致他们经常生病，然后只要出现一点不适，他们就服用抗生素。现在，有害菌将占据主导地位，难闻的气味将永远不会消失。"

"人类没有意识到他们闻到的是我们释放的气体。我们排泄什么取决于我们吃了什么，我们吃的就是人类吃的！从一开始，就是他们的食物制造了臭味！"

"All that sugary food they eat, placing their bodies under stress, making them get sick often, so they take antibiotics for every little complaint. Now the wrong bacteria will dominate, and the bad smell will never go away."

"People do not realise that what they smell is the gasses we give off. What we excrete depends on what we eat, and what we eat is what people eat! It is their food that causes the stink in the first place!"

"那么，人们应该停止吃洋葱和大蒜吗？"小细菌问道。

"不不，那没必要。我们可以应对它，因为我们细菌比森林更多样化，即使我们闻起来没有森林那么清新……"

"你很难让人类相信我们是他们口中的森林。他们只会认为我们看上去和闻起来就像嘴里的垃圾堆。"

"So, should people stop eating onions and garlic?" Small Bacterium asks.

"No, no need for that. We can deal with it, as we bacteria are more diverse than a forest, even if we do not smell as fresh as a forest…"

"You will have a hard time convincing people that we are a forest in their mouths. They will only see and smell a garbage dump in here."

……比森林更多样化……

... more diverse than a forest ...

……把自己粘在他们的牙齿上……

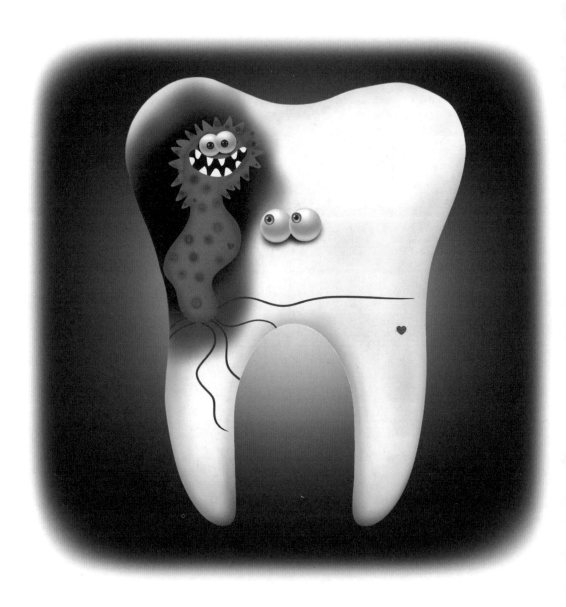

... to glue myself to their teeth ...

"你瞧,我们可不是一群只会造成蛀牙的蛀虫。我们非常有组织地生活在口腔里,在舌头、牙龈和牙齿周围建立小菌群。"大细菌解释道。

"我们无处不在,不只住在人们的嘴里。他们接触的任何东西,从硬币、纸币到厨房和浴室里的海绵,上面都会有各种各样的细菌。更不用说浴室和厕所里的那些了。"

"嗯,我喜欢把自己粘在他们的牙齿上,或者藏在他们的扁桃体下面。"

"Look here, we are not just bugs causing cavities. We are very organised, living in communities and creating little villages all around the tongue, gums and teeth," Big Bacterium explains.

"And we live everywhere, not just in people's mouths. Everything they touch, from coins and notes to kitchen and bath sponges, will have all kinds of bacteria on it. Not to even mention those in their showers and toilets."

"Well, I like to glue myself to their teeth, or hide under their tonsils."

"我们细菌都在一起生活和工作。如果人们想知道他们的口臭从何而来,他们就应该了解真菌和细菌是如何共存的,这对他们是有好处的。"

"想象一下,如果人们对足够多的人进行足够多次的取样,他们就会发现,这个星球上的大多数微生物都会出现在人的嘴里!"

……这仅仅是开始!……

"We bacteria all live and work together. If people want to understand where their smelly breath comes from, they should understand how fungi and bacteria live together – for their benefit."

"Imagine, if they sample enough people enough times they will find that most microorganisms on this planet will show up somewhere in somebody's mouth!"

... AND IT HAS ONLY JUST BEGUN!...

……这仅仅是开始！……

... AND IT HAS ONLY JUST BEGUN! ...

Did You Know?
你知道吗？

The human oral microbiome consists of more than 800 species of bacteria that colonise oral mucosa, while 1,300 species are found in the gingival crevice, and dental plaque comprises of nearly 1,000 species.

在人类口腔微生物组中，分布在口腔黏膜上的细菌有800多种，而在牙龈缝中发现的细菌有1300种，牙菌斑则含有近1000种细菌。

Gas-emitting bacteria on the tongue and below the gum line are largely responsible for bad breath, particularly when the mouth dries out after a night's sleep or a long, dehydrating plane flight.

舌头和牙龈线下释放气体的细菌是造成口臭的主要原因，尤其是经过一夜睡眠或令人脱水的长时间飞行，口腔变得干燥的情况下。

The most effective strategy for dealing with bad breath is nurturing helpful bacteria in the mouth instead of destroying the offending germs and their by-products.

应对口臭最有效的策略是培养口腔中的有益菌,而不是消灭有害菌及其副产品。

Instead of singling out one type, microbiologists focus on entire communities of microbes on the tongue, gum and teeth to figure out why some people have a sweeter-smelling oral cavity than others.

微生物学家关注的不是某一类微生物,而是舌头、牙龈和牙齿上的整个微生物群落,以发现为什么有些人的口腔气味比其他人更清新。

Sulphur compounds that easily vaporise are among the most smelly chemicals in bad breath, especially hydrogen sulphide, which smells like rotten eggs, and methyl mercaptan, which smells like rotten cabbage.

易蒸发的硫化物是口气中最臭的化学物质之一，尤其是硫化氢，闻起来像臭鸡蛋，而甲硫醇闻起来则像腐烂的卷心菜。

There are nine sites in the mouth where different bacteria build their colonies: the tongue, palate, tonsils, sub- and supra-gingival plaque on teeth, the keratinised gingiva, the buccal mucosa, the throat, and the saliva.

不同的细菌在口腔中有9个聚集的地方：舌头、上颚、扁桃体、龈下菌斑、龈上菌斑、角质化的牙龈、口腔黏膜、喉咙和唾液。

A diet rich in nitrates, provided by eating a wide variety of leafy green vegetables, may lower blood pressure a little, but will generate a different nutrient base for microorganisms in the mouth and create fresh breath.

富含硝酸盐的饮食,如食用各种绿叶蔬菜,或许只能稍微降低血压,但会为口腔中的微生物提供不同的营养基础,进而使口气清新。

When kissing, it takes only ten seconds for no less than 80 million bacteria to be exchanged through the saliva. This effect is transitory, and each individual quickly returns to his or her own bacterial equilibrium.

接吻时,只需10秒钟,就会有不少于8 000万个细菌通过唾液进行交换。这种效应是短暂的,每个人的口腔很快就会回到自己的细菌平衡状态。

Are bad smells caused by feeding the wrong food to the wrong bacteria?

难闻的气味是由给错误的细菌喂食错误的食物引起的吗?

Babies are born without any bacteria in their mouths?

婴儿出生时口腔里没有任何细菌?

Sugar causes stress to the body?

糖会给身体带来压力吗?

Is the mouth a forest of bacteria?

口腔是充满细菌的森林吗?

Check your mouth by scraping your tongue and seeing what comes off it. Take a sniff and decide if you like the smell or not. Now scrape again, and see if there is any change. Be careful not to do any damage to your tongue. Scrape until nothing is coming off anymore. Now rinse your mouth with water. Check one hour later, and one day later. Do you see a difference? Now eat a lot of raw vegetables. Does the smell change? Share what you have found out with friends and family members. What you have learnt may very well help someone else who has being suffering from bad breath to also now have fresh breath.

检查一下你的口腔。刮一下舌头,看看都刮出了什么。闻一闻,你喜欢这个味道吗?再刮一遍,看看有没有变化。小心不要刮伤舌头。刮到什么都刮不出为止。现在用水漱口。一小时后重复查看,一天后再查看。你发现区别了吗?现在多吃些新鲜蔬菜。气味改变了吗?与朋友、家人分享你的发现。你学到的东西很可能帮助那些有口臭的人拥有清新口气。

TEACHER AND PARENT GUIDE

学科知识
Academic Knowledge

生物学	革兰氏阳性和革兰氏阴性（有外细胞壁层）细菌；口腔内温度适中，是微生物的水分和营养来源；生物膜的形成与维护；微生物包括厌氧菌和真菌；生命早期，比如出生过程中获得的微生物，有助于塑造免疫系统、消化系统和大脑；牙菌斑是由微生物、多糖和糖蛋白组成的生物膜。
化学	硫化氢闻起来像臭鸡蛋，甲硫醇闻起来像腐烂的蔬菜；生物膜可能变成酸性，导致牙齿脱矿；好氧菌引发的氧化还原反应；漱口水包含洗必泰、二氧化氯或氯化十六烷基吡啶。
物理	牙菌斑变成牙石；出料板后面的塑料元件燃烧，闻起来像臭鸡蛋，表明有危险电弧。
工程学	向天然气中添加硫醇，使其产生异味，便于发现燃气泄漏；用沸水、碳酸氢钠和蒸馏醋来清除异味。
经济学	经济的标准化，即一种产品适用于世界各地，这就是所谓的规模经济。
伦理学	需要适应不同的国家、文化和地区。
历史	公元前7000年，印度河流域文明就开始应用牙科技术了；希波克拉底和亚里士多德都写过关于牙科医学的文章，特别是关于治疗蛀牙的；第一本完全关于牙科医学的书写于1530年；1728年，被誉为现代牙科之父的法国外科医生皮埃尔·福歇尔出版了《外科牙医》，介绍了补牙、使用假牙以及由糖产生的酸导致蛀牙的理论。
地理	在经济合作与发展组织(OECD)的成员国中，只有5个国家提供免费的全面牙科保健服务，分别是奥地利、墨西哥、波兰、西班牙和土耳其。
数学	微生物生长的动力学模型，如流量平衡模型和全细胞粗粒度模型。
生活方式	口腔健康问题和由此导致的口腔微生物的入侵，将影响心脏健康和认知功能；口腔微生物组，包括细菌，对人体免疫系统产生抵抗力，导致蛀牙和牙周病；口腔卫生是日常生活的一部分；过快和过于频繁地使用抗生素会降低免疫系统的反应能力；一般口臭和慢性口臭。
社会学	治标不治本。
心理学	因为害怕自己有健康问题，就不谈论问题，也不咨询医生。
系统论	口腔内细菌的多样性大于热带雨林中的动植物多样性；细菌产生的废物取决于细菌的类型和细菌食物的类型。

教师与家长指南

情感智慧
Emotional Intelligence

大细菌

大细菌抱怨口腔中的臭味，并直言不讳地表达了自己的想法。他很有信心，准备在辩论中起带头作用。他知道如何增强戏剧性以获得关注。他通过数字游戏来制造焦虑。他不准备为难闻的气味承担责任，而是把责任直接推给那些总是想要快速、简单、"可预测"的解决方案的人类。他有一种逻辑，能把细菌摄入的食物和他们发出的恶臭联系起来。他举了一些生动的例子，把口腔比作森林，而生活在这里的细菌和真菌的多样性，构成了口腔的生态系统。

小细菌

小细菌相信自己完全有能力反驳大细菌的论点。起初，她想激起大细菌的情感共鸣。她准备好为难闻的气味承担集体责任。她展现出自己的逻辑思维能力，清楚地指出漱口水中的化学物质既能杀死有害菌，也能杀死有益菌。她结合糖、压力和抗生素促进了占主导地位并导致难闻气味的细菌的生长这一事实，更好地说明了细菌应承担的责任。她还提到人类应当为自己的选择而承担责任。她想要寻找解决方案，并提出了一些显而易见的问题。她遵循大细菌的逻辑，也认识到说服人类改变是非常困难的。最后，她详述了真菌的作用以及口腔生态系统作为一个整体需要被更好地理解这一事实。

艺术
The Arts

让我们创造一个生动的视觉形象来展现细菌和真菌在口中繁衍生息的样子。画一张张开的嘴，在上面画满细菌和真菌。你的艺术作品可能有些超现实主义，但这也许可以让你的朋友和家人重视自己的口腔健康。

TEACHER AND PARENT GUIDE

思维拓展
Systems: Making the Connections

口腔健康不仅仅是没有疾病。口腔健康的关键是多种微生物，包括细菌和真菌，以及它们之间良好的生态平衡，而这取决于我们吃的食物。良好的口腔健康不仅仅是牙齿健康，还包括我们的牙龈、上颚、口腔黏膜、喉咙、舌头、嘴唇、唾液腺、咀嚼肌、神经和上下颌骨的骨骼。众所周知，口腔健康是身体其他部位健康的关键，因此它也是整体健康的关键组成部分。口腔菌群是人体中发现的最复杂、最多样化的微生物群落之一，仅次于肠道菌群。口腔是一个潮湿、营养丰富的环境，许多微生物在这里茁壮成长，形成生物膜。环境、糖摄入量、激素、吸烟等因素都可以导致微生物群落的长期变化。这些变化会提高蛀牙和口腔软组织发炎的风险。为了保持口腔健康，保持宿主与微生物之间的动态平衡至关重要。虽然口腔细菌已经得到了很好的研究，但居住在口腔中的真菌却经常被忽视。我们也忽略了这样一个事实，即在这些群落之间存在着共生关系，有许多相互作用。这些相互作用涉及物理、化学和代谢方面。代谢方面的相互作用涉及糖、碳、乳酸和氧。真菌无处不在、数量众多，所有的动植物都与之共同进化。与真菌一起生活既有益又有害。在健康人的口腔中已经发现了100多种真菌。真菌要在口腔中生存和生长，就必须与细菌、人类宿主保持共生关系。细菌和真菌具有多种代谢功能，包括分解代谢和合成代谢，这使它们能够快速适应变化的环境。根据环境条件的不同，一些真菌可能会发生突变。了解这些由代谢引起的变化，以及细菌和真菌之间的相互作用是很重要的，因为这些相互作用是建立健康的口腔生态的关键。

动手能力
Capacity to Implement

重点是细菌和真菌之间的关系。生物学家起初只把万物分为矿物、植物和动物。直到几十年前，所有的生命形式才被重新分为五个界：原核生物（如细菌）、原生生物、真菌、植物和动物。在人们认识到细菌和真菌的存在之前，这些生物之间的相互作用并没有得到很好的研究。现在是时候进一步了解细菌和真菌如何共同生存和发展，以及这些关系如何影响我们的健康。

教师与家长指南

故事灵感来自
This Fable Is Inspired by

杰西卡·L·马克·韦尔奇
Jessica L. Mark Welch

杰西卡·L·马克·韦尔奇于1989年在美国哈佛大学拉德克利夫学院获得生物学学士学位。1992年，她在德国柏林自由大学获得化学和生物学硕士学位。2001年，杰西卡获得哈佛大学分子与细胞生物学博士学位。她在马萨诸塞州伍兹霍尔海洋生物实验室接受博士后培训。她是芝加哥大学海洋生物实验室的助理科学家。在那里，她研究细菌及其在人类健康和构建健康的生态系统功能方面的关键作用。她还在波尔斯基创业与创新中心工作。

图书在版编目(CIP)数据

冈特生态童书.第八辑:全36册:汉英对照 /
(比)冈特·鲍利著;(哥伦)凯瑟琳娜·巴赫绘;
何家振等译.—上海:上海远东出版社,2021
ISBN 978-7-5476-1773-1

Ⅰ.①冈… Ⅱ.①冈…②凯…③何… Ⅲ.①生态环
境–环境保护–儿童读物—汉、英 Ⅳ.①X171.1-49

中国版本图书馆CIP数据核字(2021)第249940号

策　　划	张　蓉
责任编辑	祁东城
封面设计	魏　来　李　廉

冈特生态童书

我的舌头脏吗?
[比]冈特·鲍利　著
[哥伦]凯瑟琳娜·巴赫　绘
李原原　译

记得要和身边的小朋友分享环保知识哦！
八喜冰淇淋祝你成为环保小使者！